Thermal Energy Storage with Phase Change Materials

Thermal Energy Storage with Phase Change Materials

Edited by
Mohammed Farid
Amar Auckaili
Gohar Gholamibozanjani

CRC Press
Taylor & Francis Group
Boca Raton London New York

CRC Press is an imprint of the
Taylor & Francis Group, an **informa** business

First edition published 2021
by CRC Press
6000 Broken Sound Parkway NW, Suite 300, Boca Raton, FL 33487-2742

and by CRC Press
2 Park Square, Milton Park, Abingdon, Oxon, OX14 4RN

© 2021 Taylor & Francis Group, LLC

CRC Press is an imprint of Taylor & Francis Group, LLC

Library of Congress Cataloging-in-Publication Data
Names: Farid, Mohammed M., editor. | Auckaili, Amar M., editor. |
Gholambozanjani, Gohar, editor.
Title: Thermal energy storage with phase change materials : research
contributions of Professor Mohammed Mehdi Farid in four decades / edited
by Mohammad M. Farid, Amar M. Auckaili, Gohar Gholambozanjani.
Description: First edition. | Boca Raton : CRC Press, 2021. | Includes
bibliographical references and index. |
Summary: "This book focuses on latent heat storage, which is one of the most efficient ways of storing thermal energy. Unlike the sensible heat storage method, latent heat storage method provides much higher storage density, with a smaller difference between storing and releasing temperatures"—Provided by publisher.
Identifiers: LCCN 2021002749 (print) | LCCN 2021002750 (ebook) | ISBN
9780367559410 (hardback) | ISBN 9780367567699 (ebook)
Subjects: LCSH: Heat storage. | Materials—Thermal properties. |
Phase rule and equilibrium.
Classification: LCC TJ260 .T4933 2021 (print) | LCC TJ260 (ebook) |
DDC 621.47/120284—dc23
LC record available at https://lccn.loc.gov/2021002749
LC ebook record available at https://lccn.loc.gov/2021002750

ISBN: 978-0-367-55941-0 (hbk)
ISBN: 978-0-367-56770-5 (pbk)
ISBN: 978-0-367-56769-9 (ebk)

Typeset in Times
by codeMantra

Contents

Chapter 2.3 The Role of Natural Convection during Melting and
                      Solidification of PCM in a Vertical Cylinder 104

*Mohammed Farid, Yongsik Kim, Takuya Honda, and
Atsushi Kanzawa*

Chapter 2.4 Thermal Performance of a Heat Storage Module Using PCMs
                      with Different Melting Temperatures: Mathematical Modeling 123

Mohammed Farid and Atsushi Kanzawa

Waqar A. Qureshi, Nirmal-Kumar C. Nair, and
Mohammed M. Farid

Albert Castell and Mohammed Farid

Preface

Prof. Mohammed Farid, one of the leading thermal energy storage researchers, has offered his combined expertise and extensive knowledge to help in compiling most of his work in this book. Both Dr. Amar Auckaili and Gohar Gholamibozanjani have done their PhD in the field and made significant contribution in the development of energy storage with phase change.

With four chapters that cover the fundamental aspects of thermal energy storage and their practical applications, this book is an ideal research reference for postgraduate and undergraduate students and for those who work in academia and industry and are interested in energy storage.

Prof. Farid, through his long journey of 40 years of research and development on energy storage system, has observed significant work being repeated over these years, probably due to the lack of access to some of the old yet innovative work published on Energy Storage. By compiling his published work in this book, we hope that we have been successful in presenting a clear picture on the historical development in all aspects of energy storage with phase change, starting from fundamentals and mathematical modelling to various applications with a focus on energy saving in buildings and electricity peak load shifting.

Editors

Mohammed Farid, PhD, (MF) is a world leader in energy storage and thermal managements with an almost 40 years of experience in phase change materials development, encapsulation and in the application of these materials in buildings, cold stores and many others. More than 50% of his publications were on energy storage and renewable energy resources, which are comprehensively covered in this book.

Prof. Farid earned an MSc and a PhD in Chemical Engineering at the University of Swansea/Wales, UK, and a BSc in Chemical Engineering at the University of Baghdad, Iraq. He was the founder of the Department of Chemical Engineering at the University of Basra, Iraq in 1982 and was the Head of the department until 1991. He has also worked as a Full Professor at Jordan University of Science and Technology, the University Science Malaysia, and the University of Auckland, New Zealand, where he is currently employed as a Full Professor (Personal Chair). He is a fellow of the Institution of Chemical Engineers, London, and an active member of several international societies. He has published more than 400 papers in international high-quality journals and refereed conferences, 7 patents, 5 books and 11 chapters in books.

Prof. Farid is a member of the editorial board of several journals and has been a member of the advisory committee of many international conferences. He has received several international awards and was invited as a keynote speaker to numerous international conferences. In the Department of Chemical and Materials Engineering, he is leading the research in energy and food engineering. His research is focused on the development of innovative applications of thermal energy storage using phase change materials. In the early 1980s, he developed the effective heat capacity method, a widely used tool to mathematically solve problems in heat transfer with phase change. He has also developed a unified theory that describes most drying methods, such as freeze drying, spray drying, and frying and freezing.

Prof. Farid leads national and international research projects with teams working in energy storage and was the first New Zealand citizen to be awarded a Marie Curie Fellowship to transfer his knowledge to Europe. He has been invited to offer a number of keynote talks at international conferences in the area of energy storage and was granted several international awards, with the recent being the Fredrik Setterwall Professional Development Award by the International Energy Agency ECES in April 2018 for his worldwide contribution in the development of energy management through storage. He has substantial industrial experience aimed toward implementation of his inventions. He led national and international projects with more than NZ\$5 million such as the one received from MBIE of New Zealand, Qatar Foundation, the European commission and NZ Royal Society.

Amar Auckaili, PhD, is a professional lecturer and chartered engineer (MIChemE). He was awarded PhD in chemical and materials engineering from the University of Auckland in 2006. Shortly after, he joined the industry as a lead engineer in the fifth biggest Aluminium smelter, Emirates Global Aluminium. In the industry, he managed to supervise many vital projects and scored several professional certificates. In 2018, Dr. Amar re-joined the University of Auckland as a research fellow for one year before moving to a teaching role in 2019. Recently, he received two best teacher awards in the consecutive years 2019 and 2020. Both awards and professional certificates have added a stamp of approval and responsibility to his experience and career to date.

Gohar Gholamibozanjani, PhD, is a senior researcher at the University of Auckland. She got her PhD from the Department of Chemical and Materials Engineering at the University of Auckland, New Zealand (2016–2020). She also earned an MSc and a BSc degree in the Chemical Engineering Department at the Amirkabir University of Technology, Iran (2009–2012), and Shahid Bahonar University of Kerman, Iran (2005–2009), respectively. During PhD, her research focus was on Thermal Energy Storage systems – especially phase change materials (PCM) – and their passive/active applications in buildings. Using smart control strategies in PCM-enhanced buildings, she managed to be among the finalists of Callaghan C-Prize for environmental innovation, 2019–2020, winning a prize package valued in excess of NZ\$35,000.

Contributors

Said Al-Hallaj
Center for Electrochemical Science and
Engineering
Illinois Institute of Technology
Chicago, Illinois

I. M. AlNashef
Department of Chemical and
Environmental Engineering
Masdar Institute for Science and
Technology
Abu Dhabi, United Arab Emirates

Reza Barzin
Department of Chemical and Materials
Engineering
The University of Auckland
Auckland, New Zealand

Sam Behzadi
Department of Chemical and Materials
Engineering
The University of Auckland
Auckland, New Zealand

Luisa F. Cabeza
GREiA Innovació Concurrent
INSPIRES Research Centre, University
of Lleida
Lleida, Spain

Albert Castell
GREiA Innovació Concurrent
Universitat de Lleida
Lleida, Spain

John J. J. Chen
Department of Chemical and Materials
Engineering
The University of Auckland
Auckland, New Zealand

Paul Devaux
Department of Chemical and Materials
Engineering
The University of Auckland
Auckland, New Zealand

Mohammed Farid
Department of Chemical and
Materials Engineering, School of
Engineering
The University of Auckland
Auckland, New Zealand
and
Department of Chemical
Engineering
College of Engineering, University of
Basrah
Basrah, Iraq
and
Department of Chemical
Engineering
Jordan University of Science and
Technology
Irbid, Jordan

Cèsar Fernández
GREiA Innovació Concurrent
INSPIRES Research Centre, University
of Lleida
Lleida, Spain

Gohar Gholamibozanjani
Department of Chemical and Materials
Engineering
The University of Auckland
Auckland, New Zealand

Alvaro de Gracia
GREiA Innovació Concurrent
INSPIRES Research Centre, University
   of Lleida
Lleida, Spain
and
CIRIAF – Interuniversity Research
   Centre
University of Perugia
Perugia, Italy

M. A. Hashim
Department of Chemical Engineering
University of Malaya
Kuala Lumpur, Malaysia

Takuya Honda
Department of Chemical
   Engineering
Tokyo Institute of Technology
Ookayama, Meguro-ku,
   Tokyo, Japan

Atsushi Kanzawa
Department of Chemical Engineering
Tokyo Institute of Technology
Ookayama, Meguro-ku, Tokyo, Japan

Ali Nasser Khalaf
Department of Chemical Engineering
University of Basrah
Basrah, Iraq

Amar M. Khudhair
Department of Chemical and Materials
   Engineering, School of Engineering
The University of Auckland
Auckland, New Zealand

Yongsik Kim
Department of Chemical Engineering
Tokyo Institute of Technology
Ookayama, Meguro-ku, Tokyo, Japan

R. J. T. Lin
Department of Mechanical Engineering
The University of Auckland
Auckland, New Zealand

François Maréchal
Industrial Process and Energy Systems
   Engineering
Swiss Federal Institute of Technology
   Lausanne
Ecole Polytechnique Fédérale de
   Lausanne
Lausanne, Switzerland

F. S. Mjalli
Petroleum and Chemical Engineering
   Department
Sultan Qaboos University
Muscat, Oman

Ahmed K. Mohamed
Department of Chemical Engineering
College of Engineering, University of
   Basrah
Basrah, Iraq

Nirmal-Kumar C. Nair
Power Systems Group, Electrical &
   Computer Engineering
The University of Auckland
Auckland, New Zealand

Waqar A. Qureshi
Power Systems Group, Electrical &
   Computer Engineering
The University of Auckland
Auckland, New Zealand

Siddique Ali K. Razack
Center for Electrochemical Science and
   Engineering
Illinois Institute of Technology
Chicago, Illinois

K. Shahbaz
Department of Chemical and Materials
    Engineering
The University of Auckland
Auckland, New Zealand

Pongphat Sittisart
Department of Chemical and Materials
    Engineering, School of Engineering
The University of Auckland
Auckland, New Zealand

Joan Tarragona
GREiA Innovació Concurrent
INSPIRES Research Centre, University
    of Lleida
Lleida, Spain
and
CIRIAF – Interuniversity Research
    Centre on Pollution and Environment
    Mauro Felli
University of Perugia
Perugia, Italy

Martin Vautherot
Industrial Process and Energy Systems
    Engineering
Swiss Federal Institute of Technology
    Lausanne
Ecole Polytechnique Fédérale de
    Lausanne
Lausanne, Switzerland

Brent R. Young
Department of Chemical and Materials
    Engineering
The University of Auckland
Auckland, New Zealand

1 Phase Change Material Selection and Performance

INTRODUCTION

Latent heat storage is one of the most efficient ways of storing thermal energy. Unlike the sensible heat storage method, latent heat storage provides much higher storage density, with a smaller difference between storing and releasing temperatures. Phase change materials (PCMs), which melt and solidify at close to ambient temperature, offer a great advantage in reducing the energy needed for heating and air-conditioning of buildings [1]. Energy analysis of PCM-integrated buildings strongly depends on the melting point, latent heat, location, and the amount of the PCM incorporated into the building as well as the climatic conditions and the design of the building. Thermal comfort is usually within the range of 22°C–27°C in summer and 18°C–24°C in winter. Therefore, for building applications, the melting temperature of PCM should lie between 18°C and 27°C depending on the type of building, season, relative humidity, clothing worn, activity levels, and other factors.

PCM should melt congruently with minimum subcooling and be chemically stable, low cost, nontoxic, and noncorrosive. Materials that have been studied during the last 40 years are hydrated salts, paraffin waxes, fatty acids, and eutectics of organic and nonorganic compounds. Farid et al. [2] also introduced a new type III deep eutectic solvents based on choline chloride as a quaternary ammonium salt and calcium chloride hexahydrate as a hydrated salt. Hydrated salts have larger energy storage density and higher thermal conductivity but experience supercooling and phase segregation, and hence, their application requires the use of some nucleating and thickening agents. Paraffin waxes are cheap and have moderate thermal energy storage density but low thermal conductivity and, hence, to be applied, require a large surface area.

The main concern in using paraffin in building constructions is its flammability, as it can easily catch fire if not properly protected. Several methods have been attempted to reduce paraffin flammability. Farid et al. [3] showed that the incorporation of fire retardants into shape-stabilized PCM reduced its flammability. In an event of fire, fire retardant will delay the combustion of PCM and other building materials, allowing more time to evacuate the building.

For the PCMs to be used in any application, they must be encapsulated to prevent them from leaking out, and hence, significant research has been carried out by Farid in the application of microencapsulated or form-stable PCM where PCM is contained within the structure of polymer [4–11].

The long-term stability of PCMs is also a focal point in the efficient energy utilization in buildings. Farid et al. studied the long-term thermal performance of organic PCMs when exposed to a constant temperature above their melting point. The results showed that paraffin-based PCMs experienced significant irreversible physical change with time if it was not encapsulated, while no significant change was observed for the mixed ester compounds [12].

In addition to building application, Farid et al. have shown that PCM can be used for cold storage applications [13–20], thermal management of Li-ion battery [21–24], improving the efficiency of photovoltaic cells [25] and heat recovery system in compressed air energy storage system [26].

REFERENCES

1. Farid MM, Khudhair AM, Razack SAK, Al-Hallaj S. A review on phase change energy storage: Materials and applications. *Energy Convers Manag* 2004;45:1597–615. doi:10.1016/j.enconman.2003.09.015.
2. Shahbaz K, Alnashef IM, Lin RJT, Hashim MA, Mjalli FS, Farid MM. A novel calcium chloride hexahydrate-based deep eutectic solvent as a phase change materials. *Sol Energy Mater Sol Cells* 2016;155:147–54. doi:10.1016/j.solmat.2016.06.004.
3. Sittisart P, Farid MM. Fire retardants for phase change materials. *Appl Energy* 2011;88:3140–5. doi:10.1016/j.apenergy.2011.02.005.
4. Jamekhorshid A, Sadrameli SM, Farid M. A review of microencapsulation methods of phase change materials (PCMs) as a thermal energy storage (TES) medium. *Renew Sustain Energy Rev* 2014;31:531–42. doi:10.1016/j.rser.2013.12.033.
5. Rahman A, Adschiri T, Farid M. Microindentation of microencapsulated phase change materials. *Adv Mater Res* 2011;275:85–8.
6. Al-Shannaq R, Farid M, Al-Muhtaseb S, Kurdi J. Emulsion stability and cross-linking of PMMA microcapsules containing phase change materials. *Sol Energy Mater Sol Cells* 2015;132:311–8.
7. Al-Shannaq R, Kurdi J, Al-Muhtaseb S, Farid M. Innovative method of metal coating of microcapsules containing phase change materials. *Sol Energy* 2016;129:54–64.
8. Giro-Paloma J, Al-Shannaq R, Fernández AI, Farid MM. Preparation and characterization of microencapsulated phase change materials for use in building applications. *Materials (Basel)* 2016;9:11.
9. Al-Shannaq R, Farid MM. A novel graphite-PCM composite sphere with enhanced thermo-physical properties. *Appl Therm Eng* 2018;142:401–9.
10. Farid M, Al Shannaq R, Shaheen A-M, Kurdi J. Method for low temperature microencapsulation of phase change materials 2018.
11. Saputro EA, Al-Shannaq R, Farid MM. Performance of metal and non-metal coated phase change materials microcapsules when used in compressed air energy storage system. *Appl Therm Eng* 2019;157:113715.
12. Behzadi S, Farid MM. Long term thermal stability of organic PCMs. *Appl Energy* 2014;122:11–6.
13. Oró E, de Gracia A, Castell A, Farid MM, Cabeza LF. Review on phase change materials (PCMs) for cold thermal energy storage applications. *Appl Energy* 2012;99:513–33. doi:10.1016/J.APENERGY.2012.03.058.
14. Gin B, Farid MM, Bansal P. Modeling of phase change material implemented into cold storage application. *HVAC&R Res* 2011;17:257–67.

15. Gin B, Farid MM, Bansal PK. Effect of door opening and defrost cycle on a freezer with phase change panels. *Energy Convers Manag* 2010;51:2698–706.
16. Oro E, Miro L, Farid MM, Cabeza LF. Thermal analysis of a low temperature storage unit using phase change materials without refrigeration system. *Int J Refrig* 2012;35:1709–14.
17. Oró E, Miró L, Barreneche C, Martorell I, Farid MM, Cabeza LF. Corrosion of metal and polymer containers for use in PCM cold storage. *Appl Energy* 2013;109:449–53.
18. Oró E, Cabeza LF, Farid MM. Experimental and numerical analysis of a chilly bin incorporating phase change material. *Appl Therm Eng* 2013;58:61–7.
19. Kozak Y, Farid M, Ziskind G. Experimental and comprehensive theoretical study of cold storage packages containing PCM. *Appl Therm Eng* 2017;115:899–912.
20. Al-Shannaq R, Young B, Farid M. Cold energy storage in a packed bed of novel graphite/PCM composite spheres. *Energy* 2019;171:296–305.
21. Khateeb SA, Amiruddin S, Farid M, Selman JR, Al-Hallaj S. Thermal management of Li-ion battery with phase change material for electric scooters: experimental validation. *J Power Sources* 2005;142:345–53.
22. Khateeb SA, Farid MM, Selman JR, Al-Hallaj S. Mechanical--electrochemical modeling of Li-ion battery designed for an electric scooter. J Power Sources 2006;158:673–8.
23. Navarro L, de Gracia A, Colclough S, Browne M, McCormack SJ, Griffiths P, et al. Thermal energy storage in building integrated thermal systems: A review. Part 1. active storage systems. *Renew Energy* 2016;88:526–47. doi:10.1016/j.renene.2015.11.040.
24. Sabbah R, Farid MM, Al-Hallaj S. Micro-channel heat sink with slurry of water with micro-encapsulated phase change material: 3D-numerical study. *Appl Therm Eng* 2009;29:445–54.
25. Atkin P, Farid MM. Improving the efficiency of photovoltaic cells using PCM infused graphite and aluminium fins. *Sol Energy* 2015;114:217–28.
26. Saputro EA, Farid MM. A novel approach of heat recovery system in compressed air energy storage (CAES). *Energy Convers Manag* 2018;178:217–25.

1.1 A Review on Phase Change Energy Storage
Materials and Applications

Mohammed Farid and Amar M. Khudhair
The University of Auckland

Siddique Ali K. Razack and Said Al-Hallaj
Illinois Institute of Technology

1.1.1 INTRODUCTION

Energy storage plays important roles in conserving available energy and improving its utilization since many energy sources are intermittent in nature. Short-term storage of only a few hours is essential in most applications; however, long-term storage of a few months may be required in some applications.

Solar energy is available only during the day, and hence, its application requires an efficient thermal energy storage so that the excess heat collected during the daytime may be stored for later use during the night. Similar problems arise in heat recovery systems where the waste heat availability and utilization periods are different, requiring some thermal energy storage. Also, electrical energy consumption varies significantly during the day and night, especially in extremely cold and hot climate countries where the major part of the variation is due to domestic space heating and air conditioning. Such variation leads to an off-peak period, usually after midnight until early morning. Accordingly, power stations have to be designed for capacities sufficient to meet the peak load. Otherwise, very efficient power distribution would be required. Better power generation management can be achieved if some of the peak load could be shifted to the off-peak load period, which can be achieved by thermal storage of heat or coolness. Hence, the successful application of load shifting and solar energy depends largely on the method of energy storage used.

The most commonly used method of thermal energy storage in all the abovementioned applications is the sensible heat method. In solar heating systems, water is still used for heat storage in liquid-based systems, while a rock bed is used for air-based systems. The design of sensible heat storage units is well described in textbooks [1,2]. In the application of load leveling, heat is usually stored in a refractory bricks storage heater, known as a night storage heater [3]. These units are capable of providing space heating during the day from the stored heat during the night; however, they are heavy and bulky in size.

The latent heat method of storage has attracted a large number of applications, as will be discussed in this review article. This method of heat energy storage provides much higher energy storage density with a smaller temperature swing when compared with the sensible heat storage method. However, practical difficulties usually arise in applying the latent heat method due to low thermal conductivity, density change, stability of properties under extended cycling and sometimes phase segregation and subcooling of the phase change materials (PCMs). In this article, latent heat refers to the latent heat of melting, as other phase changes, such as evaporation, are not practical due to the large volume change associated with it.

1.1.2 PHASE CHANGE MATERIALS

1.1.2.1 CLASSIFICATION AND PROPERTIES OF PCMs

Materials to be used for phase change thermal energy storage must have a large latent heat and high thermal conductivity. They should have a melting temperature lying in the practical range of operation, melt congruently with minimum subcooling and be chemically stable, low in cost, nontoxic and noncorrosive. Materials that have been studied during the last 40 years are hydrated salts, paraffin waxes, fatty acids and eutectics of organic and nonorganic compounds.

Depending on the applications, PCMs should first be selected based on their melting temperature. Materials that melt below 15°C are used for storing coolness in air-conditioning applications, while materials that melt above 90°C are used for absorption refrigeration. All other materials that melt between these two temperatures can be applied for solar heating and for heat load-leveling applications. These materials represent the class of materials that have been studied most.

Comprehensive lists of most possible materials that may be used for latent heat storage are shown in Figure 1.1.1(a–e), as reported by Abhat [4]. Readers who are interested in such information are referred to the papers of Lorsch et al. [5], Lane et al. [6], and Humphries and Griggs [7] who have reported a large number of possible candidates for latent heat storage covering a wide range of temperatures.

Commercial paraffin waxes are cheap with moderate thermal storage densities (~200 kJ/kg or 150 MJ/m³) and a wide range of melting temperatures (Figure 1.1.1a). They undergo negligible subcooling and are chemically inert and stable with no phase segregation. However, they have low thermal conductivity (0.2 W/m °C), which limits their application. Metallic fillers, metal matrix structures, finned tubes, and aluminum shavings were used to improve their thermal conductivity [3]. Pure paraffin waxes are very expensive, and therefore, only technical-grade paraffins can be used. Commercial paraffin waxes, which melt around 55°C, have been studied most [8–10].

Farid et al. [11] have employed three commercial waxes having melting temperatures of 44°C, 53°C, and 64°C with latent heats of 167, 200, and 210 kJ/kg, respectively, in the same storage unit to improve its performance. P-116 is a commercial paraffin wax that has been used by a large number of investigators. It has a melting temperature of about 47°C and a latent heat of melting of about 210 kJ/kg. More recent information can be found in the papers of Himran et al. [12], Faith [13], and Hasnain [3].

Feldman and Shapiro [14] have analyzed the thermal properties of fatty acids (capric, lauric, palmitic, and stearic acids) and their binary mixtures. The results

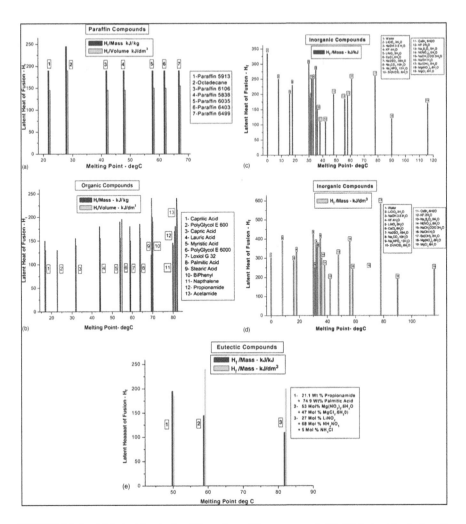

FIGURE 1.1.1   (a) Latent heat of melting of paraffin compounds [4]. (b) Latent heat of melting of non-paraffin organic compounds [4]. (c) Latent heat of melting/mass of inorganic compounds [4]. (d) Latent heat of melting/volume of inorganic compounds [4]. (e) Latent heat of melting of eutectic compounds [4].

have shown that they are attractive candidates for latent heat thermal energy storage in space heating applications. The melting range of the fatty acids was found to vary from 30°C to 65°C, while their latent heat of transition was observed to vary from 153 to 182 kJ/kg. These properties are of prime importance in the design of a latent heat thermal energy storage system.

Hasan [15] conducted an experimental investigation of palmitic acid as a PCM for energy storage. The parametric study of phase change transition included transition time, temperature range and propagation of the solid–liquid interface as well as the heat flow rate characteristics of the employed circular tube storage system. Dimaano and Escoto [16] have evaluated a mixture of capric and lauric fatty acids for low-temperature storage. The melting point of the mixture was about 14°C, and its latent heat of melting ranged between 113 and 133 kJ/kg depending on composition. Some materials have only been used to study the performance of the storage units and are unlikely to be applied in practice. Examples of that are dimethyl sulfoxide, which has a melting point of 16.5°C and a latent heat of only 86 kJ/kg [17], and maleic anhydride, which has a melting temperature of 52°C and a latent heat of 145 kJ/kg [18].

Hydrated salts (Figure 1.11c and d) are attractive materials for use in thermal energy storage due to their high volumetric storage density (~350 MJ/m³), relatively high thermal conductivity (0.5 W/m°C) and moderate costs compared to paraffin waxes, with a few exceptions. Glauber salt ($Na_2SO_4 \cdot H_2O$), which contains 44% Na_2SO_4 and 56% H_2O by weight, has been studied as early as 1952 [19,20]. It has a melting temperature of about 32.4°C, a high latent heat of 254 kJ/kg (377 MJ/m³) and is one of the cheapest materials that can be used for thermal energy storage. However, the problems of phase segregation and subcooling have limited their application. Biswas [19] suggested the use of the extra water principle to prevent the formation of the heavy anhydrous salt. Although this makes the system stable with cycling, it reduces the storage density and requires the system to be operated with a large temperature swing. The use of some thickening agents, such as Bentonite clay, with the Glauber salt has been suggested to overcome the problem of phase segregation. Unfortunately, this will reduce the rates of crystallization and heat transfer to the salt due to the lower thermal conductivity of the mixture. Borax has been long suggested by Telkes [20] as a nucleating agent to minimize subcooling. However, this required some thickening agents to prevent settling of the high-density borax. Most of the other hydrated salts share the same problems.

Table 1.1.1 shows a comparison between the sensible heat storage using a rock bed, water tank, and the latent heat storage using organic and nonorganic compounds. The advantage of the latent heat over the sensible heat is clear from the comparison of the volume and mass of the storage unit required for storing a certain amount of heat. It is also clear from Table 1.1.1 and Figure 1.1.1d that inorganic compounds, such as hydrated salts, have a volumetric thermal storage density, which is higher than that of most organic compounds due to their higher latent heat and density.

TABLE 1.1.1

Comparison between the Different Methods of Heat Storage [3]

Property	Rock	Water	Organic PCM	Inorganic PCM
Density, kg/m³	2240	1000	800	1600
Specific heat, kJ/kg	1.0	4.2	2.0	2.0
Latent heat, kJ/kg	–	–	190	230
Latent heat, kJ/m³	–	–	152	368
Storage mass for 10⁶J, kg	67,000	16,000	5300	4350
Storage volume for 10⁶J, m³	30	16	6.6	2.7
Relative storage mass	15	4	1.25	1.0
Relative storage volume	11	6	2.5	1.0

1.1.2.2 PHASE SEGREGATION AND SUBCOOLING PROBLEMS

The high storage density of salt hydrate materials is difficult to maintain and usually decreases with cycling. This is because most hydrated salts melt congruently with the formation of the lower hydrated salt, making the process irreversible and leading to the continuous decline in their storage efficiency. Subcooling is another serious problem associated with all hydrated salts. To overcome these problems, a number of investigators [21–25] have used hydrated salts in direct contact heat transfer between an immiscible heat transfer fluid and the hydrated salt solution, as will be described later. The agitation caused by the heat transfer fluid has minimized the subcooling and prevented phase segregation. The hydrated salts studied were $CaCl_2 \cdot 6H_2O$, $Na_2SO_4 \cdot 10H_2O$, $Na_2HPO_4 \cdot 12H_2O$, $NaCO_3 \cdot 10H_2O$, and $Na_2 S_2O_4 \cdot 5H_2O$.

The extra water principle was used to prevent clogging of the fluidized bed and the formation of anhydrous salts, which would reduce the storage density of the hydrated salts. The crystallization temperature of these hydrated salts is between 30°C and 50°C, which makes them very suitable for solar energy heating applications. Based on the phase diagram of these salts, it is clear that the heat discharge process employing these salts occurs with continuously decreasing crystallization temperature due to the dilution of the liquid phase. This is not desirable in most applications, which, together with the difficulties associated with the operation of such storage systems, may have limited their application.

Despite the problems associated with the application of hydrated salts in non-agitated thermal storage systems, a number of firms have put significant efforts in developing nucleating agents and stabilizers for some of the hydrated salts. There are at least two well-known companies that have developed commercial PCMs encapsulated in spherical plastic spheres. The products have been tested for a large number of cycles and are found stable. The reported latent heat for the commercial $CaCl_2 \cdot 6H_2O$ is 267 MJ/m³ [26], which is lower than that of the pure salt due to the extra water and the nucleating and thickening agents used. The volumetric storage density based on 1 m³ of a tank is 160 J. Calcium chloride was selected, even though its latent heat was lower than other hydrated salts, probably because it was easier to stabilize, showing minimum phase segregation. Ryu et al. [27] have performed an extensive study on suitable thickening and nucleating agents, which can be used for a number of hydrated salts. Table 1.1.2 summarizes their important findings, showing a significant reduction in the degree of subcooling by applying suitable nucleating and thickening agents.

Among the PCMs that have been proven useful in heat storage applications are calcium chloride hexahydrate $CaCl_2 \cdot 6H_2O$, magnesium chloride hexahydrate $MgCl_2 \cdot 6H_2O$, and magnesium nitrate hexahydrate, $Mg(NO_3)_2 \cdot 6H_2O$. As we have noted, most salt hydrates have the disadvantage that during extraction of the stored heat, the material subcools before freezing. This reduces the utility of the materials and, if too severe, can completely prevent heat recovery. Many factors determine whether a particular additive will promote nucleation, for example, crystal structure, solubility, and hydrate stability. Candidate isomorphous and isotopic nucleating additives, with crystal structures that fit well with the PCM structure, were selected from tables of crystallographic data. Epitaxial nucleators, with fewer obvious lattice structure features that promote nucleation, were selected mostly by intuition. Based

TABLE 1.1.2

Subcooling Range of Thickened PCMs with Different Nucleating Agents [28]

PCM	Thickener	T_m (°C)	Nucleating Agent (Size, µm)	Subcooling (°C) w/o Nucleator	w/Nucleator
$Na_2SO_4 \cdot 10H_2O$	SAP	32	Borax (20×50–200×250)	15–18	3–4
$Na_2HPO_4 \cdot 12H_2O$	SAP	36	Borax (20×50–200×250)	20	6–9
			Carbon (1.5–6.7)		0–1
			TiO_2 (2–200)		0–1
			Copper (1.5–2.5)		0.5–1
			Aluminum (8.5–20)		3–10
$CH_3COONa \cdot 3H_2O$	CMC	46	Na_2SO_4	20	4–6
			$SrSO_4$		0–2
			Carbon (1.5–6.7)		4–7
$Na_2S_2O_3 \cdot 5H_2O$	CMC	57	K2SO4	30	0–3
			$Na_2P_2O_7 \cdot 10H_2O$		0–2

on laboratory test results, promising materials were developed into formulations. Subsequently, attempts were made to correlate crystal structure and hydrate stability with nucleating efficacy and to speculate about active nucleating structures [28].

1.1.2.3 STABILITY OF THERMAL PROPERTIES UNDER EXTENDED CYCLING

The most important criteria that have limited the widespread use of latent heat stores are the useful life of PCMs-container systems and the number of cycles they can withstand without any degradation in their properties. Insufficient long-term stability of the storage materials is due to two factors: poor stability of the materials properties and/or corrosion between the PCM and the container [29].

The development of PCM containers must be directed towards the demonstration of physical and thermal stability, as the PCMs must be able to undergo repetitive cycles of heating and cooling. The purpose of these thermal cycling tests is to determine whether these thermal exposures will result in migration of the PCM or may affect the thermal properties of the PCM.

Kimura and Kai [30] have used NaCl to improve the stability of $CaCl_2 \cdot 6H_2O$, containing slightly more water than the stoichiometric composition. The salt was found to be very stable following more than 1000 heating–cooling cycles. Gibbs and Hasnain [31] confirmed that paraffins have excellent thermal stability as neither the cycles nor contact with metals degrades their thermal behavior. Most investigations on corrosion tests using PCMs were performed with salt hydrates [32,33]. To develop the PCM gypsum wallboard and then achieve the best energy performance, thermal cycling tests have been conducted for 24 wt% PCM-impregnated wallboards. The samples showed no tendency of the PCM (paraffin) to migrate within the wallboard, and there was no observable deterioration in the thermal energy storage capacity [34].

TABLE 1.1.3

Measured Thermophysical Data of Some PCMs [36]

Compound	Melting Temp. (°C)	Heat of Fusion (kJ/kg)	Thermal Conductivity (W/m K)	Density (kg/m³)
Inorganics				
MgCl₂·6H₂O	117	168.6	0.570 (liquid, 120°C)	1450 (liquid, 120°C)
			0.694 (solid, 90°C)	1569 (solid, 20°C)
Mg(NO₃)₂·6H₂O	89	162.8	0.490 (liquid, 95°C)	1550 (liquid, 94°C)
			0.611 (solid, 37°C)	1636 (solid, 25°C)
Ba(OH)₂·8H₂O	48	265.7	0.653 (liquid, 85.7°C)	1937 (liquid, 84°C)
			1.225 (solid, 23°C)	2070 (solid, 24°C)
CaCl₂·6H₂O	29	190.8	0.540 (liquid, 38.7°C)	1562 (liquid, 32°C)
			0.1088 (solid, 23°C)	1802 (solid, 24°C)
Organics				
Paraffin wax	64	173.6	0.167 (liquid, 63.5°C)	790 (liquid, 65°C)
			0.346 (solid, 33.6°C)	916 (solid, 24°C)
Polyglycol E600	22	127.2	0.189 (liquid, 38.6°C)	1126 (liquid, 25°C)
			–	1232 (solid, 4°C)
Fatty acids				
Palmitic acid	64	185.4	0.162 (liquid, 68.4°C)	850 (liquid, 65°C)
			–	989 (solid, 24°C)
Capric acid	32	152.7	0.153 (liquid, 38.5°C)	878 (liquid, 45°C)
			–	1004 (solid, 24°C)
Caprylic acid	16	148.5	0.149 (liquid, 38.6°C)	901 (liquid, 30°C)
			–	981 (solid, 13°C)
Aromatics				
Naphthalene	80	147.7	0.132 (liquid, 83.8°C)	976 (liquid, 84°C)
			0.341 (solid, 49.9C)	1145 (solid, 20°C)

The density of a PCM is important because it affects its storage effectiveness per unit volume. Salt hydrates are generally denser than paraffins and, hence, are even more effective on a per volume basis. The change of volume with the transition, which is in the order of 10%, could represent a minor problem [35]. Table 1.1.3 presents the experimental thermophysical properties of both the liquid and solid states for several PCMs as reported by Lane [36].

1.1.2.4 HEAT TRANSFER ENHANCEMENT METHODS

One major issue that needs to be addressed is that most PCMs have an unacceptably low thermal conductivity, and hence, heat transfer enhancement techniques are required for any latent heat thermal storage (LHTS) application [37]. Various methods are proposed to enhance the heat transfer in a latent heat thermal store. Metallic fillers, metal matrix structures, finned tubes, and aluminum shavings were used to improve paraffins' thermal conductivity [3]. The use of finned tubes in thermal storage systems with different configurations has been reported by Morcos [38], Sadasuke and Naokatsu [39], Costa et al. [40], and Padmanabhan and Murthy [41]. The use of thin aluminum plates filled with PCM was developed by Bauer and Wirtz [42]. Another method used is to embed the PCM in a metal matrix structure [43]. Mehling et al. [44], Fukai et al. [45], and Py et al. [46] proposed that the PCM should be embedded inside a graphite matrix to increase the thermal heat conductivity in the PCM without much reduction in energy storage.

1.1.3   ENCAPSULATION OF PCMs

There are many advantages of microencapsulated PCMs, such as increasing heat transfer area, reducing PCMs reactivity towards the outside environment and controlling the changes in the storage material volume as phase change occurs. Lane [47,48] has identified over 200 potential phase change heat storage materials melting from 10°C to 90°C to be used for encapsulation. Microencapsulation of $CaCl_2 \cdot 6H_2O$ in polyester resin was particularly successful, and the developments of wall and floor panels were studied. Macroencapsulation of $CaCl_2 \cdot 6H_2O$ in plastic film containers appears promising for heating systems using air as the heat transfer medium. He has assessed the technical and economic feasibility of using encapsulated PCMs for thermal energy storage in solar-driven residential heating applications, and has developed means of encapsulating a group of promising phase change heat storage materials in metal or plastic containers. After considering a number of heating and cooling schemes employing phase change heat storage, a forced hot air, central storage design using $CaCl_2 \cdot 6H_2O$ encapsulated in plastic pipes was adapted.

The encapsulation of PCMs into the micropores of an ordered polymer film was investigated by Stark [49]. Paraffin wax and high-density polyethylene (HDPE) wax were infiltrated successfully into extruded films of the ordered polymer by a solvent exchange technique to yield microcomposites with PCM levels of the order of 40 volume percent. These microcomposite films exhibit excellent mechanical stability under cyclic freeze–thaw conditions.

Royon et al. [50] have developed a new material for low-temperature storage. They contained the water as a PCM within a three-dimensional network of polyacrylamide during the polymerization process, as shown in Figure 1.1.2. The final

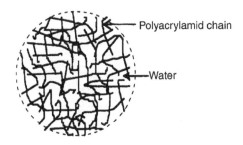

FIGURE 1.1.2   The network of the polyacrylamide containing water used for low-temperature phase change storage [50].

material remains well shaped, requiring no support or even coating, so it can be used directly. Nevertheless, the potential use of microencapsulated PCMs in various thermal control applications is limited, to some extent, by their cost.

Recently, Hong and Xin-shi [51] have employed a compound PCM, which consists of paraffin as a dispersed PCM and a HDPE as a supporting material. This new generation PCM is very suitable for application in direct contact heat exchangers. The 75% of paraffin and 25% of HDPE mixture provided a PCM that has a latent heat of 157 kJ/kg compared to 199 kJ/kg for the paraffin used and with a transition temperature of approximately 57°C, which is close to that of paraffin.

Ammonium alum and ammonium nitrate in the weight ratio of 1:1 form a eutectic that melts at 53°C and crystallizes at 48°C. Its enthalpy, as measured by drop calorimetry, was found to be 287 kJ/kg in the temperature range of 25°C–67°C, which is 1.67 times greater than that of water (172.2 kJ/kg) and 8.75 times greater than that of rock (32.8 kJ/kg). Upon several heating/cooling cycles, phase separation was observed. However, by adding 5% attapulgite clay to this eutectic mixture, phase separation was prevented. This eutectic was encapsulated in 0.0254 m diameter HDPE hollow balls and subjected to about 1100 heating/cooling cycles in the temperature range between 25°C and 65°C. At the end of these cycles, the decrease in enthalpy was found to be 5% [52].

1.1.4 MAJOR APPLICATIONS OF PCMs

The application of energy storage with phase change is not limited to solar energy heating and cooling but has also been considered in other applications as discussed in the following sections.

1.1.4.1 INDIRECT CONTACT LATENT HEAT STORAGE OF SOLAR ENERGY

Extensive efforts have been made to apply the latent heat storage method to solar energy systems where heat is required to be stored during the day for use at night. The studies varied from those related to the fundamental aspects of heat transfer to those in which the PCM is tested in full size heat storage units.

Most PCMs have low thermal conductivity that limits heat transfer rates during their applications. Hence, the PCM must be encapsulated in such a way as to prevent the large drop in heat transfer rates during its melting and solidification. The PCM

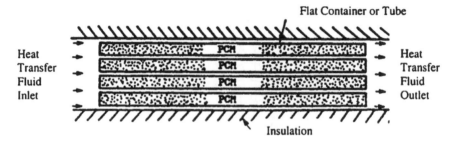

FIGURE 1.1.3 Schematic of a latent heat storage unit using flat containers for encapsulating the PCM [56].

is usually contained in a number of thin flat containers, similar to plate-type heat exchangers [10,17,53–56], as shown in Figure 1.1.3.

Alternatively, it may be contained in small diameter tubes [9,11,43,57–59] with the heat transfer fluid flowing along or across the tubes. The PCM may also be contained in the shell of a shell and tube heat exchanger [59–62].

To improve heat transfer rates, a honeycomb structure, partially filled with the PCM, was employed [56]. This arrangement has also eliminated the large stresses that may be induced by the volume expansion of the PCM. Kamimoto and Tani [63] and Velraj et al. [37] have suggested the use of high thermal conductivity materials with the PCM to increase its apparent thermal conductivity. They, as well as others [64–66], have tested the idea of using finned tubes in which the PCM was placed between the fins. Although a significant improvement in heat transfer rate was found, the high cost of the finned tubes may make their use uneconomical. It is to be noted that such arrangements may improve heat transfer rates significantly only when a liquid is used as a heat transfer fluid. In air-based systems, the heat transfer coefficients of both the air and the PCM sides are low.

A large improvement in the heat transfer rate was obtained by encapsulating the PCM in small plastic spheres [67,68] to form a packed bed storage unit. However, the expected high pressure drop through the bed and its initial cost may be major drawbacks of such units.

Most of the PCMs undergo large changes in volume (10%) during melting. This may cause high stresses on the heat exchanger walls. Volume contraction during solidification may not only reduce heat transfer area but also separate the PCM from the heat transfer surface, increasing the heat transfer resistance dramatically. The problem is usually minimized by proper selection of the containers, which should be partially filled with the PCM. Spherical encapsulation can be a good solution to this problem.

To improve the performance of phase change storage units, Farid [69] has suggested the use of more than one PCM with different melting temperatures in a thin flat container, as shown in Figure 1.1.4. The same idea has been applied later by Farid and Kanzawa [9] and Farid et al. [11] for a unit consisted of vertical tubes filled with three types of waxes having different melting temperatures, as shown in Figure 1.1.5. During heat charge, the air flows first across the PCM with the highest melting temperature to ensure continuous melting of most of the PCMs. The direction of the air must be reversed during heat discharge. Both the theoretical and experimental measurements showed an improvement in the performance of the phase change storage units using this type of arrangement.

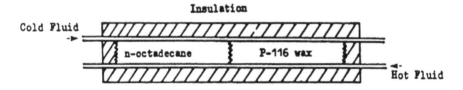

FIGURE 1.1.4 Simplified sketch of a thermal storage unit employing two types of PCM [69].

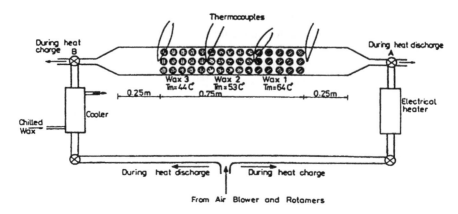

FIGURE 1.1.5 Experimental air-based PCM storage system using three waxes with different melting temperatures [10].

Velraj et al. [70] have reviewed the attempts that have been made on heat transfer enhancement. They also tested different methods of heat transfer enhancement, which included the use of longitudinal internal finned tubes in a shell and tube heat storage unit, dispersing the PCM with high thermal conductivity particles and using metallic packing, such as Leasing rings, placed inside the tubes contain the PCM. Velraj et al. [37] have previously studied the effects of using inner fins in the tubes containing the PCM, while Lacroix [59] has studied the effects of external annular fins in a shell and tube heat storage unit with the PCM placed in the shell. Both studies showed significant improvement in the heat transfer rate due to these enhancement arrangements. Ismail et al. [66] have also studied heat transfer enhancement by using axial fins on the external tube immersed vertically in a PCM. Their experimental measurements and theoretical predictions obtained from their numerical analysis for different fin heights and thicknesses showed that the use of fins enhances the heat transfer rate significantly.

1.1.4.2 THERMAL STORAGE WITH DIRECT CONTACT HEAT EXCHANGERS

The low thermal conductivity of most PCMs requires the use of large and expensive heat transfer surfaces, which is considered as the major drawback in applying them. Direct contact heat transfer arrangements eliminate the requirement of these expensive metallic surfaces. The method has been applied in two ways.

1.1.4.2.1 Solid–Solid Transition with Direct Contact Heat Transfer

A solid–solid phase change method of heat storage can be a good replacement for the solid–liquid phase change in some applications. They can be applied in a direct contact heat exchanger, eliminating the need for an expensive heat exchanger to contain them. Some materials, such as high density polyethylene (HDPE), have a solid–solid transition latent heat close to those of melting, yet retain their shape. Abe et al. [71] and Kamimoto et al. [72] have constructed a 30 kWh direct contact latent heat storage unit using form stable high density polyethylene thin rods and ethylene glycol as a heat transfer fluid. Both thermal storage density and heat transfer rate were high to the extent that the air outlet temperature from the unit during heat discharge remained almost constant at about 127°C until all the polyethylene rods lost their latent heat. A transition temperature of 127°C and a high latent heat of 300 kJ/kg were achieved by inducing cross-linking in the polyethylene using electron beam irradiation. Such a unit has not been commercialized, probably due to the cost of the cross-linking required in manufacturing the HDPE.

A series of bis (*n*-alkyl ammonium) tetrachlorometallates (II) as solid–solid PCMs were synthesized by Li et al. [73]. These potential PCMs have transition temperatures ranging from 37°C to 93°C and latent heats between 28 and 86 kJ/mol, which correspond to 122 and 340 kJ/kg, respectively. Both the transition temperature and the latent heat were found to increase with the increase in the number of carbon atoms in the metal chains.

1.1.4.2.2   Direct Contact Heat Transfer between Hydrated Salts and an Immiscible Fluid

The idea of using direct contact heat transfer between hydrated salts and an immiscible fluid was first introduced by Etherington [74] using disodium phosphate dodecahydrate as a PCM and a light mineral oil as a heat transfer fluid, bubbled through the salt solution. The agitation caused by the bubbles reduced both the subcooling and phase segregation, the two serious problems associated with crystallization of most hydrated salts.

Later, Costello et al. [21] and Edie and Melsheimer [22] studied the performance of storage units with different hydrated salts: $Na_2SO_4 \cdot 10H_2O$, $CaCl_2 \cdot 6H_2O$ and $Na_2HPO_4 \cdot 12H_2O$ using varsol as a heat transfer fluid. Similar studies have been conducted by Hallet and Keyser [75] for hydrated sodium thiosulfate, sodium acetate, and sodium sulfate and by Fouda et al. [23] for sodium sulfate and its mixture with sodium phosphate. More recently [24,25], the performance of direct contact heat transfer with phase change has been thoroughly investigated using four hydrated salts ($Na_2CO_3 \cdot 10H_2O$, $Na_2SO_4 \cdot 10H_2O$, $Na_2HPO_4 \cdot 12H_2O$ and $Na_2S_2O_3 \cdot 5H_2O$). Two storage units connected in series were used, containing a combination of $Na_2CO_3 \cdot 10H_2O$ and $Na_2S_2O_3 \cdot 5H_2O$ as the low and high crystallization temperature salts, respectively. Using this arrangement during both heat charge and discharge increased the heat transfer. Hot kerosene is bubbled through the column containing the thiosulfate first, then to the second column containing the carbonate. This has been done to provide the high temperature difference between the kerosene and the PCM, required for the high heat transfer rate. The flow of kerosene is reversed during the heat discharge cycle, as shown in Figure 1.1.6.

Direct contact heat transfer with phase change using hydrated salts provides very high heat transfer rates. The high volumetric heat transfer coefficient, which can be achieved in such a system, is mainly due to the large surface area of the droplets of the immiscible heat transfer fluid used to transfer the heat from or to the heat storage unit. The high latent heat of crystallization of most hydrated salts compensates for the extra water used to avoid precipitation of the anhydrous salt. However, it is important in future to test the operation of large units of practical sizes and for sufficiently

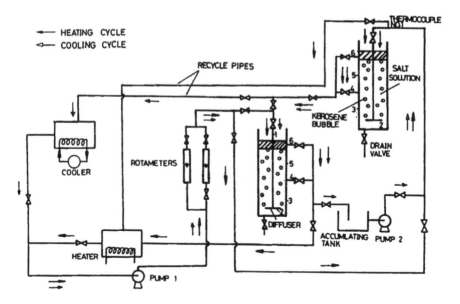

FIGURE 1.1.6   Sketch of a direct contact heat storage system using two hydrated salts [25].

long periods. More work is required on this heat storage system, as no recent work on the subject is shown in literature.

1.1.4.3   PHASE CHANGE THERMAL STORAGE FOR SHIFTING THE PEAK HEATING LOAD

Electricity consumption varies during the day and night according to the demand by industrial, commercial, and residential activities. The variation in electricity demand sometimes leads to a differential pricing system in peak and off-peak periods, usually after midnight until early morning. The shift of electricity usage from peak periods to off-peak periods will provide significant economic benefit. The development of an energy storage system may be one of the solutions to the problem when electricity supply and demand are out of phase. Energy storage systems will enable the surplus energy to be stored until such time as it is released when needed.

Winter storage heating is a direct and simple application of energy storage and has been used in many countries. The most common domestic storage heater uses ceramic bricks and structural cement [3], which is heated with electrical heating wires or heat transfer fluids (such as hot water) during the night. During the day, the heat is extracted from the heater by natural convection and radiation or by forced convection using an electric fan. Farid and Husian [10] have introduced a new concept to the design of these storage heaters by replacing the ceramic bricks with a paraffin wax encapsulated in thin metal containers. During heat charge, the wax stores a larger amount of heat as the latent heat of melting, which is continuously discharged during the other periods. The four individual storage units were filled with paraffin wax having a melting temperature of around 55°C. An electrical plate heater was fixed at the axis of each storage unit to provide low heat flux but sufficient to melt all the wax within 8 h. Using a phase change method of heat storage can lead to a significant weight reduction in domestic storage heaters. Such a unit has not yet been commercialized due to issues related to the unit capital cost.

1.1.4.4 BUILDING APPLICATIONS

The selection of PCMs to meet residential building specifications has received minor attention, although it is one of the most foreseeable applications of PCMs. The ability to store thermal energy is important for the effective use of solar energy in buildings. Because of the low thermal mass of lightweight building materials, they tend to have high-temperature fluctuations, which result in high heating and cooling demands. It has been demonstrated that paraffins, as mixtures of several linear alkyl hydrocarbons, may be tailored by blending to obtain the desired melting point required for a particular application [76]. Pure paraffins with exceptionally good properties have not yet been tested further due to the unavailability of inexpensive bulk sources. Successful utilization of PCM-impregnated building material requires a good means of containment or encapsulation. In buildings, a more interesting approach to smoothen the temperature variations within a space is done by using wallboards containing a PCM. The wall large heat transfer area supports large heat transfer between the wall and the space [35]. The wallboards are cheap and widely used in a variety of applications, making them very suitable for PCM encapsulation. However, the principles of latent heat storage can be applied to any appropriate building materials.

Kedl and Stovall [77] and Salyer and Sircar [78] presented the concept of paraffin wax-impregnated wallboard for passive solar application. The immersion process for filling the wallboards with wax has been successfully scaled up from small samples to full size sheets. Processes whereby this PCM could be incorporated into plasterboard either by postmanufacturing imbibing of liquid PCM into the pore space of the plasterboard or by addition in the wet stage of plasterboard manufacture have been successfully demonstrated.

Neeper [79,80] has examined the thermal dynamics of gypsum wallboards impregnated by fatty acids and paraffin waxes as PCMs that are subjected to the diurnal variation of room temperature. He found that the thermal storage provided by PCM–wallboard would be sufficient to enable the capture of large solar heating fractions. Stetiu and Feustel [81] have presented a thermal building simulation program based on a finite difference approach to evaluate the latent heat storage performance of PCM-treated wallboard. The thermal mass can be utilized to reduce the peak power demand and downsize the cooling or heating systems.

Athienitis et al. [82] conducted an extensive experimental and numerical simulation study in a full-scale outdoor test room with PCM gypsum board as inside wall lining. The PCM gypsum board used contained about 25% by weight proportion of butyl stearate (BS). An explicit finite difference model was developed to simulate the transient heat transfer process in the walls. It was shown that utilization of the PCM gypsum board may reduce the maximum room temperature by about 4°C during the daytime and can reduce the heating load at night significantly.

However, the development of reliable and practical thermal energy storage systems still faces some major hurdles, such as uncertainties, concerning the long-term thermal behavior and the small number of PCMs suitable for room-temperature applications.

1.1.5 NEW PCM TECHNOLOGICAL INNOVATIONS

Revankar [83] has devised a new method for satellite power testing using PCMs. Central to the solar power system are a series of metal cells containing a PCM that is liquid under high temperature, freezes during the hours of cold darkness, and then releases its latent heat. The heat released can then be used to generate electricity by driving thermoelectric units. Because the systems generate at least three times more power than batteries of comparable size, they are seen as a possible alternative to conventional satellite solar power systems that rely on batteries.

By having a hot converter at the start of a trip, auto emissions, such as hydrocarbons and carbon monoxide, can be reduced dramatically by up to 80%. NREL wrapped its catalytic converter in compact vacuum insulation to keep it at an efficient operating temperature for up to 24 h after the engine is shut off. A catalytic converter was developed by using PCMs to absorb, store and release heat as needed [84].

John et al. [85] have designed a novel ventilation nighttime cooling system (a novel combination of PCM and heat pipes) as an alternative to air conditioning. The system offers substantial benefits in terms of reducing or eliminating the need for air conditioning, thereby significantly reducing CO_2 emissions and saving energy in buildings.

Particles of microencapsulated PCM (3–100 μm) and/or macroencapsulated PCM (1–3 mm) can be included within textile fibers, composites, and clothing to provide greatly enhanced thermal protection in both hot and cold environments [86].

1.1.6 CONCLUSIONS

Organic and inorganic compounds are the two most common groups of PCMs. Most organic PCMs are noncorrosive and chemically stable, exhibit little or no subcooling, are compatible with most building materials and have a high latent heat per unit weight and low vapor pressure. Their disadvantages are low thermal conductivity, high changes in volume on phase change and flammability. Inorganic compounds have a high latent heat per unit volume and high thermal conductivity and are non-flammable and low in cost in comparison to organic compounds. However, they are corrosive to most metals and suffer from decomposition and subcooling, which can affect their phase change properties. The applications of inorganic PCMs require the use of nucleating and thickening agents to minimize subcooling and phase segregation. Significant efforts are continuing to discover those agents by commercial companies. The applications in which PCMs can be applied are vast, ranging from heat and coolness storage in buildings to thermal storage in satellites and protective clothing. A PCM with an easily adjustable melting point would be a necessity as the melting point is the most important criterion for selecting a PCM for passive solar applications. Many more applications are yet to be discovered.

REFERENCES

1. Duffie JA, Beckman WA. *Solar Energy Thermal Processes*. New York: John Wiley; 1980.
2. Garg HP, Mullick SC, Bhargava AK. *Solar Thermal Energy storage*. Dordrecht: Reidel Publishing Company; 1985. p. 154.
3. Hasnain SM. Review on sustainable thermal energy storage technologies, Part 1: Heat storage materials and techniques. *Energy Convers Mgmt* 1998;39:1127–38.
4. Abhat A. Low temperature latent heat thermal energy storage: heat storage materials. *Solar Energy* 1983;30:313–32.
5. Lorsch HG, Kauffman KW, Denton JC. Thermal energy storage for heating and air conditioning, Future energy production system. *Heat Mass Transfer Processes* 1976;1:69–85.
6. Lane GA, Glew DN, Clark EC, Rossow HE, Quigley SW, Drake SS, et al. Heat of fusion system for solar energy storage subsystems for the heating and cooling of building. Chalottesville, Virginia, USA, 1975.
7. Humphries WR, Griggs EI. A designing handbook for phase change thermal control and energy storage devices. NASA Technical Paper, 1977. p. 1074.
8. Farid MM, Mohamed AK. Effect of Natural convection on the process of melting and solidification of paraffin wax. *Chem Eng Commun* 1987;57:297–316.
9. Farid MM, Kanzawa A. Thermal performance of a heat storage module using PCMs with different melting temperatures-mathematical modeling. *Trans ASME, J Solar Energy Eng* 1989;111:152–7.
10. Farid MM, Husian RM. An electrical storage heater using phase change method of heat storage. *Energy Convers Mgmt* 1990;30:219–30.
11. Farid MM, Kim Y, Kanzawa A. Thermal performance of heat storage module using PCMs with different melting temperatures-experimental. *Trans ASME, J Solar Energy Eng* 1990;112:125–31.
12. Himran S, Suwono A, Mansoori GA. Characterization of alkanes and paraffin waxes for application as phase change energy storage medium. *Energy Sources* 1994;16:117–28.
13. Faith HE. Technical assessment of solar thermal energy storage technologies. *Renewable Energy* 1998;14:35–40.
14. Feldman D, Shapiro MM. Fatty acids and their mixtures as phase-change materials for thermal energy storage. *Solar Energy Mater* 1989;18:201–16.
15. Hasan A. Phase change material energy storage system employing palmitic acid. *Solar Energy* 1994;52:143–54.
16. Dimaano M, Escoto A. Preliminary assessment of a mixture of capric and lauric acid for low temperature thermal energy storage. *Energy* 1998;23:421–7.
17. Farid MM, Hamad FA, Abu-Arabi M. Phase change cool storage using dimethyl-sulfoxide. *Energy Convers Mgmt* 1998;39:819–26.
18. Zuca S, Pavel PM, Constantinescu M. Study of one dimensional solidification with free convection in infinite plate geometry. *Energy Convers Mgmt* 1999;40:261–71.
19. Biswas DR. Thermal energy storage using sodium sulphate decahydrate and water. *Solar Energy* 1977;19:99–100.
20. Telkes M. Nucleation of super saturated inorganic salt solution. *Indust Eng Chem* 1952;44:1308.
21. Costello YA, Melsheimer SS, Edie DD. Heat transfer and calorimetric studies of a direct contact-latent heat energy storage system; thermal storage and heat transfer in solar energy system. *ASME Meeting*, San Francisco, USA, 1978. pp. 10–5.

22. Edie DD, Melsheimer SS. An immiscible fluid-heat of fusion energy storage system. In: *Proceedings of Sharing the Sun: Solar Technology in the Seventies.* A Joint Conference of the American Section of the International Solar Energy Society and the Solar Energy Society of Canada Inc., Winnipeg. The American Section of the International Solar Energy Society, 1976. pp. 227–37.

23. Fouda AE, Despault GJ, Taylor JB, Capes CE. Solar storage system using salt hydrate latent heat and direct contact heat exchange––II, characteristics of pilot operating with sodium sulfate solution. *Solar Energy* 1984;32:57–65.

24. Farid MM, Yacoub K. Performance of direct contact latent heat storage unit. *Solar Energy* 1989;43:237–52.

25. Farid MM, Khalaf AN. Performance of direct contact latent heat storage units with two hydrated salts. *Solar Energy* 1994;52:179–89.

26. Cristopia Energy System Web Site [www.cristopia.com].

27. Ryu HW, Woo SW, Shin BC, Kim SD. Prevention of subcooling and stabilization of inorganic salt hydrates as latent heat storage materials. *Solar Energy Mater Solar Sells* 1992;27:161–72.

28. Lane GA. Phase change materials for energy storage nucleation to prevent subcooling. *Solar Energy Mater Solar Sells* 1991;27:135–60.

29. Zalba B, Marin J, Cabeza L, Mehling H. Review on thermal energy storage with phase change: materials, heat transfer analysis and applications. *Appl Therm Eng* 2003;23:251–83.

30. Kimura H, Kai J. Phase change stability of $CaCl_2 \cdot 6H_2O$. *Solar Energy* 1984;33:557–63.

31. Gibbs B, Hasnain S. DSC study of technical grade phase change heat storage materials for solar heating applications. In: *Proceedings of the 1995 ASME/JSME/JSEJ International Solar Energy Conference, Part 2*, New York, USA, 1995.

32. Porisini FC. Salt hydrates used for latent heat storage: corrosion of metals and reliability of thermal performance. *Solar Energy* 1988;41:193–7.

33. Cabeza L, Illa J, Roca J, Badia F, Mehling H, Hiebler S, et al. Immersion corrosion tests on melt-salt hydrate pairs used for latent heat storage in the 32–36°C temperature range. *Mater Corros* 2001;52:140–6.

34. Khudhair A, Farid M, Ozkan N, Chen J. Thermal performance and mechanical testing of gypsum wallboards with latent heat storage. In: *Proceedings of Annex 17, advanced thermal energy storage through phase change materials and chemical reactions––feasibility studies and demonstration projects* (www.fskab.com/annex17), Indore, India, 2003.

35. Dincer I, Rosen MA. *Thermal Energy Storage: System and Applications.* West Sussex: John Wiley and Sons; 2002.

36. Lane GA. Low temperature heat storage with phase change materials. *Int J Energy Res* 1980;5:155–60.

37. Velraj R, Seeniraj RV, Hafner B, Faber C, Schwarzer K. Heat transfer enhancement in a latent heat storage system. *Solar Energy* 1999;65:171–80.

38. Morcos VH. Investigation of a latent heat thermal energy storage system. *Solar Wind Technol* 1990;7:197–202.

39. Sadasuke I, Naokatsu M. Heat transfer enhancement by fins in latent heat thermal energy devices. *Solar Eng, ASME* 1991; 113:223–8.

40. Costa M, Buddhi D, Oliva A. Numerical simulation of a latent heat thermal energy storage system with enhanced heat conduction. *Energy Convers Mgmt* 1998;39:319–30.

41. Padmanabhan PV, Murthy MV. Outward phase change in a cylindrical annulus with axial fins on the inner tube. *Int J Heat Mass Transfer* 1986;29:1855–68.

42. Bauer C, Wirtz R. Thermal characteristics of a compact, passive thermal energy storage device. In: *Proceedings of the 2000 ASME IMECE*, Orlando, USA, 2000.

43. Kamimoto M, Abe Y, Kanari K, Swata S, Tani T, Ozawa T. Heat transfer in latent heat thermal storage units using pentarythritol slurry, thermal energy storage. *World Congress of Chemical Engineering*, Tokyo, Japan, 1986.

44. Mehling H, Hiebler S, Ziegler F. Latent heat storage using a PCM-graphite composite material: advantages and potential applications. In: *Proceedings of the 4th Workshop of IEA ECES IA Annex 10*, Bendiktbeuern, Germany, 1999.

45. Fukai J, Morozumi Y, Hamada Y, Miyatake O. Transient response of thermal energy storage unit using carbon fibers as thermal conductivity promoter. In: *Proceedings of the 3rd European Thermal Science Conference*, Pisa, Italy, 2000.

46. Py X, Olives S, Mauran S. Paraffin/porous graphite matrix composite as a high and constant power thermal storage material. *Int J Heat Mass Transfer* 2001;44:2727–37.

47. Lane GA. Encapsulation of heat of fusion storage materials. In: *Proceedings of 2nd Southeastern Conference on Application of Solar Energy*, Baton Rouge, Louisiana, USA, 1976. pp. 442–50.

48. Lane GA. Low temperature heat storage with phase change materials. *Int J Ambient Energy* 1980;1:155–68.

49. Stark P. PCM-impregnated polymer microcomposites for thermal energy storage. *SAE (Soc Automotive Eng) Trans* 1990;99:571–88.

50. Royon L, Guiffant G, Flaud P. Investigation of heat transfer in a polymeric phase change material for low level heat. *Energy Convers* 1997;38:517–24.

51. Hong Y, Xin-shi G. Preparation of polyethylene–paraffin compounds as a form-stable solid–liquid phase change material. *Solar Energy Mater Solar Sells* 2000;64:37–44.

52. Goswami DY, Jotshi CK. Thermal storage in ammonium alum/ammonium nitrate eutectic for solar space heating. In: *Proceedings of Solar Energy Annual Conference*, Minneapolis, Minnesota, July 15-20, USA, 1995. pp. 336–41.

53. Morrison DJ, Abdul Khalik SI. Effect of phase change energy storage on the performance of air-based and liquidbased solar heating systems. *Solar Energy* 1978;20:57–67.

54. Marshall R, Dietsche C. Comparisons of paraffin wax storage subsystem models using liquid heat transfer media. *Solar Energy* 1982;29:503–11.

55. Vakilaltojjar SM, Saman W. Analysis and modelling of a phase change storage system for air conditioning applications. *Appl Therm Eng* 2001;21:249–63.

56. Bailey JA, Mulligan JC, Liao CK, Guceri SI, Reddy MK. Research on solar energy storage subsystem utilizing the latent heat of phase change of paraffin hydrocarbons for the heating and cooling of buildings. Report to the National Science Foundation for Grant GI-4438, North Carolina State University, USA, 1976.

57. Dietz D. Thermal performance of a heat storage module using calcium chloride hexahydrate. *Trans ASME, J Solar Energy Eng* 1984:106–11.

58. Farid MM, Kim Y, Honda T, Kanzawa A. The role of natural convection during melting and solidification of PCM in a vertical cylinder. *Chem Eng Commun* 1989;84:43–60.

59. Lacroix M. Numerical simulation of a shell and tube latent heat thermal energy storage unit. *Solar Energy* 1993;50:357–67.

60. Abe Y, Takahashi Y, Kanari K, Kamimoto M, Sakamoto R, Ozawa T. Molten salt latent thermal energy storage for load following generation in nuclear power plants. In: *Proceedings of 21st Intersociety Energy Conversion Engineering Conference*, San Diego, California, USA, 1986. pp. 856–61.

61. Riahi M. Efficiency of heat storage in solar energy systems. *Energy Convers Mgmt* 1993;34:677–85.

62. Ismail KAR, Concalves MM. Thermal performance of a PCM storage unit. *Energy Convers Mgmt* 1999;40:115–38.

63. Kamimoto M, Tani T. Effect of conduction promoters and fins on heat transfer in latent storage unit. *Bul Electrotech Lab* 1980;44:278–368.

64. De Jong AG, Hoogendoorn CJ. Improved of heat transport in paraffin for latent heat storage systems. In: *Proceedings of TNO Symposium on Thermal Storage of Solar Energy*, Amsterdam, Holland, 1980. pp. 99–110.
65. Eftekhar J, Haji-Sheikh A, Lou DYS. Heat transfer in paraffin wax thermal storage system. *J Solar Energy Eng* 1984;106:299–306.
66. Ismail KAR, Alves CLF, Modesto MS. Numerical and experimental study on the solid-ification of PCM around a vertical axially finned isothermal cylinder. *Appl Therm Eng* 2001;21:53–77.
67. Wood RJ, Gladwell SD, Callahar PWO, Probert SD. Low temperature thermal energy storage using packed beds of encapsulated phase-change materials. In: *Proceedings of the International Conference on Energy Storage*, Brighton, UK, 1981. pp. 145–58.
68. Saitoh T, Hirose K. High-performance of phase change thermal energy storage using spherical capsules. *Chem Eng Commun* 1986;41:39–58.
69. Farid MM. Solar energy storage with phase change. *J Solar Energy Res* 1986;4:11–29.
70. Velraj R, Seeniraj RV, Hafner B, Faber C, Schwarzer K. Experimental analysis and numerical modeling of inward solidification on a finned vertical tube for a latent heat storage unit. *Solar Energy* 1997;60:281–90.
71. Abe Y, Takahashi Y, Sakamoto R, Kanari K, Kamimoto M, Ozawa T, et al. Charge and Discharge characteristics of a direct contact latent thermal energy storage unit using form-stable high-density polyethylene. *Trans ASME, J Solar Energy Eng* 1984;106:465–74.
72. Kamimoto M, Abe Y, Swata S, Tani T, Ozawa T. Latent thermal storage unit using form-stable high density polyethylene; Part I: Performance of the storage unit. *Trans ASME, J Solar Energy Eng* 1986;108:282–9.
73. Li W, Zhang D, Zhang T, Wang T, Ruan D, Xing D, et al. Study of solid–solid phase change of $(n$-$C_n H_{2n+1} NH_3)_2 MCL_4$ for thermal energy storage. *Thermochim Acta* 1999;326:183–6.
74. Etherington TL. A dynamic heat storage system. *Heat Piping Air-Cond* 1957;29(12):147–51.
75. Hallet J, Keyser G. Power characteristics of a continuous crystallization latent heat recovery system. ASME Publ 1979;21.
76. Stoval TK, Tomlinson JJ. What are the potential benefits of including latent storage in common wallboard? *Trans ASME* 1995;117:318–25.
77. Kedl RJ, Stovall TK. *Activities in support of the wax-impregnated wallboard concept: thermal energy storage researches activity review*. New Orleans, LA: U.S. Department of Energy; 1989.
78. Salyer IO, Sircar AK. Phase change materials for heating and cooling of residen-tial buildings and other applications. In: *Proceedings of 25th Intersociety Energy Conversion Engineering Conference*, Reno, Nevada, USA, 1990. pp. 236–43.
79. Neeper DA. Potential benefits of distributed PCM thermal storage. In: Coleman MJ, editor. *Proceedings of 14th National Passive Solar Conference*. Denver, Colorado, USA: American Solar Energy Society; 1989. pp. 283–8.
80. Neeper DA. Thermal dynamics of wallboard with latent heat storage. *Solar Energy* 2000;68:393–403.
81. Stetiu C, Feustel HE. Phase change wallboard as an alternative to compressor cooling in Californian residences. In: *Proceedings of 96 ACEE Summer Study for Energy Efficient Building*, California, USA, 1996.
82. Athienitis AK, Liu C, Hawes D, Banu D, Feldman D. Investigation of the thermal per-formance of a passive solar test-room with wall latent heat storage. *Building Environ* 1997;32:405–10.

83. Revankar S. Purdue University News Bulletin, 2001. Available from: <http://news.uns. purdue.edu/UNS/html4ever/010607.Revankar.solar.html>.
84. NREL Web Site [www.ctts.nrel.gov/bent].
85. John T, David E, David R. A novel ventilation system for reducing air conditioning in buildings: Testing and theoretical modeling. *Appl Therm Eng* 2000;20:1019–37.
86. Colvin DP, Bryant YG. Protective clothing containing encapsulated phase change materials. In: *Proceedings of the 1998 ASME International Mechanical Engineering Congress and Exposition*, Anaheim, CA, USA, 1998. pp. 123–32.

1.2 Fire Retardants for Phase Change Materials

Pongphat Sittisart and Mohammed Farid
The University of Auckland

1.2.1 INTRODUCTION

Phase change materials (PCMs) are used in a wide range of applications including latent heat thermal energy storage in buildings [1]. In order for the PCMs to be used in any application, they must be encapsulated to prevent them from leaking out, and hence, significant research has been done in the application of form-stable PCM, where PCM is contained within the structure of polyethylene or any other polymer. One of the most suited PCMs are paraffin, fatty acids and their esters because it has many desirable properties such as high latent heat of fusion, varied phase change temperature, negligible supercooling and lower vapour pressure in the melt, chemically inert and stable, self-nucleating, no phase segregation and commercial availability [1].

Research conducted by Farid et al. [2,3] has shown that the application of paraffin (RT20 or RT21) in building materials can provide thermal storage benefits. This is due to its desirable physical and thermal attributes including its suitable melting temperature of 20°C–22°C, which is close to human comfort temperature. Experimental results based on gypsum wallboard impregnated with PCM (PCMGW) showed that during summer, PCM can effectively reduce diurnal daily fluctuations of indoor air temperatures and maintain indoor temperature at the desired comfort level for a longer period of time. Major benefits of thermal energy storage in winter are capturing solar radiation and reducing electrical demand charges by limiting the need to run electricity for heating during peak load periods [2].

The main concern in using paraffin in building constructions is its flammability, as it can easily catch fire if not properly protected. Several methods have been attempted to reduce paraffin flammability. One method involves encapsulating it inside a composite building block and then placing the block inside a hollow container surrounded by non-combustible concrete [4]. Some have tried impregnating PCM into gypsum wallboard [5]. Others have encapsulated the PCM in polymers forming what is known as shape-stabilized PCM, which could be formed to any shape for use in buildings structure and with home and office furniture. However, both the polymer and PCM increase fire hazard and require the use of suitable fire retardants [6]. The main issues in the application and use of PCM in buildings are cost and leakage. Incorporating PCM into building material will incur additional expenses. However, previous studies [2,3] on the use of PCMs showed a good potential in energy saving that may compensate for the extra cost of these materials. There is always a possibility for leakage, which would render PCM unprotected during

24

an event of fire. PCM would then become an additional fuel to the fire. To avoid these problems, two things must be done. First, use a cheaper supporting material for PCM, and second, reduce the flammability of PCM. This can be achieved by adding suitable fire retardants with the PCM into HDPE. In addition to this, low-cost fire retardants should be used [6]. It must be noted that PCM could be impregnated in any building materials such as gypsum boards; however, the shape-stabilized PCM products have the benefit of being flexible and could be formed to any shape for use not only on building walls and ceiling but with home furniture. It is the cost of PCM that need to be introduced before they can be applied commercially in buildings.

There are approximately four types of fire retardants. The first type, and the most common, is known as flame quencher. Halogenated alkanes can limit or extinguish nearby flame source. The second type is called heat absorber. Materials such as magnesium hydroxide and aluminium hydroxide, which absorb heat from the surrounding and decompose endothermically preventing nearby material from heating up and hence minimizing its combustion. The third type, which is relatively new, is called intumescent fire retardant (IFR). Materials such as ammonium polyphosphate (APP) + pentaerythritol (PER) or expanded graphite are common examples. IFR works by creating a voluminous char layer which prevents the underlying material from further exposure to ignition source. The last type is known as synergist, and is of two types, one that is not a fire retardant but can work with other types of fire retardants to improve their fire retardancy. Antimony oxide is known as a synergist for halogenated alkanes. The second type (will be refer as synergist system) is where two or more fire retardants work together to improve overall fire retardancy of a material. Montmorillonite clay (MMT) is a fire retardant that works together with heat absorber and IFR [7–9]. It is also shown that APP can be used together with expanded graphite to improve fire retardancy [10,11].

Zhanga et al. [12] investigated the effects of adding different combination of nanoclay, magnesium hydroxide and aluminium trihydroxide to a polymer blend consisting of ethylene–vinyl acetate and low-density polyethylene. Cone calorimeter result has shown that the combination of polymer blend, nanoclay and aluminium tri-hydroxide had the lowest peak heat release and therefore experienced the best fire retardancy. In addition, Bellayer et al. [13] also investigated the mechanism of intumescence of a polyethylene/calcium carbonate/stearic acid system. Three different grades of chalk known as chalk 3, 6 and 30 were used. The results have shown that using chalk 3 yielded the best swelling effect, which leads to the highest reduction of peak heat release rate.

Thermal stability and flammability properties of fire-retarded form-stable PCM containing different types of fire retardants, such as those describe previously, were extensively studied by Cai et al. [1,4,7–11] using paraffin as PCM with a melting temperature of 54°C–56°C. Zhang et al. [14] studied the effects of EG on PCM and the effects of iron powder on IFR [15] using paraffin with melting temperature of 51°C. They concluded that adding EG has improved both fire retardancy and thermal conductivity while adding iron powder to IFR has also increased its fire retardancy as well. Both Cai et al. and Zhang et al. have succeeded in improving the fire resistance of paraffin with high melting temperatures (51°C and 54°C–56°C) but with no test appeared to be done on low melting temperature paraffin such as RT21, which is used in building application. Paraffin (RT21) poses higher flammable risk than those having higher melting temperature because of its higher vapour pressure.

This article is focused on how to reduce fire risk of a shape-stabilized PCM with a low melting point such as paraffin RT21, which has not been done before. Different fire retardants are tested. The effect of the best fire retardant will also be tested on a shape-stabilized ester for the first time.

1.2.2   EXPERIMENTS

1.2.2.1   MATERIALS

RT21 (latent heat of 136.3 J/g) was supplied by Rubitherm. Propyl ester (latent heat of 120 J/g) was manufactured in the Department of Chemical and Materials Engineering. HDPE was supplied by Dow Chemical. EG was provided by AllCell technologies, Chicago. Magnesium hydroxide was supplied by ECP Limited. Aluminium hydroxide and PER were supplied by Sigma Aldrich. MMT, which was treated with a surface modifier (dimethyl, di-hydrogenated tallow ammonium +1% silane) to improve its mix-ability during sample preparations, was supplied by Nanocor. APP was kindly provided by SpecialChem.

1.2.2.2   PREPARATION OF FIRE-RETARDED FORM-STABLE PCM

Encapsulating PCM in plastic, known as shape-stabilized PCM, is important product for use to cover building walls and ceiling to increase thermal mass of buildings. A desired quantity of RT, HDPE and fire retardants were mixed in Brabender Plastograph for approximately 10 min. Temperature and rotational speed were set at 150°C and 50 rpm, respectively. The ratio of APP: PER was 1:1 by weight. Detailed compositions are listed in Table 1.2.1. The resulting mixtures were grinded (for TGA, DSC) and pressed into rectangular bar (for vertical burning).

TABLE 1.2.1

Samples Identification and Classification

Samples	Compositions
PCM1	60 wt.% RT21 + 40 wt.% HDPE
PCM2	60 wt.% RT21 + 20 wt.% HDPE + 20 wt.% magnesium hydroxide
PCM3	60 wt.% RT21 + 15 wt.% HDPE + 20 wt.% magnesium hydroxide + 5 wt.% MMT
PCM4	60 wt.% RT21 + 20 wt.% HDPE + 20 wt.% aluminium hydroxide
PCM5	60 wt.% RT21 + 15 wt.% HDPE + 20 wt.% aluminium hydroxide + 5 wt.% MMT
PCM6	60 wt.% RT21 + 20 wt.% HDPE + 20 wt.% EG
PCM7	60 wt.% RT21 + 15 wt.% HDPE + 20 wt.% EG + 5 wt.% MMT
PCM8	60 wt.% RT21 + 20 wt.% HDPE + 20 wt.% (APP + PER)
PCM9	60 wt.% RT21 + 15 wt.% HDPE + 20 wt.% (APP + PER) + 5 wt.% MMT
PCM10	60 wt.% RT21 + 20 wt.% HDPE + 10 wt.% APP + 10 wt.% EG

1.2.2.3 TEST FOR FIRE RETARDANCY

1.2.2.3.1 Vertical Burning Test

The vertical burning test was carried out inside a fume hood. Samples were held vertically with tongs at one end and burned from the free end. Samples were exposed to ignition source for 10 s, and then they were allowed to burn above a cotton wool until both sample and cotton wool were extinguished [16]. Observable parameters were recorded to assess fire retardancy.

1.2.2.3.2 Thermal Stability Test

Thermogravimetry analysis was carried out using TGA50 by heating the samples from 25°C to 690°C with a linear heating rate of 10°C/min under inert gas flow. Nitrogen as an inert gas flow was set to approximately 80 ml/min and sample mass in alumina pan was measured with an initial mass of 5–10 mg [1,6,7,9,10]. The results were used to assess thermal stability of samples.

1.2.2.3.3 Cone Calorimeter Test

In addition, cone calorimeter test was carried out under a heat flux of 35 kW/m^2. Samples with approximate dimensions of 6 cm ×6 cm ×2 cm were ignited by an electric spark under a selected heat flux. Combustible gas emitted during burning was collected for analysis [17]. The test was performed in the Plastic Centre of the University of Auckland, New Zealand.

1.2.2.3.4 DSC Test

Differential scanning calorimeter (DSC) was carried out using DSC50. The sample was heated from −20°C to 40°C with a linear heating rate of 3°C/min. This was done to determine whether the addition of fire retardants affects the physical properties of the PCM such as its latent heat and its melting range.

1.2.3 RESULTS AND DISCUSSION

1.2.3.1 FIRE RETARDANCY AND FIRE SPREAD USING VERTICAL BURNING TEST

The results in Table 1.2.2 clearly show that heat absorbers such as aluminium hydroxide and magnesium hydroxide had little effect in improving fire retardancy of the composite. Addition of heat absorber reduced dripping, but a very small residue was obtained after burning. However, when MMT was added together with heat absorber, the size of residue increased, material self-extinguished and there was no dripping. This showed that there is a synergist effect between heat absorber and MMT, which contributed to an improvement in fire retardancy of the samples.

Cai et al. [7] used paraffin with a melting temperature of 54°C–56°C. Organophillic montmorillonite (OMT) was made by treating pristine montmorillonite by ion exchange reaction using hexadecyl trimethyl ammonium bromide (C16) in water. This treatment was done in order to increase the miscibility between clay and polymer matrix, but the MMT used was treated differently. Cai et al. [8] used OMT together with microencapsulated red phosphorus (MRP) and magnesium hydroxide (heat

TABLE 1.2.2
Vertical Burning Test Results and Comments

	Dripping	Cotton Ignite	Cotton Residue	Sample Residue	Self-Extinguish	Comment	Fire Retardancy
PCM1	Yes	Yes	None	None	No	Significant dripping observed	Very poor
PCM2	Yes	Yes	Small	Very small	No	Three to four drips observed	Poor
PCM3	No	Yes	Medium	Small	Yes	Material broke apart due to decrease in mechanical property	Good
PCM4	Yes, material drips during ignition.	Yes	Small	Very small	No	Three to four drips observed	Poor
PCM5	No	Yes	Medium	Small	Yes	Material broke apart due to decrease in mechanical property	Good
PCM6	Small burning ash	Yes	Large	Large	Yes	Small ash drops often extinguish quickly	Good
PCM7	Small burning ash	Yes	Large	Large	Yes	Small ash drops often extinguish quickly	Good
PCM8	Yes, material drips during ignition	Yes	Medium	Medium	Yes	A lot of dripping. Char formation	Good
PCM9	Yes/No[a]	Yes/No[a]	Large	Large	Yes	Few drips, sometimes none. Char formation	Very good
PCM10	No	Yes	Very large	Very large	Yes	Material broke apart due to decrease in mechanical property	Very good

[a] Cotton did not ignite when sample did not drip.

absorber). They indicated that the results from both TGA and cone calorimeter have shown a significant improvement in thermal stability and fire retardancy of the samples, which suggested that there is a synergist effect between MRP, MH and OMT [8].

On the other hand, IFR showed a better improvement compared to heat absorber. Starting with EG (refer to PCM6 in Table 1.2.2), a small burning ashes were observed but distinguished quickly. This means that fire can spread quickly if ignition source is near, otherwise it may extinguish before it reaches an ignition source. The tested sample had large residues and can self-extinguish indicating that it has high resistance to fire. The addition of MMT (refer to PCM7 in Table 1.2.2) did not have any significant effect.

Using APP + PER (refer to PCM8 in Table 1.2.2) in intumescent system, significant dripping was observed, but the material can self-extinguish. The material formed an insulating char layer protecting the underlying material from burning, which results in medium size residues. An addition of MMT (refer to PCM9 in Table 1.2.2) to APP + PER system indicates a significant improvement in fire retardancy. The material had only few drips, sometimes none and it can also self-extinguish. There is a synergist effect between MMT and APP + PER, which results in a formation of larger char layer, as reported by Cai et al. [8].

The addition of OMT together with IFR like melamine polyphosphate (MPP) + PER [6] and APP + PER [9] has increased fire retardancy of material by increasing the amount of residue [7,9].

PCM10 (refer to Table 1.2.2) contained a synergist system of APP + EG. The material does not drip, can self-extinguish and has a very large residues. This indicates that APP + EG system has higher fire resistance than normal intumescent system like EG only (PCM8).

Cai et al. have tested EG by itself and have concluded that it was not as good as using EG with APP [11]. It was suggested that there is a synergist effect between EG and APP (refer to Section 1.2.2.3.4), resulting in an increased in residues and also a decreased in peak heat release rate which improved the overall fire retardancy of the material [1,10].

1.2.3.2 THERMAL STABILITY OF FIRE-RETARDED FORM-STABLE PCM

Figure 1.2.1 shows the TGA analysis for the shape-stabilized PCM containing no fire retardant. The figure shows three periods of mass loss. The first period is between 25°C and 220°C, the second period is between 220°C and 510°C and the last period is between 520°C and 590°C. It can be seen from Figure 1.2.1 that nothing happens until the temperature reaches about 100°C, where RT21 starts to decompose or evaporate until none remained at 220°C (60 wt.% loss). In the second period, no mass loss is observed until the temperature reaches 450°C where HDPE starts to decompose. The figure shows that HDPE is fully decomposed at 520°C leaving no residues.

Figure 1.2.2 shows that the addition of fire retardant to PCM produces similar TGA. It can be observed that up until 100°C no mass loss occurs, and then PCM starts to leave the sample until no PCM remained when the temperature reaches approximately 220°C. It is only after that where the effect of fire retardant could be observed. At 220°C, 60% weight loss indicates complete loss of RT21 with no loss of HDPE while 70% weight loss means 10% of HDPE is lost. With RT21, PCM2

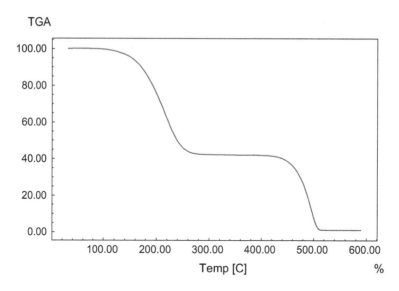

FIGURE 1.2.1   TGA result showed that PCM1 decompose according to their constituent (RT21 and HDPE decompose at 100°C and 450°C, respectively). There are no residues.

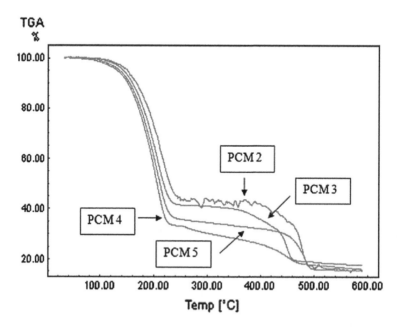

FIGURE 1.2.2   TGA graph of form-stable PCM containing heat absorber (with and without MMT).

(containing magnesium hydroxide) showed approximately 55% weight loss while PCM4 (aluminium hydroxide) showed 67% weight loss at 220°C indicating magnesium hydroxide is a better fire retardant than aluminium hydroxide. The additions of fire retardants resulted in approximately 18% residues at the end of the test, demonstrating that the material became more thermally stable.

Figure 1.2.2 also shows that an addition of MMT together with aluminium hydroxide (PCM5) slightly decreased the weight loss at 220°C from 67% to 65%. However, an opposite effect can be seen when MMT was added to magnesium hydroxide (PCM3). Adding MMT slightly increased the weight loss at 220°C from 55% to 58%. This suggested that MMT should only be used together with aluminium hydroxide.

It can be seen from Figure 1.2.3 that the weight loss at 220°C for PCM6, PCM7, PCM10, PCM8 and PCM9 were 48%, 55%, 59%, 65% and 68%, respectively. The weight loss of PCM7 (contained EG and MMT) suggests that MMT should not be added together with EG as it increased the weight loss of PCM at 220°C. PCM8 (10% APP, 10% PER) had higher weight loss than PCM10 (10% APP, 10% EG) indicating that the combination of APP+EG gives better fire retardancy. The weight loss of PCM9 suggests once again that MMT should not be used together with APP+PER. These results have shown that PCM6 had the best improvement in fire retardancy. However, as PCM6 had 10% more EG than PCM8, PCM9 and PCM10, accurate comparison cannot be made. Further investigation is necessary.

The graph also shows that PCM6 and PCM7 (containing EG only and EG+MMT, respectively) produced the highest residues (high thermal stability), while PCM8 and PCM9 (containing APP+PER and APP+PER+MMT) produced the smallest residues (low thermal stability). Addition of MMT did not show an improvement in

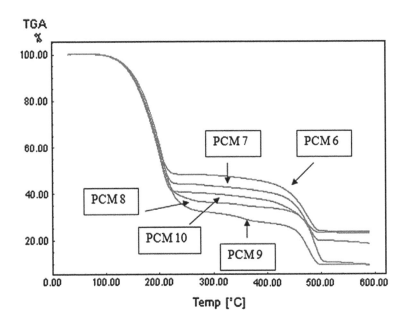

FIGURE 1.2.3   TGA graph of form-stable PCM containing IFR (with and without MMT).

thermal stability of the samples and did not increase the amount of residues formed. PCM6 and PCM7 produced about 23% residues; PCM10 produced 20% residues, while PCM8 and PCM9 produced only 10% residues.

1.2.3.3 FLAMMABILITY OF FORM-STABLE PCM

Heat release rate, in particular peak heat release rate (PHRR), has been found to be the most important parameter in the evaluation of fire safety [6,7]. Lower PHRR indicates that the material is more fire safe. Since the vertical burning test described in Section 1.2.3.1 showed that PCM10 has the best fire retardancy and due to the high cost of associated with the cone calorimeter test, the PHRR testing was conducted to PCM10 and PCM1 containing no fire retardant. It can be seen from Figure 1.2.4 that PHRR of PCM1 and PCM10 (containing APP and expanded graphite) are 1507 kW/m², and 1107 kW/m², respectively. Figure 1.2.4 shows that PCM1 contains two peaks, the reason for this could be the difference in the heat of gasification of paraffin and HDPE, and the two peaks correspond to the flammability of the paraffin and HDPE, respectively [12].

Figure 1.2.4 also shows that the addition of fire retardant (EG + APP) has reduced first peak from 1500 kW/m² to approximately 1100 kW/m² with the second peak

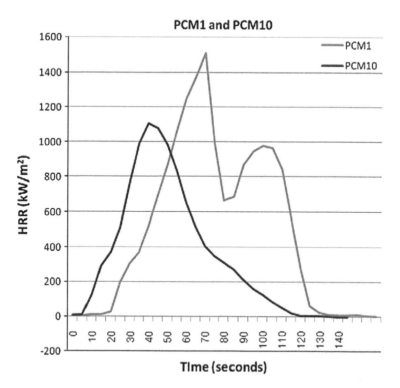

FIGURE 1.2.4   Cone calorimeter of PCM1 and PCM10. It can be seen that PCM1 has two peaks while PCM10 only has one peak. The addition of EG and APP removed the second peak as well as reduced the peak HRR from approximately 1500 to 1100 kW/m².

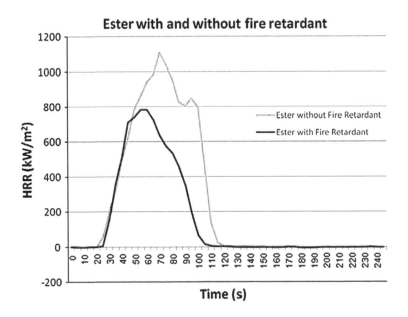

FIGURE 1.2.5 Cone calorimeter of ester/HDPE with and without fire retardant. It can be seen that peak HRR reduced from 1100 to 800 kW/m² due to an addition of fire retardant.

became smoother. This is due to the formation of char layer which protect the underlying material from further combustion. This clearly shows that an addition of fire retardant improved fire safety of PCM.

Esters could also be used as PCM since they have good latent heat of melting within a suitable melting range. Figure 1.2.5 shows that heat flux generated from tallow propyl ester/HDPE composite (peak = 1100 kW/m²) is lower than the corresponding one for paraffin (peak = 1500 kW/m²) shown in Figure 1.2.4, which is expected due to the lower vapour pressure of the ester. Figure 1.2.5 also shows how fire retardant could lower fire risk in the ester/HDPE sample (reduce the peak heat flux from 1100 to 800 kW/m²). The fire retardant used was the same one used in PCM10. To our knowledge no one has tested fire hazard of shape-stabilized PCM esters.

1.2.3.4 Latent Heat of Fire-Retarded Form-Stable PCM

The latent heat of RT21 is 136.3 J/g [18]. DSC tests were conducted to all samples, but only two are presented since the addition of fire retardant shows no effect other than the reduction in the latent heat due to the lower mass fraction of the PCM in the composites. The DSC measurements shown in Figures 1.2.4 and 1.2.5 are for RT21 in PCM9 and PCM10. When the measured latent heat is corrected based on the mass fraction of the PCM in the samples, its value was reduced from 136.3 J/g to 125.3 J/g and 127.28 J/g, respectively. This is very small reduction considering the common error involved in DSC measurements. Other samples behaved the same. Cai et al. [1] suggested that the molecular movement of paraffin with a melting temperature of 54°C–56°C was restricted by the three-dimensional network of form-stable PCM, therefore decreasing the latent heat. However, this difference is within the error associated with DSC measurements, and hence, a strong conclusion cannot be drawn.

1.2.4 CONCLUSIONS

Fire-retarded form-stable PCM was prepared using the Brabender Plastograph. The material used were based on RT21, HDPE and different fire retardants such as aluminium hydroxide, magnesium hydroxide, EG, APP, PER, MMT and IFR. Vertical burning test has shown that IFR is a better fire retardant for the PCM composite. Results indicated that PCM composite which contained IFR has better fire resistance than that containing heat absorber. TGA results showed that adding fire retardants to PCM composite has increased the amount of residues, which contributed to an improvement in thermal stability. Cone calorimeter test indicated that an addition of fire retardant had improved fire safety of PCM by reducing peak HRR. DSC results have shown that an addition of fire retardant had little effect on the latent heat of PCM. It can be concluded that fire retardants used in PCM10 is the best. It showed a "very good" rating from the vertical burning test (one of the best two), 20% residues (second best) from TGA and a clear reduction in peak HRR. Cone calorimeter tests carried out on shape-stabilized tallow propyl ester showed lower heat peak indicating that esters are safer PCMs. Finally, it can be concluded that the incorporation of fire retardant into PCM reduced its flammability (as seen from the reduction of peak HRR). In an event of fire, as the combustion of PCM releases less heat, other building materials would combust less readily therefore this would increase evacuation time.

ACKNOWLEDGEMENT

This research was supported by the University of Auckland and the Foundation for Research Science and Technology of New Zealand for supporting this project.

GLOSSARY

TGA: thermogravimetric analysis
DSC: differential scanning calorimeter
HDPE: high-density polyethylene
APP: ammonium polyphosphate
PER: pentaerythritol
EG: expanded graphite
MMT: montmorillonite
OMT: organophillic montmorillonite PCM: phase change material
PCMGW: gypsum wallboard impregnated with PCM
RT21: paraffin with melting temperature of 20°C–21°C
RPM: revolution per minute
IFR: intumescent fire retardant
PHRR: peak heat release rate

REFERENCES

1. Cai YB, Wei Q, Huang F, Gao W. Preparation and properties studies of halogenfree flame retardant form-stable phase change materials based on paraffin/ high density polyethylene composites. *Appl Energy* 2008;85:765–75.
2. Farid MM. Energy storage for efficient use in domestic and industrial applications. In: *Keynote in the International Conference on Renewable Energy: Generation and Applications (ICREGA'2010)*, March 8–10; 2010, UAE.
3. Farid MM, Khudhair AM, Razak SA, Al-Hallaj SA. Review on phase change energy storage: materials and applications. *Energy Convers Manage* 2004;45:1597–615.
4. Salyer IO. Building products incorporating phase change materials and method of making the same. 26 May 1998, patent no. 5755216.
5. Banu D, Feldman D, Haghighat F, Paris J, Hawes D. Energy storing wallboard: flammability tests. *J Mater Civil Eng* 1998;10:98–105.
6. Cai YB, Hu Y, Song L, Tang Y, Yang R, Zhang YP, et al. Flammability and thermal properties of high density polyethylene/paraffin hybrid as a form-stable phase change material. *J Appl Polym Sci* 2005;99:1320–7.
7. Cai YB, Hu Y, Song L, Kong QH, Yang R, Zhang YP, et al. Preparation and flammability of high density polyethylene/paraffin/organophilic montmorillonite hybrids as a form stable phase change material. *Energy Convers Manage* 2007;48:462–9.
8. Cai YB, Wei QF, Shao DF, Hu Y, Song L, Gao WD. Magnesium hydroxide and microencapsulated red phosphorus synergistic flame retardant form stable phase change materials based on HDPE/EVA/OMT nanocomposites/paraffin compounds. *J Energy Inst* 2009;82:28–36.
9. Cai YB, Hua Y, Song L, Lu HD, Chen ZY, Fan WC. Preparation and characterizations of HDPE–EVA alloy/OMT nanocomposites/paraffin compounds as a shape stabilized phase change thermal energy storage material. *Thermochim Acta* 2006;451:44–51.
10. Cai YB, Wei QF, Huang FL, Lin SL, Chen F, Gao WD. Thermal stability, latent heat and flame retardant properties of the thermal energy storage phase change materials based on paraffin/high density polyethylene composites. *Renew Energy* 2009;34:2117–23.
11. Cai YB, Song L, He QL, Yang DD, Hu Y. Preparation, thermal and flammability properties of a novel form-stable phase change materials based on high density polyethylene/ poly (ethylene-co-vinyl acetate)/organophilic montmorillonite nanocomposites/paraffin compounds. *Energy Convers Manage* 2008;49:2055–62.
12. Zhanga J, Hereida J, Hagena M, Bakirtzisa D, Delichatsiosa MA, Finab A, et al. Effects of nanoclay and fire retardants on fire retardancy of a polymer blend of EVA and LDPE. *Fire Saf J* 2009;44(4):504–13.
13. Bellayera S, Tavardb E, Duquesnea S, Piechaczykb A, Bourbigota S. Mechanism of intumescence of a polyethylene/calcium carbonate/stearic acid system. *Polym Degrad Stab* 2009;94(5):797–803.
14. Zhang P, Hu Y, Song L, Ni JX, Xing WY, Wang J. Effect of expanded graphite on properties of high-density polyethylene/paraffin composite with intumescent flame retardant as a shape-stabilized phase change material. *Sol Energy Mater Sol Cells* 2010;94:360–5.
15. Zhang P, Hu Y, Song L, Lu HD, Wang J, Liu QQ. Synergistic effect of iron and intumescent flame retardant on shape-stabilized phase change material. *Thermochim Acta* 2009;487:74–9.
16. Chomerics premier conductive plastic. Test report. 10 February 2006. <http://vendor. parker.com/Groups/Seal/Divisions/Chomerics/Chomerics%20Product% 20Library.nsf/05eb61a92fc220c78525695800744d42/455d5cf9ec73052f8525712 b006f87cd/$FILE/TR%201007%20EN%200206%20NEBS.pdf.>

17. SINTEF. ISO 5660-1:1993: the cone calorimeter test. Published 20 April 2005. <http://www.sintef.no/Home/Building-and-Infrastructure/SINTEF-NBL-as/Key-projects-and-topics/Fire-testing-of-railway-seats/ISO-5660–11993-THECONE-CALORIMETER-TEST/>.
18. Fang XM, Zhang ZG. A novel montmorillonite-based composite phase change material and its applications in thermal storage building materials. *Energy Build* 2006;38:377–80.

1.3 Long-Term Thermal Stability of Organic PCMs

Sam Behzadi and Mohammed Farid
The University of Auckland

1.3.1 INTRODUCTION

Maintaining sustainable supply of energy and the increasing emphasis towards the use of renewable energy are challenges that are being faced worldwide. Space heating and cooling of buildings are major contributors to total energy used globally, equating to more than 30% energy consumption in residential buildings. Therefore, priorities should be given to strategies that enhance the thermal performance of buildings to reduce the amount of energy consumed. Most buildings in the developed world use conventional building materials: bricks and mortar, concrete, steel or wood framing for structural components. Insulation must be used to prevent heat loss and gains; however, thermal mass of buildings can vary significantly. For example, a concrete wall has much higher thermal mass than a framed wood wall. In buildings, thermal mass reduces temperature swings, which increases occupant's comfort level and reduces the amount of energy needed to regulate the desired indoor temperature [1]. To create lightweight construction, most buildings are fitted with fibreglass insulation and finished with gypsum wallboards.

For more than 30 years, researchers have investigated ways to efficiently increase the thermal mass of building by incorporating phase change materials (PCMs) into the building materials. The latent heat required to cause the material to change phase is much larger than the sensible heat of all building materials. For example, the heat capacity of wallboard with 30% PCM by weight is about five times greater than the heat uptake of conventional wallboard for small temperature changes [1,2]. Consequently, the use of PCM increases the thermal mass of the building when temperatures change within the transition temperature of PCM [1,3–7]. This not only addresses the problem of intermittent and variable character of solar energy but also lowers indoor air temperature fluctuations and improves comfort. Storage using latent heat represents an attractive option for thermal energy storage. A wide range of PCMs have been investigated and used, including paraffin wax, salt hydrates and non-paraffin organic compounds. The economic feasibility of employing latent heat storage materials in a system depends, among many other factors, on the life of the storage material, i.e. there should not be major changes in the melting temperature and latent heat of fusion with time due to thermal cycling of the storage material [4,8–13]. Most PCMs suitable for use in buildings typically change phase from solid to liquid at a temperature within human comfort range of 21°C–25°C. Ideally, phase change should occur at a precise temperature, but in practice, most PCMs change phase over a range of temperatures. For organic PCMs, the range of the transition temperature depends on its purity. High-purity PCMs have narrow melting and high latent heat but very expensive for use in commercial building. Most PCMs can experience some degree of subcooling. This can be an issue in some process, and accommodations need to be made to account for this.

For latent heat storage, commercial-grade organic PCMs are preferred due to their low cost, better stability, reduced supercooling and large-scale availability [2,4,-8–13]. The thermophysical properties of commercial organic PCMs are found to be much different than those quoted in the literature for laboratory-grade PCMs. Matching of the transition temperature range of the PCM to the delivered energy temperature for a given application is the most important aspect of PCM energy storage system. Eliminating the problems of supercooling, phase separation and stability over a long period of the application are also important for the successful application of the PCM. However, to achieve this ideal situation, a stable PCM needs to be used. A comprehensive knowledge of the thermal reliability of the PCMs as a function of repeated thermal cycles is essential for assurance to long-term performance and good economics. Recent reviews on the topic suggest that renewable fatty acids and their esters are suitable PCM candidates. These materials as opposed to paraffin's have much lower vapour pressure allowing them to be more thermally stable and less volatile as identified by Sittisart and Farid [22]. More importantly, these materials have lower environmental impact and are less fire hazardous [3,8,14–22].

Even though PCMs have been applied in numerous applications since many years, their thermal stability has not been well assessed [2,23]. The available literature shows insufficient information about the long-term thermal performance of the materials especially when used in buildings. The utilisation of PCMs in buildings increases the thermal mass and potentially improves energy efficiency of buildings [24]. However, for PCM to be used, it needs to be contained within a medium for practical applications. Most of the research to date has utilised wallboards as a means of containing PCM for thermal energy storage, since they are cheap, widely used and compatible with most PCMs [4,25,26]. The knowledge of the thermophysical properties of materials used for latent heat storage is essential as it dictates how they can be utilised in specific applications. It is the objective of this article to discuss the change in the melting temperatures and latent heat of fusion of various selected PCMs when stored at temperatures well above their melting point. This is an area that has been previously overlooked and requires further research.

1.3.2 MATERIALS AND METHOD

1.3.2.1 EXPERIMENTAL PROCEDURE

1.3.2.1.1 Oven Trials

The thermal characteristic (i.e. melting point and latent heat of fusion) of RT21 and propyl ester mixtures of stearic and palmitate (80:20) with melting point in the range of 18°C–25°C were tested after an exposure to temperatures of 30°C, 35°C and 55°C [3,7,14–16,27]. These temperatures were chosen to ensure PCM remained in the liquid phase while being tested. Furthermore, this article highlights the potential fire and health hazards caused by PCM migration to indoor environment. The long-term

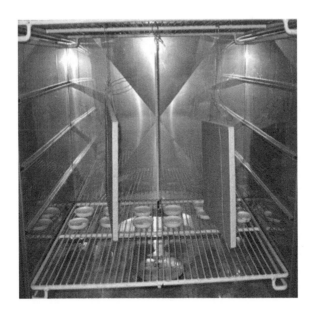

FIGURE 1.3.1A PCM samples inside the temperature-controlled oven. (Note: every 10 days, PCM samples were removed, weighed and analysed using a DSC).

thermal and physical stability of the selected PCM was initially evaluated by placing 8–10 g of each PCM sample in ceramic containers (5 cm in diameter and 1.5 cm in height) inside a temperature control oven and monitored for weight loss (refer to Figure 1.3.1a). To eliminate human error and improve experimental accuracy, two identical samples of the mentioned PCM were placed in the oven for a period of 120–200 days and were monitored periodically for physical changes. In addition, samples of similar weights of octadecane and hexadecane were also placed in the oven and used as control samples.

In addition, two gypsum panels (i.e. GIB board), impregnated by direct immersion either with RT21 or with ester, were also tested for weight loss in the oven. As indicated previously, majority of the research to date uses gypsum panel as a storage medium for storing PCM since they are more exposed to internal temperature changes. Therefore, it was crucial to ascertain their ability as a primary storage for PCM. The most flexible process whereby the PCM can be incorporated into gypsum wallboards is the immersion process, which does not interfere with the manufacturing processes of the gypsum wallboards. It involved dipping the boards in molten PCM. The process was easy to control and simple to setup. Impregnation of the gypsum wallboard with 24%–28% by weight of RT21 or ester mix was achieved by immersing ordinary gypsum wallboard for 60 min in a bath that was filled with molten PCM at 70°C–80°C [28]. These PCMs were purchased from Sigma and Rubitherm GmbH.

1.3.2.1.2 Naturally Exposed

Furthermore, a gypsum board (300 mm × 370 mm × 13 mm) impregnated with 27 wt% of RT21 was prepared and installed in testing facility at the University

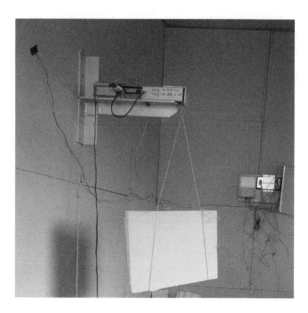

FIGURE 1.3.1B PCM-impregnated gypsum sample inside the temperature-controlled Huts at the University of Auckland. (Note: weight changes were monitored continuously through a digital strain gauge).

of Auckland Tamaki Campus for a real-life weight loss analysis (Figure 1.3.1b). These are full-scale size facility consisting of two identical-shaped test offices and a computer office built at the Tamaki Campus of the University of Auckland (New Zealand) with a view to conduct long-term thermal performance involving monitoring and modelling work. The facilities known as cabins are a lightweight construction built according to New Zealand building code. Each cabin in the test facility is a single-storey design of a typical lightweight construction. Each measures $2.6 \times 2.6 \times 2.6$ m giving floor area of 5.76 m^2 each. Their wooden frames were made of 9.8×6.3 cm dressed pine timber. The interior coverings were sets of either gypsum wallboards panels or gypsum panel impregnated with RT21 mounted on the wooden frame. The exterior walls were 1.25 cm thick sheets of plywood. The wall cavities were filled with fibreglass thermal insulations. The insulation is installed with no gaps and no folds so as to achieve high thermal resistance to heat flow. The thickness of the insulation is 9.4 cm for both the walls and ceilings. The test facility faces north to maximise sun exposure, as this is a key aspect of energy-efficient building design. Each office was supplied with one window on the north side. The test cabin was not equipped with any outside shading or overhangs with a view of making the conditions as severe as possible. The test building was situated in a large open area with a 4.2 m distance between the neighboring test cabins and was free from any shading. The research units were also equipped with a heat pump that used during winter and summer period for heating and cooling. These facilities are also equipped with the appropriate temperature measuring device to monitor the internal and external temperature changes [28].

The amount of impregnation was ascertain by Khudhair and Farid [28] whom determined the optimum amount to be 24–28 wt% PCM. The panel was connected to a digital scale and allowed to hang freely while it was monitored for weight loss. This paraffinic PCM has a melting range of 18°C–23°C and a latent heat of fusion of 134 kJ/kg⁻¹. These cabins are also equipped with identical heat pumps that allow the cabins to be cooled to 23°C during summer and heated to 22°C during the winter periods. For further detail on these test facilities, refer to previous publication from this research group [28].

1.3.2.1.3   Cyclic Trials

In addition, cyclic experiments were performed in a controlled heating/cooling water bath for the mentioned PCMs at temperatures between 10 and 40°C. These experiments were conducted by using 20 g of PCM in a 50 ml sealed glass containers with type K thermocouple immersed in the PCM and connected to a data logger and PC to measure the PCM temperature during cycling.

1.3.2.2   Analytical Method

Thermal analysis of the PCM samples was conducted using a differential scanning calorimetry (Shimadzu DSC-60). In the DSC testing, the sample mass was limited to 5.5–6.5 mg to minimise the error caused by the temperature distribution within the sample. As identified by previous work, the heating rate was limited to less than 3°C/min. The DSC was calibrated using pure n-C18 standard of a known latent heat of 245 J/g and a melting point of 29°C.

1.3.2.3   Materials

In this project, RT21 was purchased from Rubitherm Technologies GmbH. The propyl esters were manufactured in the laboratory based on the esterification procedures developed at the University of Auckland using stearic and palmitic acid purchased from Sigma chemicals. Similarly propanol, and other reagents such as octadecane and hexadecane were purchased from Sigma Chemicals.

1.3.3   RESULTS AND DISCUSSION

The focus of the current research was to examine the long-term thermal characteristics of PCM used in building envelope such as wallboards. Many substances have been studied as potential PCM, but only a few of them are commercialised [27]. Most of the commercially available PCM offered by companies, such as BASF, DuPont, Rubitherm GmbH, Microtek and PureTemp, are a mixture of organic paraffin or renewable fatty acid and esters. These materials are vulnerable to natural degradation and evaporation with time if left exposed. The associated chemical changes have potential application issues and unwanted health risks. Hence, this study examined the long-term performance of RT21 a commercial paraffin mixture and propyl ester mix consisting of propyl stearate propyl palmitate. These mixtures were subjected to constant temperatures above their boiling point, and their thermal and physical

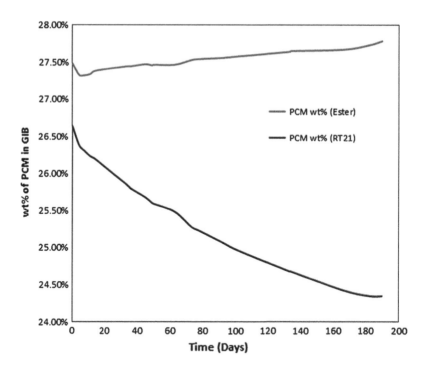

FIGURE 1.3.2 Difference in weight loss for two gypsum samples impregnated with RT21 and ester mix at 30°C inside the oven.

changes are examined. Figure 1.3.2 illustrates the physical changes associated with the two gypsum panels impregnated with the mentioned PCM mixtures at a fixed temperature of 30°C. During the experimental period, RT21 experienced an average weight loss of 10%, while the propyl ester mix experienced a slight weight gain which is believed to be caused by moister absorption. As described by Oliveira et al. [29], alcohol esters at storage temperatures of 30°C–50°C can potentially absorb 3–5 g of water per kg of alcohol ester.

The results shown in Figure 1.3.2 were used to calculate the rate of mass loss from both samples and presented in Figure 1.3.3. The rate of mass loss from ester is almost zero, and at times it appears that samples gained weight as indicated by the negative zone on graph 3. The gains are associated by changes in relative humidity. On the other hand, the paraffin samples showed a constant weight loss.

As indicated in both Figures 1.3.2 and 1.3.3, initially the samples experienced a high mass loss before gradually slowing down. It should be noted that RT21 continued to lose weight after 200 days of storage at 30°C.

As stated earlier, the experimental cabins at the University of Auckland's research facility had its internal wallboard impregnated with RT21, and it was decided to use this to monitor the PCM weight loss. Therefore, a freshly impregnated panel was placed inside the test cabin, and its weight was monitored using a digital scale. As illustrated in Figure 1.3.4, the panel experienced a 13 wt% weight loss over a period of 20 months. The weight loss was less than the corresponding oven trials due to the

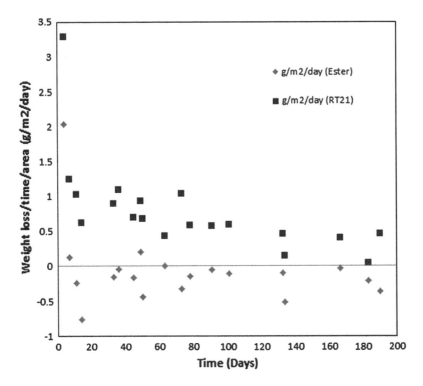

FIGURE 1.3.3 Weight loss per unit area time for the for two gypsum samples impregnated with RT21 and ester mix at 30°C inside the oven.

temperatures inside the test cabin being lower. However, at present, more than 30% of the impregnated PCM would have evaporated over the initial 5 years, thus changing the thermophysical characteristics of the PCM, potentially increasing its melting point, and making it inadequate for human comfort temperature.

As stated earlier, the PCM samples in the oven were analysed using a differential scanning calorimeter (DSC). As illustrated in Figure 1.3.5, during the experiment majority of the lighter hydrocarbons that make up the RT21 mix must have evaporated leaving behind a heavier fraction with higher latent heat of fusion and higher melting point. As evident by the DSC test results obtained, there was no change in the ester mix at 30°C. The average latent heat of fusion for ester remained at 110 J/g, while the RT21 changed from 135 to 165 J/g. Similarly, there was no change in the peak melting point of the ester, whereas the RT21 peak melting point changed from 21°C to 28°C.

The previously mentioned experiments at 30°C were subsequently performed at higher temperatures to establish the maximum potential weight loss from RT21 and the propyl ester mix. As illustrated in Figures 1.3.6 and 1.3.7, there was no evidence of weight loss from the ester mix at the selected experimental temperatures, while the RT21 paraffin mix experienced a significant weight loss. While inside the oven at 55°C, the RT21 experienced an average weight loss of 15%–40%. It is expected that during this period, most of the lighter hydrocarbons that make up the RT21 mix

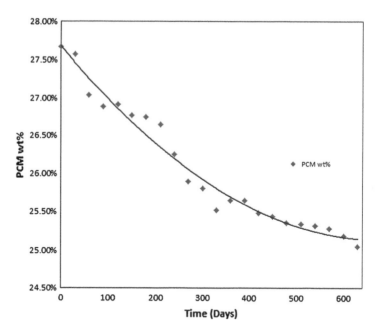

FIGURE 1.3.4 Average weight loss for gypsum sample impregnated with RT21 inside the experimental cabins at University of Auckland, Tamaki Campus.

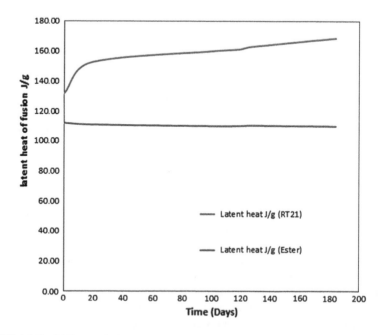

FIGURE 1.3.5 Difference in latent heat of fusion for RT21 and ester mix at 30°C for 185 days inside the oven.

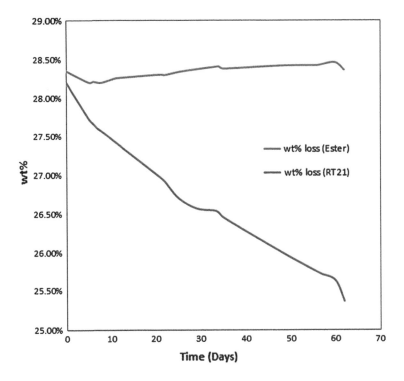

FIGURE 1.3.6 Difference in weight loss for two GIB samples impregnated with RT21 and ester mix at 35°C inside the oven.

have evaporated leaving behind a hydrocarbon fraction containing compounds such as octadecane with a peak melting point of 29°C. Therefore, RT21 and other light paraffin mix are not suitable for direct impregnation as they are vulnerable to mass loss, which results in irreversible changes making them inadequate as heat storage materials in buildings, unless properly encapsulated.

Figure 1.3.8 illustrates the changes in latent heat of fusion for the two PCM samples tested in the oven at an elevated temperature of 55°C. This temperature was chosen to give an accelerated effect of weight loss and thermal change. As shown, RT21 experienced a significant change in latent heat of fusion resulting in an increased from 134 to 170 J/g. This is accompanied with the increase in its peak melting point from 21°C to 29°C.

These weight loss can be related to the vapour pressure of the various hydrocarbons that make up both PCMs. As illustrated in Table 1.3.1, the propyl esters that make up the propyl ester mixture have almost zero vapour pressure between 20°C and 23°C (i.e. comfort living temperature). However, the paraffins such as tetradecane and hexadecane have higher vapour pressures. Hence, if not encapsulated, these paraffins will evaporate over time. Only high melting temperature paraffins have similar properties to that of esters.

As illustrated in Table 1.3.1, the commercial paraffin's containing materials such as dodecane, teradecan and hexadecane are all volatile at the comfort-level

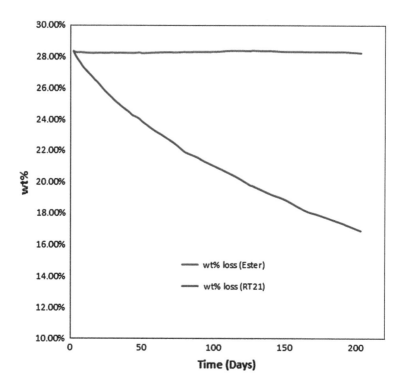

FIGURE 1.3.7 Difference in weight loss for two GIB samples impregnated with RT21 and ester mix at 55°C inside the oven.

temperatures and have significant associated health risks if not properly encapsulated. These materials, as stated earlier, could be a major fire hazard. However, as illustrated by Sittisart and Farid (2011), these materials can be contained in matrix combined with various fire retardants that reduced their health and fire hazards [22]. The results also indicated that esters are less fire hazards than paraffin-based PCMs and hence would require less fire retardant. However, the research failed to illustrate the cost increase associated with the addition of these fire retardants and the potential environmental impact caused by the inclusion of these materials after the building has been demolished. This is an area that will require further study and refinement to ascertain the overall benefit of using volatile PCMs in the building envelope.

An important criterion for PCM selection is its thermal cycle performance. Therefore, experiments were conducted to ascertain the cyclic properties of the two PCM studied in this project. For comparative purposes, the cyclic properties of *n*-octadecane were also included. As illustrated in Figure 1.3.9, both samples performed similarly with no notable difference. This similarity is due to the containment of the samples, which prevented potential mass loss through encapsulation. As expected, RT21 performed in accordance to its manufacturing specification. For this reason, paraffins can only be used as PCM when they are encapsulated. This procedure makes PCMs more expensive and restricts the recovery of these materials

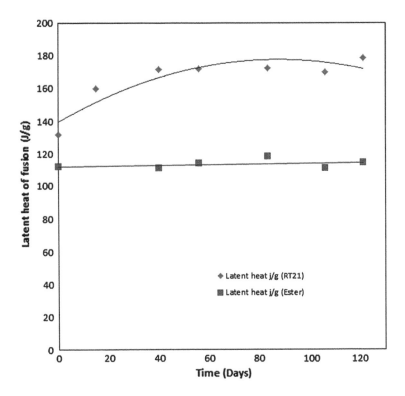

FIGURE 1.3.8 Difference in latent heat of fusion for RT21 and ester mix at 55°C for 120 days inside the oven.

TABLE 1.3.1
Vapour Pressure of Esters and Hydrocarbon Compounds [30,31]

Compound Name	M_w (g/mol)	T_{min} (°C)	T_{max} (°C)	Vapour Pressure $@T_{min}$ (mmHg)	Vapour Pressure $@T_{max}$ (mmHg)	Vapour Pressure @20°C	Vapour Pressure @23°C
Propyl stearate	326.563	232.72	556	1	3000	0	0
Propyl palmitate	298.51	198.08	504.04	1	3000	0	0
Dodecane	170.338	−9.58	385.05	0.0051	13,681.11	0.09769	0.12618
Tetradecane	198.392	5.86	419.25	0.0014	12,158.49	0.007127	0.00976
Hexadecane	226.446	18.19	447.45	0.007	10,643.37	0.000819	0.00116
Octadecane	254.5	61.5	473.85	0.0075	9652	6.02884E−05	9.16435E−05

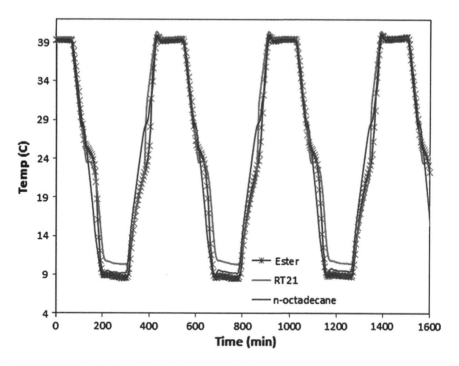

FIGURE 1.3.9 Cyclic thermal performance of PCMs.

once the building is destroyed. However, esters do not have these limitations and can easily be incorporated into the building envelope. More importantly, these esters can be produced from renewable sources making them more environmentally friendly.

1.3.4 CONCLUSIONS

This article presents the potential changes in the thermal characteristics of organic PCMs such as RT21 and esters with time when heated above their melting points. It was found that the paraffin-based PCM undergoes an irreversible physical change which alters its thermal characteristics. Paraffin mixtures have higher vapour pressures than most esters, which makes them more susceptible to weight loss and change in physical properties due to encapsulation of light compounds. On the other hand, the more stable ester mixture experiences no change in mass and was thermally stable during the entire heating period.

ACKNOWLEDGEMENTS

This research was supported by the University of Auckland and the Foundation for Research Science and Technology of New Zealand (MBIE).

REFERENCES

1. Roth K, Westphalen D, Brodrick J. PCM technology for building materials. *ASHRAE J* 2007;49(7):129–31.
2. Zhou D, Zhao CY, Tian Y. Review on thermal energy storage with phase change materials (PCMs) in building applications. *Appl Energy* 2012;92:593–605.
3. Farid MM et al. A review on phase change energy storage: materials and applications. *Energy Convers Manage* 2004;45(9–10):1597–615.
4. Hawes DW, Feldman D, Banu D. Latent heat storage in building materials. *Energy Build* 1993;20(1):77–86.
5. Kuznik F, Virgone J. Experimental investigation of wallboard containing phase change material: data for validation of numerical modeling. *Energy Build* 2009;41(5):561–70.
6. Castellon C et al. Experimental study of PCM inclusion in different building envelopes. *J Solar Energy Eng, Trans ASME* 2009;131(4):0410061–66.
7. Castellon C et al. Improve thermal comfort in concrete buildings by using phase change material. In: *Energy Sustainability Conference*, June 27, 2007 – June 30, 2007. Long Beach, CA, United States: American Society of Mechanical Engineers; 2007.
8. Hasnain SM. Review on sustainable thermal energy storage technologies, Part I: heat storage materials and techniques. *Energy Convers Manage.* 1998;39(11):1127–38.
9. Hawes DW, Banu D, Feldman D. Latent heat storage in concrete. *Solar Energy Mater* 1989;19(3–5):335–48.
10. Hawes DW, Banu D, Feldman D. Latent heat storage in concrete. II. *Solar Energy Mater* 1990;21(1):61–80.
11. Hawes DW, Feldman D. Absorption of phase change materials in concrete. *Sol Energy Mater Sol Cells* 1992;27(2):91–101.
12. He B, Martin V, Setterwall F. Phase transition temperature ranges and storage density of paraffin wax phase change materials. *Energy* 2004;29(11):1785–804.
13. Shapiro MM et al. PCM wallboard tempers passive-solar temperature swings. *Sunworld* 1988;12(2):46–47, 55.
14. Feldman D, Banu D, Hawes D. Low chain esters of stearic acid as phase change materials for thermal energy storage in buildings. *Sol Energy Mater Sol Cells* 1995;36(3):311–22.
15. Feldman D et al. Energy storage building materials with organic PCM's. In: *Proceedings of the 28th Intersociety Energy Conversion Engineering Conference*, August 8, 1993 – August 13, 1993. Atlanta, GA, USA: Publ by SAE; 1993.
16. Feldman D, Khan MA, Banu D. Energy storage composite with an organic PCM. *Solar Energy Mater* 1989;18(6):333–41.
17. Inoue T et al. Solid–liquid phase behavior of binary fatty acid mixtures: 2. Mixtures of oleic acid with lauric acid, myristic acid, and palmitic acid. *Chem Phys Lipids* 2004;127(2):161–73.
18. Inoue T et al. Solid–liquid phase behavior of binary fatty acid mixtures: 3. Mixtures of oleic acid with capric acid (decanoic acid) and caprylic acid (octanoic acid). *Chem Phys Lipids* 2004;132(2):225–34.
19. Inoue T et al. Solid–liquid phase behavior of binary fatty acid mixtures: 1. Oleic acid/ stearic acid and oleic acid/behenic acid mixtures. *Chem Phys Lipids* 2004;127(2):143–52.
20. Khudhair AM, Farid MM. A review on energy conservation in building applications with thermal storage by latent heat using phase change materials. *Energy Convers Manage* 2004;45(2):263–75.
21. Suppes GJ, Goff MJ, Lopes S. Latent heat characteristics of fatty acid derivatives pursuant phase change material applications. *Chem Eng Sci* 2003;58(9): 1751–63.
22. Sittisart P, Farid MM. Fire retardants for phase change materials. *Appl Energy* 2011;88(9):3140–5.

23. Giro-Paloma J et al. Physico-chemical and mechanical properties of microencapsulated phase change material. *Appl Energy* 2013;109:441–8.
24. Smith MC. Microencapsulated phase change materials for thermal energy storage: development, evaluation, and application, in Chemical and Materials Engineering, The University of Auckland, 2009.
25. Banu D et al. Energy-storing wallboard: flammability tests. *J Mater Civ Eng* 1998;10(2):98–105.
26. Sharma A et al. Review on thermal energy storage with phase change materials and applications. *Renew Sustain Energy Rev* 2009;13(2):318–45.
27. Cabeza LF et al. Materials used as PCM in thermal energy storage in buildings: a review. *Renew Sustain Energy Rev* 2011;15(3):1675–95.
28. Khudhair AM, Farid MM. Use of phase change materials for thermal comfort and electrical energy peak load shifting: experimental investigations. In: *Solar World Congress*. Beijing, China: Solar Energy and Human Settlement; 2007.
29. Oliveira MB et al. Prediction of water solubility in biodiesel with the CPA equation of state. *Ind Eng Chem Res* 2008;47(12):4278–85.
30. Yaws CL. *Knovel (Firm), Yaws Handbook of Thermodynamic Properties for Hydrocarbons and Chemicals*. New York: Knovel; 2009.
31. Yaws CL. *Knovel (Firm), Yaws' Thermophysical Properties of Chemicals and Hydrocarbons*. Norwich, NY: Knovel; 2009.

1.4 A Novel Calcium Chloride Hexahydrate-Based Deep Eutectic Solvent as a Phase Change Material

K. Shahbaz
University of Auckland

I. M. AlNashef
Masdar Institute for Science and Technology

R. J. T. Lin
University of Auckland

M. A. Hashim
University of Malaya

F. S. Mjalli
Sultan Qaboos University

Mohammed Farid
University of Auckland

1.4.1 INTRODUCTION

In 1990s when the concept of green chemistry was introduced, ionic liquids (ILs) received a significant attention because of their favorable features such as being environmentally friendly, nonvolatile, easily designable, high solvating, nonflammable, thermally stable, and having simple regeneration processes [1]. However, some ILs are very toxic and poorly biodegradable [2,3]. Furthermore, the synthesizing process of ILs is complicated, expensive, and non-environmentally friendly since it demands a large amount of salts and solvents to allow the anions to exchange completely [4,5].

Recently, a new generation of solvents, namely deep eutectic solvents (DESs), has emerged to overcome some of the principal disadvantages of ILs. The DESs serve as a low-cost alternative to ILs because DESs have similar common properties of ILs [6,7].

The DESs can be easily acquired by mixing two or more components, which are capable of forming a eutectic mixture through hydrogen bonding or metal halide bond interactions [2,8]. The compositions of the DESs can manifest four types of DESs. A quaternary salt can be mixed either with a metal salt-containing chloride ion (Type I), a metal chloride hydrated salt (Type II) or with a hydrogen bond donor (Type III) that can consist of alcohol, amide or carboxylic acid [9]. Moreover, mixing a metal chloride salt or a metal chloride hydrated salt with a hydrogen bond donor (HBD) can form type IV DES [10,11].

Generally, the physicochemical properties of DESs can be further adapted to tailor particular application by altering the characteristics of constituting components. In the last decade, DESs as affordable green solvents have been successfully demonstrated in various applications such as electrochemical processes [12], biological catalysis [13], synthesis of solar cells [14], separation and purification processes [15–17], nanotechnology [18], and many other potential applications [9]. In addition, the potential of DESs usage in some new applications can be examined due to their unique features.

Developing renewable sources of energy plays an important role in reducing the environmental impact related to energy use. Designing thermal energy storage systems is one of the options that can be followed for energy saving and for preserving it for later use. Sensible heat storage (SHS), latent heat storage (LHS), and thermochemical storage (TCS) are the three main thermal energy storage methods [19]. Among these, LHS is the most attractive method for storing thermal energy in a wide range of applications due to its ability to provide energy storage with high capacity and less temperature fluctuation during the phase change process [20]. Generally, the LHS method is employed to improve the use of energy and conserve it in a number of applications. It is commonly used to improve thermal performance of buildings since it can fulfill the demand for thermal comfort and energy conservation in the buildings [21–23]. In addition, its use covers many other applications [20,24].

Phase change material (PCM) is the main representative of the LHS method. The theory of PCM use in buildings is quite simple. When the indoor temperature of a building increases, the PCM is changed from solid to liquid, absorbing heat being an endothermic process. Likewise, decreasing the temperature in the building solidifies the PCM, which releases the heat it absorbed. The use of PCMs in buildings enhances the energy storage capacity of the building envelope and can justify the use of renewable and nonrenewable energies more efficiently [25].

An appropriate PCM for building applications should meet a number of criteria related to the desired thermophysical, kinetic and chemical properties, as well as economics. For the thermal properties, the PCM must have a high latent heat and high thermal conductivity [26]. Moreover, they should have a melting temperature slightly above 20°C to meet the need for thermal comfort of building. According to the American Society of Heating, Refrigerating and Air-Conditioning Engineers (ASHRAE), the suggested comfort room temperature is 21.0°C–23.0°C in winter and 23.5°C–25.5°C in summer [22]. When it comes to physical properties, the PCM should have a high density that allows more heat to be stored into a limited space and with small volume changes. In terms of the kinetic properties, the PCM must have minimum supercooling and must have sufficient crystallization rate to form crystals when the heat energy is released. Under the chemical properties, the PCM must have long life time, be compatible with building materials, noncorrosive, nontoxic, nonflammable and show no degradation after numerous cycles. From an economical point of view, PCMs should be readily available in large quantities at low cost [27].

PCMs used in buildings and other applications can be classified into three different types, namely, organic, inorganic and eutectic. For example, paraffin, fatty acids and fatty acid ester lie under the organic class. The main advantages of organic PCMs are their moderate heat of fusion at low temperature, small volume changes and do not undergo supercooling. But they usually have poor thermal conductivity and expensive for use in buildings [22]. However, the thermal conductivity of the organic PCMs can be improved through microencapsulation using different methods [28].

Metal salts and hydrated salts are a few examples of inorganic PCMs. They have high latent heat of fusion and high thermal storage density. They are cheap compared to organic PCMs, nonflammable, and have high thermal conductivity. The main drawbacks of the inorganic PCMs are supercooling and phase segregation. Based on reports in the published literature, the supercooling problem of these PCMs could be eliminated by adding nucleating agent or some impurities, mechanical stirring and encapsulation [24,29]. Adding thickening agents, using excess water, mechanical stirring, and PCM encapsulation are available techniques to rectify the phase segregation problem [20,29,30]. Among the inorganic PCMs, calcium chloride hexahydarte ($CaCl_2$ $6H_2O$) with a large latent heat of fusion (170–190 kJ/Kg) and a low melting temperature (29°C–30°C) has been shown as a potential candidate useful in heat storage applications [31]. This hydrated salt is cheap and nontoxic. However, apart from the two serious drawbacks for inorganic PCMs, supercooling and phase segregation, its melting temperature (29°C) is also too high for use in buildings, which requires much lower melting point in accordance with the human comfort temperature (20°C–25.5°C). It is therefore necessary for the eutectic PCMs to be developed for tailored properties from the synergistic effects of various materials.

A eutectic PCM could be a combination of different materials including organic–organic, organic–inorganic and inorganic–inorganic, which can produce a mixture that has a specifically designed property such as melting point and latent heat of fusion [32]. Hence, there is no limit to the number of eutectic PCMs that can be synthesized from available organic and inorganic chemicals. For instance, a new eutectic PCM can be synthesized by mixing an inorganic compound that has a great energy storage capacity with an organic compound to adjust the melting temperature for a particular application.

Carlsson et al. [33] investigated the phase diagram of $CaCl_2$-H_2O system and also the effect of strontium chloride hexahydrate ($SrCl_2$ $6H_2O$) on the system to define compositions and temperature in which the formation of calcium chloride tetrahydrate ($CaCl_2$ $4H_2O$) is prevented. They also showed that the presence of $SrCl_2$ $6H_2O$ had a significant effect on the phase diagram. Adding $SrCl_2$ $6H_2O$ can decrease the solubility of $CaCl_2$ $6H_2O$ and also shift the peritectic point towards a higher concentration of $CaCl_2$ $6H_2O$. They also found that 2 wt% of $SrCl_2$ $6H_2O$ can avoid the formation of tetrahydrate. Later, Feilchenfeld et al. [34] reported the melting temperature of $CaCl_2$ $6H_2O$ with $SrCl_2$ $6H_2O$ as a nucleating agent and Cab-o-SilR as a thickening agent through 1000 thermal cycles without any signs of phase segregation. Hiroshi et al. [35] revealed that $CaCl_2$ $6H_2O$ with a slight excess water was stable over 1000 thermal cycles (in a heat cycle of 18°C–35°C) in a long vertical glass tube. In addition, it was also reported that NaCl had a nucleating effect to improve the phase change repeatability. Brandstetter [36] stated that $CaCl_2$ $6H_2O$ can be stabilized against supercooling and incongruent melting using different nucleating agents such as strontium or barium salts and sodium chloride and the formation of $CaCl_2$ $4H_2O$ was inhibited entirely using the extra water principle. Paris et al. [37] studied $Ba(OH)_2$ $8H_2O$ and $SrCl_2$ $6H_2O$ as nucleating agents for $CaCl_2$ $6H_2O$, and the results showed that 3 wt% $SrCl_2$ $6H_2O$ revealed a stable thermal behavior with no supercooling. Kaneff and Brandstetter [38] in their patent pointed out that the tetrahydrate formation problem reduces heat storage capacity of $CaCl_2$ $6H_2O$ and for solving the problem they proposed the fumed silica as a thickening agent as well as the use of extra water in excess of the stoichiometric composition of $CaCl_2$ $6H_2O$. Moreover, they also suggested other additives such as sodium chloride and potassium chloride to reduce the melting temperature of $CaCl_2$ $6H_2O$.

Gao and Deng [29] introduced an inorganic PCM-based $CaCl_2$ $6H_2O$ for thermal energy storage with a phase change temperature of 22.6°C. They revealed that calcium chloride solution containing 5 wt% of $Mg(NO_3)_2$ $6H_2O$ produced a stable PCM with latent heat of 160 kJ/Kg. The results indicated that, by adding 2 wt% $SrCl_2$ $6H_2O$, the degree of supercooling of the PCM was reduced and no phase segregation was observed. Recently, Li et al. [39] prepared a eutectic hydrated salt based on $CaCl_2$ $6H_2O$ and $MgCl_2$ $6H_2O$ as a new PCM with the phase change temperature and latent heat of 21.41°C and 102.3 kJ/kg. Their results exhibited that 3 wt% $SrCl_2$ $6H_2O$ and 1 wt% $SrCO_3$ were effective nucleating agents, and 0.5 wt% hydroxyethyl cellulose made the PCM stable over 50 thermal cycles. As shown above, $CaCl_2$ $6H_2O$ has long been considered as a low-temperature PCM for thermal energy storage and many attempts were made to manipulate its properties and solve the problems associated with its use, yet it has not been applied in real buildings, probably due to the problems mentioned above.

The aim of this article is to introduce a new type III DESs prepared from choline chloride (ChCl) and $CaCl_2$ $6H_2O$. Different compositions of ChCl salt and hydrated salt were prepared and characterized. Furthermore, the prepared DESs were investigated as potential PCMs for building applications by measuring their properties such as freezing temperature, latent heat of fusion, degree of supercooling and thermal stability. Most previous researchers used extra water to ensure the long-term stability of $CaCl_2$ $6H_2O$. From the phase diagram of $CaCl_2$-H_2O system [33], it is clear that adding water above the degree of hexahydrate can reduce the concentration of $CaCl_2$. Consequently, the melting temperature of composition decreases and no tetrahydrate is formed. However, any loss of water due to improper encapsulation would initiate the problems of phase separation. In this work, using extra water was not proposed, and the thickening agents were examined for the stability of the synthesized DESs as PCMs.

1.4.2 EXPERIMENTAL DETAILS

1.4.2.1 CHEMICALS

Choline chloride ($C_5H_{14}ClNO$), calcium chloride hexahydrate ($CaCl_2$ $6H_2O$), strontium chloride hexahydrate ($SrCl_2$ $6H_2O$), hydroxyethyl cellulose and fumed silica (SiO_2)

with an average particles size of 0.2–0.3 μm were purchased from Sigma-Aldrich (St. Louis, USA). All chemicals were of synthesis grade (purity > 98%) and were used without further purifications.

1.4.2.2 PREPARATION OF DESs

The DESs in this work were prepared based on the method described in previous work [5]. The mixture of ChCl as quaternary ammonium salt (QAS) and $CaCl_2$ $6H_2O$ as hydrated salt at different mole fractions were mixed mechanically into 150 ml Schott bottle at 400 rpm using magnetic stirrer. The synthesis temperature was maintained at 90°C. The DES is said to be fully prepared when the mixture turns into homogeneous liquid phase without any solid precipitation. To prevent moisture absorption, the DESs' bottles were sealed with Parafilm. Table 1.4.1 shows the $ChCl$:$CaCl_2$ $6H_2O$ combinations used in this work along with their mole fractions. Abbreviations were given to each DES for convenient identification.

1.4.2.3 CHARACTERIZATION OF DESs

The freezing temperature of DESs was determined using the temperature history method. The freezing performance of DESs was conducted using a programmable cycling water bath (PolyScience, US). Ten milliliters of each DES was added in a glass vial. Subsequently, a type K thermocouple was inserted at the center of DES, and the vial was placed in a water bath at 40°C. After 15 min, the sample was cooled to −15°C. The thermocouple was connected to a temperature data logger (Pico TC-08, UK) to record the temperature of DES at time interval of 5 s. The thermal cycling test was also carried out using the above procedure for 100 melting and freezing cycles with the following temperature program: held at 40°C for 15 min then cooling from 40°C to 5°C, held at 5°C for 15 min and heating from 5°C to 40°C.

The latent heat of the synthesized DESs was measured using a Shimadzu Differential Scanning Calorimetry instrument (DSC-60). A small amount of sample was placed in an aluminum pan with lid and subjected to the DSC furnace. The DSC measurements were carried out at a heating rate of 1°C/min. Octadecane and tetradecane were used to calibrate the DSC instrument, which allowed latent heat to be measured with uncertainty better than ±2 kJ/kg.

Densities of synthesized DESs were determined using an Anton Paar DMA 500 vibrating-tube density meter. The accuracy of the density meter was checked using distillated water and air. The measured values were compared with the corresponding values in the density tables, showing measurements accuracy of ±0.0001 g/cm³. Density measurements were performed at temperatures from 15°C to 40°C with 5°C intervals and three replicates for each reading.

Viscosity measurements were conducted at different temperatures using a controlled stress AR-G2 rheometer (TA Instrument Ltd, US). The rheometer was fitted with a conical concentric cylinder geometry and was calibrated using an oil standard. Viscosity values were taken between 15°C and 40°C at 5°C intervals. The temperature was controlled by an external water bath and circulator. The uncertainty in the viscosity measurements was ±0.02 Pa s.

1.4.3 RESULTS AND DISCUSSION

In this work, ChCl as quaternary ammonium salt and $CaCl_2$ $6H_2O$ as hydrated salt were selected to prepare the new type III DESs. This was accomplished by mixing the quaternary salt and the hydrated salt at different mole fractions. Accordingly, five DESs based on $CaCl_2$ $6H_2O$ were prepared from these mixing experiments (Table 1.4.1), and their fundamental physicochemical properties including density, viscosity, freezing temperature, and latent heat were measured. It should be pointed out that to obtain the eutectic point and study the influence of ChCl on freezing temperature of the DESs, five mole fractions of ChCl (0.09, 0.11, 0.14, 0.20, and 0.34) were selected in this work. All synthesized DESs at the selected mole fractions were homogeneous colorless liquids at room temperature; however, at higher mole fractions of ChCl (>0.34), prepared mixtures were solid at ambient conditions.

Density is an important and useful physical property for the DESs. The knowledge of this property plays an important role in many engineering application designs of DESs. The densities of all synthesized DESs (DES1–DES5) were measured at temperatures from 15°C to 40°C in 5°C intervals and at atmospheric pressure. The results of the measured densities of DES1–DES5 are presented in Figure 1.4.1. As expected, the liquid densities of all DES exhibit a temperature-dependent behavior and decreased linearly with the increase in temperature due to thermal expansion of the DESs. The DESs' densities also increased lineally with the decrease in mole fraction of ChCl. The following liner equation was found to fit the measured DESs' densities as a function of temperature:

$$\rho = aT + b \qquad (1.4.1)$$

where ρ and T are density (g/cm^3) and temperature (°C), respectively, while a and b are empirical constants as tabulated in Table 1.4.2 (a depends on the type of DES and b depends on mole fractions of constituting components).

The viscosity is another important property in the early design stage of any processes employing new DESs. Viscosity data for many types of DESs were reported in the literature as part of new DESs characterization studies [40,41]. In this work, viscosities of all $CaCl_2$ $6H_2O$-based DESs (DES1–DES5) were measured experimentally at a temperature range of 15°C–40°C, and the values were plotted against temperature in Figure 1.4.2. As expected, the viscosity of the DESs decreased with

TABLE 1.4.1

Compositions and Abbreviations of Synthesized DESs

QAS	Hydrated Salt	x_{QAS}	$x_{Hydrated\ salt}$	Abbreviation
Choline chloride	Calcium chloride hexahydrate	0.34	0.66	DES1
		0.20	0.80	DES2
		0.14	0.86	DES3
		0.11	0.89	DES4
		0.09	0.91	DES5

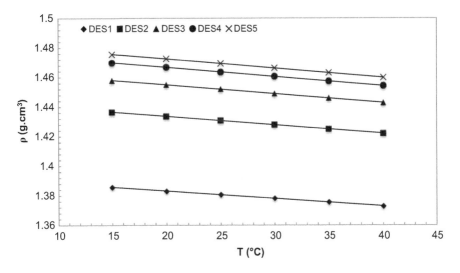

FIGURE 1.4.1   Density of calcium chloride hexahydrate-based DESs as a function of temperature.

TABLE 1.4.2
Values of Fitting Constants for Density and Viscosity

DES	a	b	μ_o (Pa s)	E_o (Pa m³/mol)
1	-5×10^{-4}	1.3936	1×10^{-9}	5349.10
2	-6×10^{-4}	1.4454	2×10^{-7}	3736.20
3	-6×10^{-4}	1.4672	1×10^{-7}	3700.00
4	-6×10^{-4}	1.4797	3×10^{-7}	3458.10
5	-6×10^{-4}	1.4853	2×10^{-7}	3471.10

temperature. In addition, the viscosity of DES1 was much higher than that of DES2, DES3, DES4 and DES5. This could be attributed to stronger hydrogen bonding between ChCl and CaCl$_2$ 6H$_2$O in DES1 as shown by its low freezing temperature (Table 1.4.3). As a result, the mole fraction of ChCl has an obvious effect on the viscosity of DESs. The DES with lower mole percentage of ChCl has lower viscosity at any temperature. The viscosity values of each DES were fitted with a high accuracy by an Arrhenius-like equation [40], given below:

$$\mu = \mu_0 e^{\frac{E_0}{RT}} \tag{1.4.2}$$

where μ is the viscosity (Pa s), μ_o is a constant (Pa s), E_o is the activation energy (Pa m³/mol), R is the gas constant (Pa m³/mol/K) and T is the temperature in Kelvin. The values of μ_o and E_o are given in Table 1.4.2.

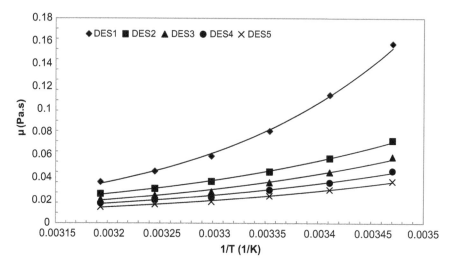

FIGURE 1.4.2 Viscosity of calcium chloride hexahydrate-based DESs as a function of temperature $(1/T)$.

TABLE 1.4.3
Freezing Temperatures and Latent Heat of Fusion for the Synthesized DESs

DES	Freezing Temperature (°C)	Latent Heat (kJ/kg)
DES1	2.70	–
DES2	16.80	100.05
DES3	20.65	127.22
DES4	23.05	135.21
DES5	24.10	146.57

The melting/freezing temperature is of utmost importance in defining the liquidity range of the newly prepared DESs. Moreover, this property plays a key role in the selection of PCM for a particular application. The freezing temperatures of all studied DESs were measured using the temperature history method described in the earlier section. The reason for using this method in this work is that it offers true and precise information about melting/freezing temperature of a sample as compared to DSC, especially in the presence in supercooling which is very common for both ILs and DESs. This method is able to provide a very slow heating/ cooling rate in comparison with DSC analysis. In addition, it gives a better simulation in the real use of PCMs in building applications. Table 1.4.3 presents the measured freezing temperatures for DES1–DES5. As can be seen, the freezing temperatures of all synthesized DESs were less than the freezing temperatures of their constituting components, which is a main character of DESs. Results also show that the freezing temperature of DES is a function of mole fraction of ChCl. The freezing temperature gradually decreased with increasing ChCl mole fraction and eutectic temperature occurred at DES1 with a freezing temperature of 2.7°C. As abovementioned, the ideal phase change temperature of PCMs used in building application should be in accordance with the human comfort temperature. DES3, DES4 and DES5 satisfied this expectation with a phase change temperature of 20.65°C, 23.05°C and 24.10°C, respectively.

The latent heat of PCM is another important criterion used for the selection of a proper PCM in building application. The measured latent heats of fusion for resulting DESs are also listed in Table 1.4.3. From the obtained results, it can be concluded that latent heat of DESs depends on the mole fractions of its constituents (ChCl and $CaCl_2$ $6H_2O$). Moreover, all DESs have a latent heat above 100 kJ/kg except for DES1. The values of the latent heat of DES3, DES4 and DES5 revealed that they have a good potential for use as PCMs in building applications. The DES1 as eutectic point was characterized by having a low density, high viscosity and low freezing temperature as compared to other DESs. DSC analysis of DES1 showed no melting peak, but a glass transition was observed instead. Such behavior is typical for DESs because the high viscosity of DES hinders nucleation and crystal growth. Furthermore, the strength of hydrogen bonds and other intermolecular interactions between ChCl and $CaCl_2$ $6H_2O$ decrease the freezing temperature deeper and impress the trends in phase transition temperature. From Table 1.4.3, it is clear that the ability of hydrogen bonding and other forces in DESs increased with increasing mole fraction of ChCl. Accordingly, the stability of the DESs was correlated with intensity of hydrogen bonds.

From above results, DES3 and DES4 were nominated as amendable PCMs and examined for their supercooling behavior and thermal stability over 100 thermal cycles. Figure 1.4.3 shows the cooling curves for $CaCl_2$ $6H_2O$, DES3 and DES4 with a cooling rate of 0.95°C/min. It is obvious from Figure 1.4.3 that they all subcooled

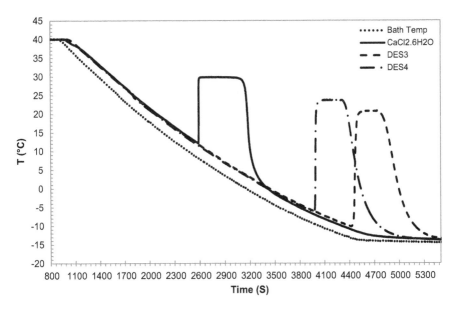

FIGURE 1.4.3 Cooling curves of calcium chloride hexahydrate, DES3 and DES4 (with a cooling rate of 0.95°C/min).

heavily before freezing. The degree of supercooling of $CaCl_2\ 6H_2O$ is about 18°C and the freezing temperature is 29°C. The reported value in literature [39] for the supercooling degree of $CaCl_2\ 6H_2O$ was about 20°C, which is close to the value obtained in this work. The supercooling degree of DES3 and DES4 was found even higher. The supercooling of DESs is high and can prevent the heat recovery in an application. Therefore, DESs should be modified before using them as PCMs.

To overcome this problem, $SrCl_2\ 6H_2O$ as a nucleating agent was employed to provide crystal nucleon in the DESs. $SrCl_2\ 6H_2O$ was added into DES3 and DES4 at different concentrations (0.5, 1 and 2 wt%) to determine the effective nucleation rate. The experimental results indicated that 2% by weight of $SrCl_2\ 6H_2O$ had a very highly efficient nucleation rate for DES3 and DES4. The degree of supercooling of DES3 and DES4 during the 100 cooling processes was recorded and presented in Table 1.4.4. The maximum degree of supercooling of DES3 and DES4 was found to be 2.54°C and 2.33°C, respectively, and the average supercooling degree for DES3 was 1.47°C and for DES4 was 1.23°C. The cooling curves of DES3 and DES4 with 2 wt% of $SrCl_2\ 6H_2O$ at first cooling cycle are shown in Figure 1.4.4. As can be seen from Figures 1.4.3 and 1.4.4, the subcooling was significantly reduced by the nucleating agent used.

To investigate the stability of DES3 and DES4, they were melted and frozen 100 times. Phase segregation was observed after eight cycles for DES3 and five cycles for DES4. In addition, the phase change temperature of the DESs increased as cycling was continued.

Finally, the freezing temperature reached 28°C after 20 cycles and then stayed constant for both DESs for the remaining cycling test. It was expected that calcium chloride tetrahydrate is formed in the DESs after few cycles since the chemical stability of studied DESs was influenced by concentration of ChCl. One promising

TABLE 1.4.4

Degree of Supercooling of DES3 and DES4 with 2 wt% Nucleating Agent through the 100 Cycles

No. of Cycles	Supercooling(°C)	
	DES3	DES4
1	1.23	0.57
10	0.88	0.89
20	0.94	1.04
30	1.22	1.69
40	1.8	1
50	0.4	1.49
60	2.12	1.47
70	2.54	2.33
80	1.64	0.75
90	1.71	1.03
100	1.77	1.33

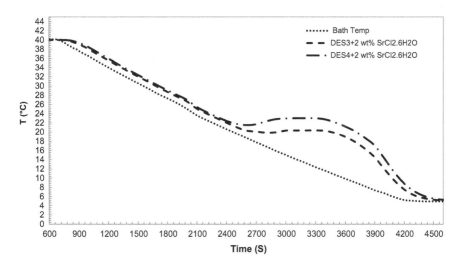

FIGURE 1.4.4   Cooling curves of DES3 and DES4 with 2 wt% nucleating agent (with a cooling rate of 0.6°C/min).

method to tackle this problem in hydrated salt PCMs is to add a thickening agent. It has also been found that fumed silica and hydroxyethyl cellulose can be used as a thickener in a PCM based $CaCl_2 \, 6H_2O$ to process long-term stability [38,39]. Thus, both fumed silica and hydroxyethyl cellulose were employed in this work to examine their applicability as a thickening agent for the $CaCl_2 \, 6H_2O$-based DESs.

The hydroxyethyl cellulose was mixed with DES3 and DES4 at different concentrations (0.5, 1 and 2 wt%) to reveal the optimum condition. The cycling test showed that hydroxyethyl cellulose at selected concentrations could not prevent phase segregation in DES3 and DES4 over 100 cycles, and tetrahydrate formation was observed after 10 cycles. On the other hand, the fumed silica was also added to DES3 and DES4 at different weight percentages (0.5, 1 and 2 wt%). The phase change temperatures of DES3 and DES4 with fumed silica for the 100 cycles were presented in Figures 1.4.5 and 1.4.6, respectively. The additions of 0.5 and 1 wt% of this thickener were not able to stabilize the DES3 and DES4 through the 100 cycles, and the phase change temperature of the DESs gradually increased to 27.8°C. However, it can be noted that there is a noticeable improvement in the stability of DES3 and DES4 by adding 2 wt% of fumed silica as a thickener. The freezing point of both DES3 and DES4 remained unchanged during cycling indicating excellent stability.

The DSC results of DES3 and DES4 before and after 100 cycles are shown in Figures 1.4.7 and 1.4.8, respectively. The latent heat of DES3 was reduced from 123.02 to 112.69 kJ/kg after 100 cycles. The deviation of latent heat for DES4 was also found to be 7.8% after 100 cycles (Figure 1.4.8). From the obtained results, it can be concluded that ChCl: $CaCl_2 \, 6H_2O$ DESs in the molar ratios of 1:6 (DES3) and 1:8 (DES4) have the potential to be utilized as PCMs in building applications with large latent heat (127.2–135.2 kJ/kg), proper phase change temperature (20.65°C–23.05°C), and low cost. Based on rough estimates of $CaCl_2 \, 6H_2O$ (0.25 USD/Kg) and ChCl (1 USD/kg), the cost of DES3 and DES4 is about 0.35 USD/kg. Although DES3 and

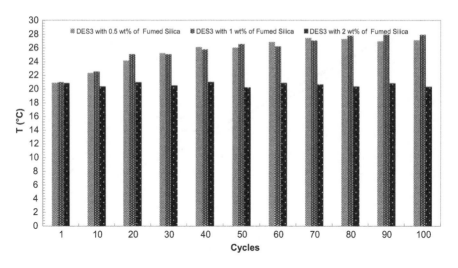

FIGURE 1.4.5   Phase change temperature of DES3 with thickening agent during 100 cycles.

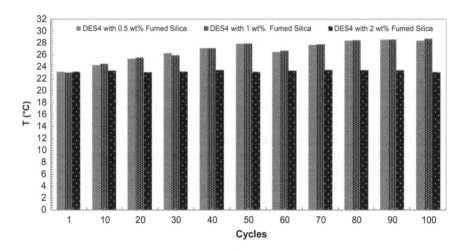

FIGURE 1.4.6   Phase change temperature of DES4 with thickening agent during 100 cycles.

DES4 suffered from the supercooling and phase segregation, the use of 2 wt% of $SrCl_2$ $6H_2O$ and fumed silica were successful to mend these problems. It should also be noted that these new PCMs can be macro-encapsulated and be used in many applications, since microencapsulation of hydrated salts was proven to be still difficult [28].

1.4.4   CONCLUSION

In this work, five new type III DESs were prepared using ChCl as the quaternary ammonium salt and $CaCl_2$ $6H_2O$ as the hydrated salt, and their important physicochemical properties were also measured. It was found that the mole fraction of ChCl

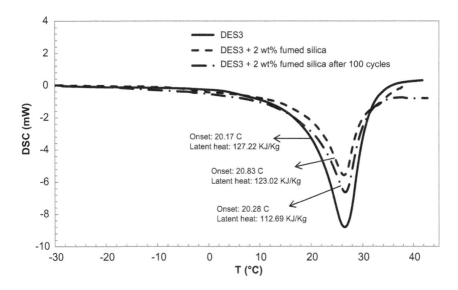

FIGURE 1.4.7 DSC melting curves for DES3, DES3 with thickening agent, and DES3 with thickening agent after 100 cycles.

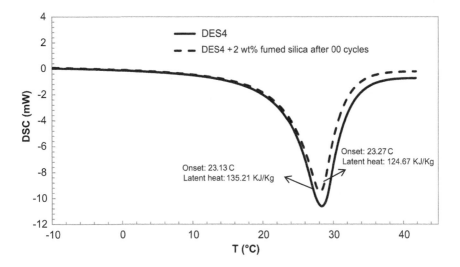

FIGURE 1.4.8 DSC melting curves for DES4 and DES4 with thickening agent after 100 cycles.

has noticeable effects on the freezing temperature, latent heat, density, viscosity, and chemical stability of the synthesized DESs (DES1–DES5). Besides, the potential of using DESs as PCMs in building application was investigated. DES3 and DES4 with a large latent heat of 127.2 and 135.2 kJ/kg and also a proper phase change temperature of 20.65°C and 23.05°C, respectively, were selected for thermal cycling test.

The cycling results showed that the supercooling degrees of both DESs were huge and phase segregation occurred after a few cycles. Hence, both DESs were required to be modified by using nucleating and thickening agents before utilizing them as PCMs. The experimental results specified that adding 2 wt% of $SrCl_2$ $6H_2O$ as a nucleating agent reduced the degree of supercooling with an average supercooling degree of 1.47°C for DES3 and 1.23°C for DES4 over 100 cooling cycles. In addition, the hydroxyethyl cellulose and fumed silica were employed as thickening agents to tackle the phase segregation problem. The hydroxyethyl cellulose was not able to prevent tetrahydrate formation with different concentrations (0.5, 1 and 2 wt%) investigated. On the contrary, when 2 wt% of fumed silica was used, the cycling results indicated an excellent stability for both DES3 and DES4. These results successfully present the newly prepared DESs as potential PCMs for building applications.

ACKNOWLEDGMENTS

The authors gratefully acknowledge the University of Auckland, and this work is funded by the University of Auckland Research Grant (FRDF-3707421), Masdar Institute for Science and Technology, United Arab Emirates and University of Malaya, Malaysia for their support to this research.

APPENDIX A. SUPPLEMENTARY MATERIAL

Supplementary data associated with this article can be found online at http://dx.doi.org/10.1016/j.solmat.2016.06.004.

REFERENCES

1. A.A.P. Kumar, T. Banerjee, Thiophene separation with ionic liquids for desulphurization: a quantum chemical approach, *Fluid Phase Equilibria* 278 (2009) 1–8.
2. Q. Zhang, K.D.O. Vigier, S. Royer, F. Jérôme, Deep eutectic solvents: syntheses, properties and applications, *Chem. Soc. Rev.* 41 (2012) 7108–7146.
3. L. Bahadori, M.H. Chakrabarti, F.S. Mjalli, I.M. AlNashef, N.S.A. Manan, M.A. Hashim, Physicochemical properties of ammonium-based deep eutectic solvents and their electrochemical evaluation using organometallic reference redox systems, *Electrochim. Acta* 113 (2013) 205–211.
4. M.H. Chakrabarti, F.S. Mjalli, I.M. AlNashef, M.A. Hashim, M.A. Hussain, L. Bahadori, C.T.J. Low, Prospects of applying ionic liquids and deep eutectic solvents for renewable energy storage by means of redox flow batteries, *Renew. Sustain. Energy Rev.* 30 (2014) 254–270.
5. K. Shahbaz, F.G. Bagh, F. Mjalli, I. AlNashef, M. Hashim, Prediction of refractive index and density of deep eutectic solvents using atomic contributions, *Fluid Phase Equilibria* 354 (2013) 304–311.
6. A. Boisset, J. Jacquemin, M. Anouti, Physical properties of a new Deep Eutectic Solvent based on lithium bis [(trifluoromethyl) sulfonyl] imide and N-methylacetamide as superionic suitable electrolyte for lithium ion batteries and electric double layer capacitors, *Electrochim. Acta* 102 (2013) 120–126.
7. R.B. Leron, A. Caparanga, M.-H. Li, Carbon dioxide solubility in a deep eutectic solvent based on choline chloride and urea at T = 303.15–343.15 K and moderate pressures, *J. Taiwan Inst. Chem. Eng.* 44 (2013) 879–885.

8. K. Shahbaz, S. Baroutian, F.S. Mjalli, M.A. Hashim, I.M. AlNashef, Prediction of glycerol removal from biodiesel using ammonium and phosphunium based deep eutectic solvents using artificial intelligence techniques, *Chemom. Intell. Lab. Syst.* 118 (2012) 193–199.

9. B. Tang, H. Zhang, K.H. Row, Application of deep eutectic solvents in the extraction and separation of target compounds from various samples, *J. Sep. Sci.* 38 (2015) 1053–1064.

10. H.M. Abood, A.P. Abbott, A.D. Ballantyne, K.S. Ryder, Do all ionic liquids need organic cations? Characterisation of [AlCl$_2$ n Amide]$+$AlCl^{4-} and comparison with imidazolium based systems, *Chem. Commun.* 47 (2011) 3523–3525.

11. A.P. Abbott, A.A. Al-Barzinjy, P.D. Abbott, G. Frisch, R.C. Harris, J. Hartley, K.S. Ryder, Speciation, physical and electrolytic properties of eutectic mixtures based on CrCl$_3$ 6H$_2$O and urea, *Phys. Chem. Chem. Phys.* 16 (2014) 9047–9055.

12. A.P. Abbott, G. Capper, K.J. McKenzie, K.S. Ryder, Electrodeposition of zinc–tin alloys from deep eutectic solvents based on choline chloride, *J. Electroanal. Chem.* 599 (2007) 288–294.

13. E. Durand, J. Lecomte, B. Baréa, G. Piombo, E. Dubreucq, P. Villeneuve, Evaluation of deep eutectic solvents as new media for Candida antarctica B lipase catalyzed reactions, *Process Biochem.* 47 (2012) 2081–2089.

14. M. Steichen, M. Thomassey, S. Siebentritt, P.J. Dale, Controlled electrodeposition of Cu–Ga from a deep eutectic solvent for low cost fabrication of CuGaSe$_2$ thin film solar cells, *Phys. Chem. Chem. Phys.* 13 (2011) 4292–4302.

15. M.A. Kareem, F.S. Mjalli, M.A. Hashim, M.K. Hadj-Kali, F.S.G. Bagh, I.M. Alnashef, Phase equilibria of toluene/heptane with tetrabutylphosphonium bromide based deep eutectic solvents for the potential use in the separation of aromatics from naphtha, *Fluid Phase Equilibria* 333 (2012) 47–54.

16. K. Shahbaz, F. Mjalli, M. Hashim, I. AlNashef, Eutectic solvents for the removal of residual palm oil-based biodiesel catalyst, *Sep. Purif. Technol.* 81 (2011) 216–222.

17. K. Shahbaz, F.S. Mjalli, M.A. Hashim, I.M. AlNashef, Using deep eutectic solvents based on methyl triphenyl phosphunium bromide for the removal of glycerol from palm-oil-based biodiesel, *Energy Fuels* 25 (2011) 2671–2678.

18. A. Abo-Hamad, M. Hayyan, M.A. AlSaadi, M.A. Hashim, Potential applications of deep eutectic solvents in nanotechnology, *Chem. Eng. J.* 273 (2015) 551–567.

19. A. Pasupathy, R. Velraj, R. Seeniraj, Phase change material-based building architecture for thermal management in residential and commercial establishments, *Renew. Sustain. Energy Rev.* 12 (2008) 39–64.

20. M.M. Farid, A.M. Khudhair, S.A.K. Razack, S. Al-Hallaj, A review on phase change energy storage: materials and applications, *Energy Convers. Manag.* 45 (2004) 1597–1615.

21. S. Wi, J. Seo, S.-G. Jeong, S.J. Chang, Y. Kang, S. Kim, Thermal properties of shape-stabilized phase change materials using fatty acid ester and exfoliated graphite nanoplatelets for saving energy in buildings, *Sol. Energy Mater. Sol. Cells* 143 (2015) 168–173.

22. T. Kousksou, P. Bruel, A. Jamil, T. El Rhafiki, Y. Zeraouli, Energy storage: applications and challenges, *Sol. Energy Mater. Sol. Cells* 120 (Part A) (2014) 59–80.

23. F. Chen, M. Wolcott, Polyethylene/paraffin binary composites for phase change material energy storage in building: a morphology, thermal properties, and paraffin leakage study, *Sol. Energy Mater. Sol. Cells* 137 (2015) 79–85.

24. A. Sharma, V. Tyagi, C. Chen, D. Buddhi, Review on thermal energy storage with phase change materials and applications, *Renew. Sustain. Energy Rev.* 13 (2009) 318–345.

25. F. Kuznik, D. David, K. Johannes, J.-J. Roux, A review on phase change materials integrated in building walls, *Renew. Sustain. Energy Rev.* 15 (2011) 379–391.

26. S. Behzadi, M.M. Farid, Experimental and numerical investigations on the effect of using phase change materials for energy conservation in residential buildings, *HVACR Res.* 17 (2011) 366–376.

27. S. Riffat, B. Mempouo, W. Fang, Phase change material developments: a review, *Int. J. Ambient Energy* 36 (2015) 102–115.

28. A. Jamekhorshid, S.M. Sadrameli, M. Farid, A review of microencapsulation methods of phase change materials (PCMs) as a thermal energy storage (TES) medium, *Renew. Sustain. Energy Rev.* 31 (2014) 531–542.

29. D. Gao, T. Deng, Energy storage: preparations and physicochemical properties of solid-liquid Phase change materials for thermal energy storage, Mater. *Process. Energy Commun. Curr. Res. Technol. Dev.* (2013) 32–44.

30. V. Tyagi, D. Buddhi, Thermal cycle testing of calcium chloride hexahydrate as a possible PCM for latent heat storage, *Sol. Energy Mater. Sol. Cells* 92 (2008) 891–899.

31. W. Su, J. Darkwa, G. Kokogiannakis, Review of solid–liquid phase change materials and their encapsulation technologies, *Renew. Sustain. Energy Rev.* 48 (2015) 373–391.

32. H. Kumano, T. Hirata, T. Kudoh, Effects of poly-vinyl alcohol on supercooling phenomena of water, *Int. J. Refrig.* 32 (2009) 454–461.

33. B. Carlsson, H. Stymne, G. Wettermark, An incongruent heat-of-fusion system — $CaCl_2$ $6H_2O$—Made congruent through modification of the chemical composition of the system, *Sol. Energy* 23 (1979) 343–350.

34. H. Feilchenfeld, J. Fuchs, S. Sarig, A calorimetric investigation of the stability of stagnant calcium chloride hexahydrate melt, *Sol. Energy* 32 (1984) 779–784.

35. H. Kimura, J. Kai, Phase change stability of $CaCl_2$ $6H_2O$, *Sol. Energy* 33 (1984) 557–563.

36. A. Brandstetter, On the stability of calcium chloride hexahydrate in thermal storage systems, *Sol. Energy* 41 (1988) 183–191.

37. J. Paris, R. Jolly, Observations sur le comportement a la fusion-solidification du chlorure de calcium hexahydrate, *Thermochim. Acta* 152 (1989) 271–278.

38. S. Kaneff, A. Brandstetter, Calcium chloride hexahydrate formulations for low temperature heat storage application, in: *Proceedings of the Google Patents*, 1990.

39. G. Li, B. Zhang, X. Li, Y. Zhou, Q. Sun, Q. Yun, The preparation, characterization and modification of a new phase change material: $CaCl_2$ $6H_2O$–$MgCl_2$ $6H_2O$ eutectic hydrate salt, *Sol. Energy Mater. Sol. Cells* 126 (2014) 51–55.

40. F.S. Ghareh Bagh, K. Shahbaz, F.S. Mjalli, M.A. Hashim, I.M. AlNashef, Zinc (II) chloride-based deep eutectic solvents for application as electrolytes: preparation and characterization, *J. Mol. Liq.* 204 (2015) 76–83.

41. M.A. Kareem, F.S. Mjalli, M.A. Hashim, I.M. AlNashef, Phosphonium-Based Ionic Liquids Analogues and Their Physical Properties, *J. Chem. Eng. Data* 55 (2010) 4632–4637.

2 Mathematical Analysis of Phase Change Processes

INTRODUCTION

Farid initiated his work on thermal energy storage in the early 1980s to establish a comprehensive mathematical analysis of the moving interface associated with the use of phase change materials (PCMs). He has developed the effective specific heat capacity method to describe heat transfer during phase change, which has simplified the analysis significantly and is used nowadays by a large number of researchers [1].

Studying the effect of natural convection on the process of melting and solidification of paraffin PCM, contained in a cylindrical cell, was established in his early work [2]. Detailed temperature distribution and solid–liquid interface positions were measured during melting and solidification of the PCM. This initial work was followed shortly by a more comprehensive work to study the role of natural convection during melting and solidification of two PCMs in vertical rectangular geometry [3]. The theoretical model, developed in the work, was used for the prediction of both melting and solidification experiments, considering the effect of initial sub-cooling or superheating that might occur during the process. The agreement of the model was good not only with the experimental results of that work but also with the published work of the others.

In a further study, the performance of a heat storage unit consisting of a number of vertical cylindrical capsules filled with PCMs having different melting points, with air flowing across them for heat exchange, was analysed [4], which was probably the first attempt of using the PCM cascade system. The mathematical model developed in this work was based on solving the heat conduction equation in both melt and solid phases in cylindrical coordinates, considering the radial temperature distribution in both phases. The study showed a significant improvement in the rate of heat transfer during heat charge and discharge when PCMs with different melting temperatures were used.

The performance of a direct contact latent heat storage unit, which consists of two columns with different hydrated salts, was investigated experimentally in detail [5]. Hydrated salts have larger energy storage density and higher thermal conductivity but experience supercooling and phase segregation, and hence, their application requires the use of some nucleating and thickening agents. The use of direct contact heat exchange has eliminated phase segregation and supercoiling. Results showed "that" significant improvement in heat transfer rates could be further achieved by using two separate columns containing hydrated salts having different crystallisation temperatures (cascade system). This cascade system is particularly suitable for solar energy applications in which the collector temperature may vary significantly during the day. Modelling was achieved by building a heat transfer model between the bubble phase (heat transfer fluid) and the continuous phase (hydrated salts).

REFERENCES

1. Farid MM. A new approach in the calculation of heat transfer with phase change. *9th Int. Congr. Energy Environ.* Miami, 1989, pp. 1–19.
2. Farid MM, Mohamed AK. Effect of natural convection on the process of melting and solidification of paraffin wax. *Chem Eng Commun* 1987;57:297–316. doi:10.1080/00986448708960492.
3. Farid M, Kim Y, Honda T, Kanzawa A. The role of natural convection during melting and solidification of PCM in a vertical cylinder. *Chem Eng Commun* 1989;84:43–60. doi:10.1080/00986448908940334.
4. Farid MM, Kanzawa A. Thermal performance of a heat storage module using pcm's with different melting temperatures: Mathematical modeling. *J Sol Energy Eng Trans ASME* 1989;111:152–7. doi:10.1115/1.3268301.
5. Farid MM, Khalaf AN. Performance of direct contact latent heat storage units with two hydrated salts. *Sol Energy* 1994;52:179–89. doi:10.1016/0038-092X(94)90067-1.

2.1 A New Approach in the Calculation of Heat Transfer with Phase Change

Mohammed Farid
University of Basrah

2.1.1 INTRODUCTION

Heat transfer with phase change is a subject with wide applications such as thermal storage, casting of metals, manufacturing of ice and food preservation. In thermal storage, the heat obtained from the solar collector is stored as latent heat of fusion, which may be utilized later. The phase change material (PCM), used to store the heat, is usually encapsulated in cylindrical or flat geometries. The PCM exchanges its heat with a heat transfer fluid passing along or across the capsules containing it.

Modelling of the phase-change storage unit [1] requires the evaluation of the temperature distribution and interface location in every capsule in the storage unit. Extensive work has been carried out on heat transfer with phase change in single capsules of different geometries subjected to different boundary conditions. The analysis requires the solution of the unsteady heat conduction equation

$$\partial^2 T/\partial x^2 = \frac{1}{\alpha}\partial T / \partial t \qquad (2.1.1)$$

in the melt and solid phases. The existence of a moving interface complicates the solution significantly when the problem is two-dimensional.

The first well-known solution to the problem of heat transfer with phase change was given by Neumann [2]. The analytical solution was limited to one-dimensional heat transfer with a semi-infinite body of PCM subjected to a step-change in the temperature at the boundary. Later, more general solutions using the approximate integral method of solution [3] were reported. The explicit finite difference method of Murray and Landis [4] was found suitable for the analysis of one-dimensional heat transfer with phase change [5–8]. In the method, the melt and solid regions were divided into equal space increments having width varying with time. Boger and Westwater [5] and Farid and Mohamed [6] have modified the variable space network (VSN) method of Murray and Landis to take into account the effect of natural convection in the melt. Thermal conductivity of the melt was replaced by an effective value that includes the effect of natural convection. Farid and Mohamed [6] have reported the following correlation for the effect of natural convection in a thick slab of paraffin wax heated from the bottom:

$$k_e/k_l = 0.159\,\mathrm{Ra}^{0.34} \tag{2.1.2}$$

Farid and Hussain [8] have developed a similar correlation for a thin vertical slab of wax subjected to a constant heat flux at one side and convection boundary at the other side:

$$k_e/k_l = 0.1\,\mathrm{Ra}^{0.25} \tag{2.1.3}$$

The VSN method was found successful for one-dimensional problems. However, in any storage unit, the one-dimensional heat transfer within the capsule never exists except for very thin capsules. The main features of the VSN method may be summarized as follows:

1. The method is suitable for calculation for any type of capsules and usually requires only a small number of space divisions.
2. It provides an accurate prediction of the interface locations and hence natural convection in the melt phase may be predicted with a reasonable accuracy.
3. The method is based on melting and solidification of the PCM at a specified temperature. Hence, it should be applied with care when the PCM undergoes phase change with a wide temperature range, which is true for most commercial waxes.
4. Application of the method requires five computational steps: (a) sensible heating of the solid PCM, (b) approximate estimate for the time required for the formation of thin melt layer, (c) major melting of PCM, (d) approximate time to melt the last thin layer of the solid and (e) sensible heating of the melt. Similar steps are required in solidification.
5. If the temperature at the boundary of the PCM undergoes periodic changes, then more than one interface may form which may complicate the application of the method. This may occur in the applications of solar energy.

The limitations of the VSN method show the importance of finding a more simple solution, but with minimum assumptions so that it may be applied for practical systems. Abe et al. [9] and Kamimoto et al. [10] have studied the performance of a storage unit consisting of bundles of thin polyethylene rods as a PCM. The solid–solid transition of the polyethylene occurs within a wide range of temperature, and hence the latent heat was accounted for by using an effective heat capacity. The use of thin rods validates the assumption of no radial temperature distribution within the rods.

In this article, the previous method for solid-solid transition is modified and applied to cases where true melting and solidification occur. Unlike solid–solid transition, melting and solidification may occur within a narrow temperature range. They require the interface to be traced continuously for the evaluation of the effect of natural convection which may be significant.

2.1.2 ANALYSIS

In the previous analysis of heat transfer with phase change, melting and solidification were assumed to occur with the formation of a sharp interface. Hence, the unsteady heat conduction equation was solved in both phases with an interface where all the latent heat effects occur.

In the present analysis, the phase change is assumed to develop within a finite region having a width that depends on the nature of the PCM. This phase-change region moves in the direction of heat flow. The heat conduction equation is assumed to describe heat transfer in the solid, melt and phase-change regions. The only variation is in the value of the thermal diffusivity (α). Density and thermal conductivity of the PCM in the phase-change region are taken as their average values in the melt and solid regions. The latent heat effect was accounted for, by using an effective heat capacity (C_e) which undergoes large variation in its values within the phase-change region. Such variation may be of the order of one hundredth or less depending on the melting range of the PCM. The temperature dependency of the effective heat capacity of the PCM may be obtained from the differential scanning calorimetric (DSC) measurements. However, it is found in this work that such accurate information is unnecessary if the approximate melting range of the PCM is known. Hence, any function may be used for the variation of C_e with temperature, which satisfies the following integral:

$$\int_{T_{m1}}^{T_{m2}} C_e \, dT = \text{Latent heat of fusion} \tag{2.1.4}$$

To study the effect of melting (or solidification) range, a simple delta function with different widths is used to represent the variation of C with temperature, as shown in Figure 2.1.1. The area under these delta functions is always equal to the latent heat of fusion of the PCM.

The effect of natural convection in the melt phase cannot be ignored. Previous investigators [5–8] have included this effect using correlations similar to those given by Equations (2.1.2) and (2.1.3). The use of such correlations requires the knowledge of a sharp interface. For this purpose, the interface, in the present article, is defined as the position where C_e is maximum, even though no true sharp interface exists.

The method of analysis adopted in this work, for the calculation of heat transfer with phase change, has been tested against the following sets of experimental measurements.

1. Melting of the thin vertical slab of PCM [8] subjected to constant heat flux at one side and convection boundary at the other.

   The wax (PCM1), with the physical properties shown in Table 2.1.1, was enclosed in a 48 mm thick container (with 0.85 m height and 0.52 m width). The heating element was fixed vertically to divide the container into two symmetrical sections receiving equal heat with a constant rate as sketched in Figure 2.1.2. The enclosed air adjacent to the container walls was continuously heated by the heat lost from the PCM. The measured air temperature

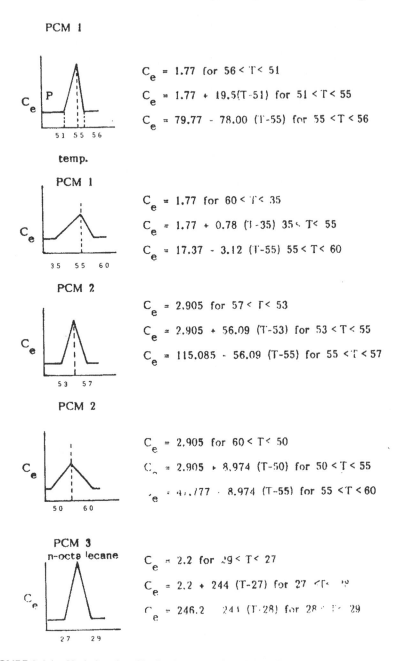

PCM 1

C_e = 1.77 for $56 < T < 51$

C_e = 1.77 + 19.5(T-51) for $51 < T < 55$

C_e = 79.77 - 78.00 (T-55) for $55 < T < 56$

temp.

PCM 1

C_e = 1.77 for $60 < T < 35$

C_e = 1.77 + 0.78 (T-35) $35 < T < 55$

C_e = 17.37 - 3.12 (T-55) $55 < T < 60$

PCM 2

C_e = 2.905 for $57 < T < 53$

C_e = 2.905 + 56.09 (T-53) for $53 < T < 55$

C_e = 115.085 - 56.09 (T-55) for $55 < T < 57$

PCM 2

C_e = 2.905 for $60 < T < 50$

C_e = 2.905 + 8.974 (T-50) for $50 < T < 55$

C_e = 47.177 - 8.974 (T-55) for $55 < T < 60$

PCM 3
n-octadecane

C_e = 2.2 for $29 < T < 27$

C_e = 2.2 + 244 (T-27) for $27 < T < 29$

C_e = 246.2 - 244 (T-28) for $28 < T < 29$

FIGURE 2.1.1   Variation the effective heat capacity of the PCMs with temperature.

TABLE 2.1.1

Physical Properties of the Waxes Used in the Analysis

Properties	PCM 1	PCM 2	PCM 3
Melting temperature, °C	55	55	28
Solid density, kg/m^3	910	785.3	814
Melt density, kg/m^3	822	769.2	768.4
Solid thermal conductivity W/m °C	0.29	0.24	0.30
Melt thermal conductivity W/m °C	0.21	0.18	0.157
Melt viscosity near melting temperature, kg/m s	8×10^{-3}	5×10^{-3}	3.9×10^{-3}
Specific heat capacity of solid and melt, kJ/kg °C	1.770	2.905	2.200
Latent heat of fusion kJ/kg	195	224.36	244.00
Thermal expansion coefficient (β), k^{-1}	8.0×10^{-3}	8.0×10^{-3}	9.0×10^{-3}

was used as a known boundary condition with the calculated air convection coefficient ($U = 10\,W/m^2\,°C$). It is to be noted that the container was divided into three sections along its height. The results reported in Figures 2.1.2–2.1.6 belong to the lowest section. The wax used was of the commercial type containing impurities that affect its melting characteristics. The differential scanning calorimetric measurements carried out for this wax (result is not shown) suggest the unsymmetrical variation of C_e with temperature as shown in Figure 2.1.1.

2. Melting of thick PCM body due to sudden rise in the temperature of its bottom face [6].

The wax (PCM 2) was contained in a 100 mm long cylindrical test cell having inside diameter of 90 mm. The PCM body was thick enough to be considered as a semi-infinite body. The wall of the cell was insulated to ensure one-dimensional heat flow as shown in the sketch in Figure 2.1.7. The wax used was of high purity and more symmetrical variation of C_e with temperature is expected as shown in Figure 2.1.1. Physical properties of the wax are given in Table 2.1.1.

3. Melting and solidification of 89 mm thick slab of n-octadecane (PCM 3) due to a sudden rise in the temperature of the bottom face [11].

The width and depth of the test cell were 146 mm and 22 mm, respectively. The pure n-octadecane is expected to undergo a symmetrical variation in C_e with temperature as shown in Figure 2.1.1. Physical properties of n-octadecane are shown in Table 2.1.1.

4. Melting of 24.54 mm thick slab of wax (n-octadecane) due to heating from above with constant heat flux, while the bottom face being insulated [12].

With such an arrangement, natural convection does not play any important role in heat transfer.

In the numerical solution of the conduction equation using explicit finite differences, the PCM body was divided into equal space divisions in the direction of heat flow. In all the four sets of experiments, heat transfer was one-dimensional, so the analysis

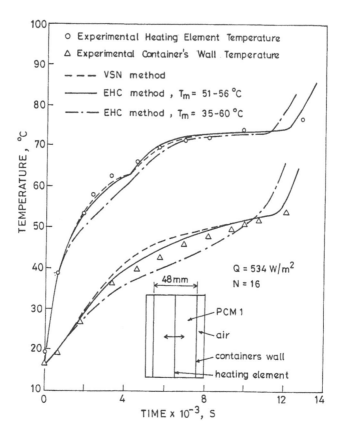

FIGURE 2.1.2 Prediction of the heating element and container's wall temperature during melting of vertical slab of PCM (8).

was limited to such condition; the extension of the analysis to multidimension is straightforward.

The metallic surface separating the PCM from the heating source does not have an important contribution to the total thermal resistance due to the high thermal conductivity of the metal compared to that of the wax. However, the heat capacity of the metallic surface was included as it is expected to be important, particularly, for the experiments of heating with constant heat flux.

2.1.3 RESULTS

Figures 2.1.2–2.1.6 exhibit the first set of experiments of heating a thin vertical slab of PCM with constant heat flux at one side and convection boundary at the other side. Figures 2.1.2 and 2.1.4 show the variation of the temperatures of the heating surface and of the container's wall for two different heat fluxes. Two delta functions were assumed to define C_e with different melting ranges. The variable space network method (VSN) and the effective heat capacity method (EHC) with $T_m = 51°C–56°C$

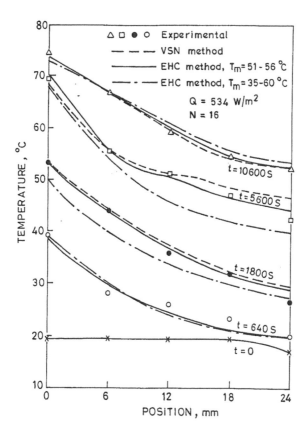

FIGURE 2.1.3 Prediction of the temperature distribution in a vertical PCM slab during melting with constant heat flux (8).

agree well with each other and with the experimental measurements. When the PCM was assumed to melt with a very wide temperature range of 35°C–60°C, the agreement between the EHC method and the experimental results was poor. This is even more evident in Figures 2.1.3 and 2.1.5, showing the temperature distribution within the PCM for the same two experiments. When the PCM temperature has not yet reached 35°C, the EHC method gives good results irrespective of the melting range assumed. The EHC method with T_m =35°C–60°C suggests that the melting starts soon after the temperature reaches 35°C. This early melting causes a decrease in the rate of temperature change with time as shown in Figures 2.1.2 and 2.1.4. After the PCM temperature has reached 51°C, some deviation is also evident, but to the less extent, between the VSN and the EHC methods assuming T_m=51°C–56°C.

The prediction of the interface position by the EHC method is not very accurate and requires an interpolation. The interface position is defined as the location where the temperature is close to 55°C (within ±0.5°C), the smooth curves, represented by the solid lines in Figure 2.1.6, are not the true interface calculated by the EHC method. The interface position changes in a stepwise manner and not continuously;

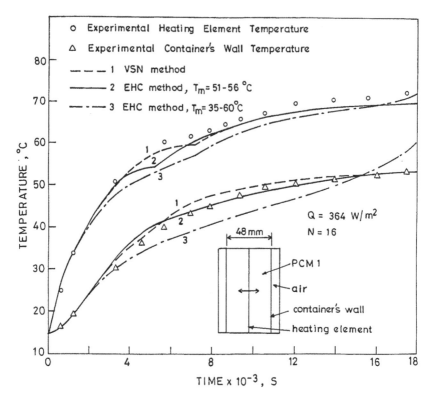

FIGURE 2.1.4 Prediction of the heating element and container's wall temperatures during melting of vertical slab of PCM (8).

however, they were smoothed for clarity. Total number of space divisions (Δx) were eight for both methods; however, increasing the number to 16 had a very little effect that could not be shown in the figures. The change in the melting range was big (from 51°C–56°C to 35°C–60°C). Smaller changes showed only a very little effect on the results. A very narrow melting range should be avoided otherwise significant amount of the latent heat will not be accounted.

The second set of experiments in Figures 2.1.7–2.1.9 was for melting of thick PCM slab due to heating from below. The temperature of the lower face of the PCIM was raised from 50°C to 59°C. Two melting ranges of 54°C–56°C and 50°C–60°C were selected as shown in Figure 2.1.1 for PCM 2. The deviation between the experiments and the prediction of the EHC method shown in Figure 2.1.7 is mainly due to the large Δx of 3.3 mm. For the wide melting range (50°C–60°C), the EHC method is expected to give a lower interface position, as discussed before. Hence, the good agreement of the method is by a mere coincidence. Figure 2.1.8 shows large deviation between the measured temperature distribution and that predicted by the EHC method ($T_m = 50°C–60°C$) after 1200 s.

As we have noted earlier, the EHC method predicts the interface in a stepwise manner. If the interface position is not required during the computation, then this will

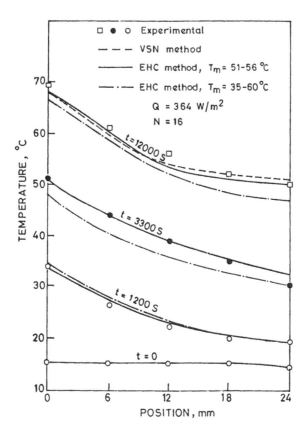

FIGURE 2.1.5 Prediction of the temperature distribution in a vertical PCM slab during melting with constant heat flux (8).

not cause any error in the calculation of heat transfer. However, the interface position is usually required for the evaluation of the effect of natural convection. If natural convection is very large, then accumulated error will be expected. Figure 2.1.9 shows the results when the PCM slab bottom face temperature was raised from 46.5°C to 73.5°C. Under this condition, severe convection in the melt is expected. However, the condition of Figure 2.1.9 has limited applications and may be considered unimportant.

The EHC method, as shown in Figure 2.1.10, predicts well the experimental results of Hale and Viskanta [11] for melting and solidification of 89 mm thick slab of n-octadecane from below. The prediction of the VSN method is also given. For the experiment of melting, natural convection was evaluated using an equation similar to Equations (2.1.2) and (2.1.3) but with different constants. The constants (0.16 and 0.305) were evaluated from a direct comparison between the VSN method and the experimental results. Only one melting range (27°C–29°C) is assumed for the n-octadecane, as shown in Figure 2.1.1. The assumption of phase change with a wider range of temperature will generate erroneous results since the initial temperature of the wax (27°C–29°C) was very close to the phase-change temperature (28°C). The

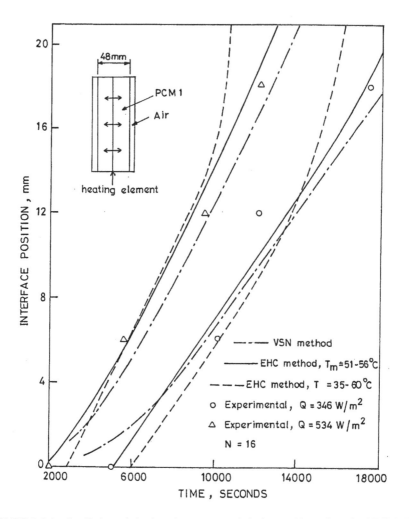

FIGURE 2.1.6 Prediction of the interface measured during melting of vertical PCM slab with constant heat flux (8).

EHC method predicts higher rates of solidification when the number of space divisions is reduced from 30 to 12. This is because that significant part of the latent heat is not accounted.

2.1.4 CONCLUSION

The variable space network method was found successful in describing the one-dimensional heat transfer with phase change. Major limitations are as follows: (a) the method is difficult to be applied for multidimensional system, (b) it is suitable for PCM which melts only with narrow melting range, and (c) the method is not suitable for systems with cyclic boundary conditions which may cause more than one solid–liquid interface within the PCM body. Further, the method requires an approximate

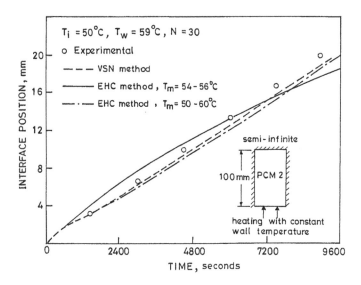

FIGURE 2.1.7 Prediction of the interface position in a PCM slab during melting from below with constant wall temperature (6).

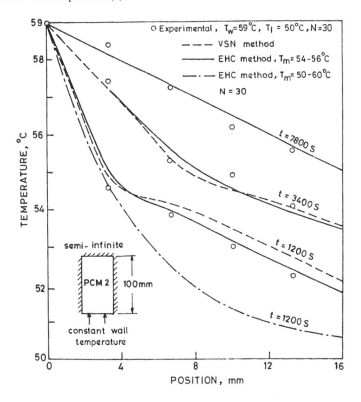

FIGURE 2.1.8 Prediction of the temperature distribution in a thick stab of PCM during melting with constant wall temperature (6).

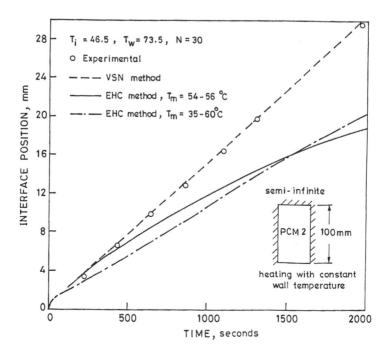

FIGURE 2.1.9   Prediction of the interface position in a PCM slab during melting from below with constant wall temperature (6).

estimate for an initial melted or solidified layer causing some error in the early estimate of heat transfer rate. The effective heat capacity method, suggested in this work, does not carry any of these limitations.

This method was found successful in describing the experimental measurements of one-dimensional heat transfer with phase change. The extension of the method to multidimensions and different geometries is straightforward. The method is not sensitive to little changes in the melting range. Very narrow melting ranges should be avoided; otherwise, part of the latent heat effect will not be accounted. Increasing the number of space divisions allows the use of narrow melting ranges.

The method is easy to program and requires less computer time than that required by the VSN method.

NOMENCLATURE

C = specific heat capacity, kJ/kg °C
C_e = effective heat capacity, kJ/kg °C
EHC = effective heat capacity
F_0 = Fourier number, α, t/H
g = acceleration due to gravity, m/s²
H = PCM thickness in the direction of heat, m
k = thermal conductivity, W/m °C

k_e = effective thermal conductivity, W/m °C
N = number of space division
Ra = Rayleigh number, $\rho_1^2 g Y^3 C (T_w - T_m)/\mu k_1$
S = dimensionless melt layer thickness Y/H
Ste = Stefan number, $(T_w - T_m) C / \lambda$
T = temperature, °C
t = time, s
U = heat loss coefficient, W/m²°C
VSN = variable space network
X = position, m
Y = melt layer thickness, m
α = thermal diffusivity, m²/s
β = volumetric thermal expansion coefficient, k⁻¹
λ = Latent heat of fusion, kJ/kg
ρ = density, kg/m³
μ = melt viscosity, kg/m s
∂ = partial derivative

Subscript

a = air
e = effective value
i = initial
l = liquid
m = melting
m_1 = starting temperature for melting
m_2 = final temperature for melting
w = wall

REFERENCES

1. Farid, M.M. and Kanzawa, A., "Thermal Performance of a Heat Storage Module Using PCM's with Different Melting Temperatures: Mathematical Modelling", *Trans. ASME, J. Sol Energy Eng.*, Vol. III, 1989, pp. 152–157.
2. Carslaw, H.S. and Jaeger, J.C., *"Conduction of Heat in Solids"* 2nd Edition, Oxford, Clarendon Press, 1959, pp. 282–286.
3. Tien, C. and Yen, Y.C., "Approximate Solution of Melting Problem with Natural Convection", *AIChE Chem. Eng. Prog. Symp. Ser.* 62, 1966, pp. 166–172.
4. Murray, W.D. and Landis, F., "Numerical and Machine Solution of Transient Heat Conduction Problems Involving Phase Change", *Trans. ASME, Ser. C, J. Heat Transfer*, 81, 1959, pp. 106–112.
5. Boger, D.V. and Westwater, J.W., "Effect of Buoyancy on the Melting and Freezing Process", *Trans. ASME, Ser. C, J. Heat Transfer*, 189, 1967, pp. 81–87.
6. Farid, M.M. and Mohamed, A.K., "Effect of Natural Convection on the Process of Melting and Solidification of paraffin Wax", *Chem. Eng. Comm.*, 57, 1987, pp. 297–316.
7. Farid, M.M., Kim, Y., Honda, T. and Kanzawa, A., "The Role of Natural Convection during Melting and Solidification of PCM in a Vertical Cylinder", *Chem. Eng. Comm.* (In Press).

8. Farid, M.M. and Husian, R.M., "An Electrical Storage Heater Using Phase-Change Method of Storage", to be published.

9. Abe, Y. et al, "Charge and Discharge Characteristics of a Direct Contact Latent Thermal Energy Storage Unit Using Form-Stable High Density Polyethylene", *Trans. ASME, J. Sol. Energy Eng.*, 106, 1984, pp. 465–474.

10. Kamimoto, M. et al., "Latent Thermal Energy Unit Using Form-Stable High Density Polyethylene; Part II; Numerical Analysis of Heat Transfer", *Trans. ASME, J Sol Energy Eng.*, 108, 1986, pp. 290–297.

11. Hale, N.W. and Viskanta, R., "Solid-Liquid Phase Change Heat Transfer and Interface Motion in Materials Cooled or Heated from above or below", *Int. J. Heat and Mass Transfer*, 23, 1980, pp. 283–292.

12. Pujudo, P.R., Stermole, J.F. and Golden, J.O., "Melting of a Finite Paraffin Slab as Applied to Phase Change Thermal Control", *J. Spacecraft Rockets*, 6, 1969, pp. 280–284.

2.2 Effect of Natural Convection on the Process of Melting and Solidification of Paraffin Wax

Mohammed Farid and Ahmed. K. Mohamed
University of Basrah

2.2.1 INTRODUCTION

Heat transfer involving phase change has become very important due to its wide practical application in manufacturing of ice, storage of solar energy, food preservation, growth of crystals, etc. Extensive work has already been carried out to study heat transfer during melting and solidification of different materials, both experimentally and theoretically.

The earliest analytical solutions to this problem given by Neumann[1,2] were limited to a one-dimensional heat flow with a semi-infinite region subjected to a step-change in temperature at the boundary. Neumann has solved Fourier equation in both phases with the stated boundary and initial conditions. For other geometries and boundary conditions, attempts have been made to solve the problem of solidification by applying numerical methods based either on finite difference formulation[3–7] or by using approximate integral rnetbods.[8,9]

In the early solutions, heat was assumed to flow by conduction in both phases, and hence the solutions were valid for only limited practical conditions of no convection. Hale and Viskanta[8] and Tien and Yen,[9] both applied the approximate mate heat balance integral method of Goodman,[13] have included the effect of convection by replacing the conduction heat flux in the liquid with a convective heat flux. Fourier equation was then solved for the solid phase with the modified boundary condition at the interface. The convective heat flux was calculated from previous correlations developed originally for natural convection in a fluid confined between two parallel plates. In studying ice–water system, Boger and Westwater[5] have solved the conduction equation in both phases using the numerical technique of Murray and Landis. The effect of convection was also included by replacing liquid thermal conductivity with an effective thermal conductivity. Although Boger and Westwater have developed their own correlation for the effective thermal conductivity, there was no significant difference between their correlation and those obtained with no phase change.

Two distinct techniques have been found in literature to identify the solid–liquid interface position. First is the photographic method that provides no idea about temperature distribution. The second is to record temperature profile throughout the test material in the cell using thermocouples. From the known fusion temperature of the test material, the interface position, in the second method may be traced easily. This method has been widely used by most of the investigators.

Literature provides very little information about temperature distribution in the liquid phase during phase-change processes involving convection. Hale and Viskanta[8] carried some measurements of temperature distribution during melting of, n-octadecane from below; however, their results do not provide a clear picture about temperature distribution in the liquid. Yin-Chao Yen[11] has made similar measurements during the melting of ice. The temperature profile in the melted water was found linear at the early stages of the experiment in which no convection occurred. However as the melted layer thickness increased, the temperature in the liquid phase became uniform due to convective mixing. Most of the other investigators assumed complete mixing in the liquid phase during the process of melting from below. However, to fully understand the role of convection in such processes, it will be necessary to carry out detailed measurements of temperature distribution in both phases. It is expected that viscous materials such as paraffin waxes will behave differently than water.

In this work, measurements have been carried out in determining the temperature distribution in both phases and the position of the interface. These results have then been compared with theoretical predictions based on both conduction and convection models. A finite difference method[5] was used for both models employing the exact boundary conditions used in this work. The method was a modification to the variable-space network method of Murray and Landis.[3] Neumann solution, based on semi-infinite condition, was also tried for comparison even though the test cell thickness was less than 100 mm.

2.2.2 APPARATUS AND PROCEDURE

A schematic diagram of the test apparatus is, shown in Figure 2.2.1. Heat is conducted vertically either upward or downward through the paraffin wax (mixture of pentacosane and hexacosane) in a 100 mm long cylindrical test cell made of a glass tube of 90 mm inside diameter. A thick layer of glass wool was wrapped around the cylinder to minimize heat transfer in the radial direction, hence implementing one-dimensional heat flow.

The glass tube was fixed between two copper boxes by four studs as shown in the diagram. Rubber gaskets were used between the glass tube and the boxes to prevent leaks. During the experiment, water was circulated through the boxes from two separate water baths capable of providing constant water temperature within ±0.5°C. The bottom box was used as a heat source or sink and was designed carefully to achieve an instantaneous, stepwise change in the temperature of its face in contact with the paraffin wax and also to maintain this temperature constant during the process of

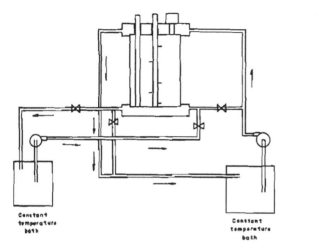
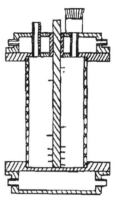

FIGURE 2.2.1 Schematic diagram of the test arrangement.

melting and solidification. Baffles were used in both boxes to increase water velocity ensuring high rates of heat transfer. Heat balance across the thin copper plate showed that the difference between plate and water temperature will never exceed 1°C.

In the upper box, water at a temperature equal to the initial temperature of the wax was circulated to prevent any heat loss from the top layer of the wax. The two boxes were similar except that the upper box contained three passages in it. The central passage was used to introduce the plastic tube containing the thermocouples. One of the other passages was connected to a small reservoir soldered to the upper box and filled with the same paraffin wax kept at a temperature near the initial temperature of the wax. This has been made to compensate for the decrease in volume of the wax in the cell during solidification, which was of the order of 2%. The flow of the wax from the reservoir to the cell was too slow to disturb the condition inside the cell seriously.

During solidification, a glass rod was inserted through the third passage right to the bottom of the cell. The rod was then pulled while the paraffin was solidifying leaving a cylindrical hollow space in the solid paraffin. This was to accommodate the increase in volume during melting.

Temperature distribution in both phases was measured using eleven 0.3 mm Iron-Constantan thermocouples, fixed axially through holes in the small diameter plastic tube. These thermocouples were displaced 90° to reduce disturbance to the measurements. Positions of the thermocouples, as measured from the lower end were, respectively, 3.3, 6.6, 9.9, 13.2, 16.5, 19.8, 29.8, 49.8, 89.8, and 100 mm. The last two thermocouples were used only to ensure the condition of no heat flux at the upper end of the cell. The temperature of the lower plate was also measured by using 0.5 mm Copper–Constantan thermocouples soldered carefully to the surface of the plate. All the 12 thermocouples were connected to a multichannel recorder capable of recording temperature directly.

In the solidification experiments, special care was taken during the process of charging to get rid of any air bubble trapped in the cell by vibrating the test cell during charging. The wax was first melted outside the cell and heated to the required initial temperature. Water at a temperature slightly above the initial temperature of the melted wax (about 1°C) was circulated through the two boxes until the whole material reached a steady uniform temperature. During this process, the other bath was set to the desired wall temperature. Suddenly, the hot water supplied to the lower box was stopped and replaced by the cold water from the other bath using the valves shown in Figure 2.2.1. The temperature at the different locations was continuously recorded until the interface reached a distance of about 30 mm from the bottom.

In the melting experiments with the material in the solid phase, it was necessary to run the apparatus for many hours to reach a steady uniform temperature. Then, the procedure was similar to that in solidification.

2.2.3   THEORY AND METHODS OF COMPUTATION

2.2.3.1   NEUMANN ANALYSIS

In this work, the semi-infinite boundary condition was approximated so that the analytical solution of Neumann[1,2,8] may be checked against a more general numerical solution.[3,5]

If convection in the liquid phase is neglected, then the temperature distribution in the solid and liquid phases is governed by the conduction equations:

$$\frac{\partial T_s}{\partial t} = \alpha_s \frac{\partial^2 T_s}{\partial y^2} \tag{2.2.1}$$

$$\frac{\partial T_l}{\partial t} = \alpha_l \frac{\partial^2 T_l}{\partial y^2} \tag{2.2.2}$$

Neumann has solved the above equations analytically with the assumptions of semi-infinite region and uniform initial temperature distribution. Physical properties of the two phases were assumed different but independent of temperature. The change in volume due to the density difference of the two phases was included by introducing ratio of the two densities, γ. However, in this work, a unity value was assumed for γ since the arrangement has been done to get rid of the effect of such change.

The nature of the problem necessitates the use of the following interface boundary conditions:

$$T_s(Y,t) = T_l(Y,t) = T_f \tag{2.2.3}$$

$$\rho_s L \frac{dY}{dt} = K_s \left. \frac{\partial T_s}{\partial y} \right|_{y=Y} - K_l \left. \frac{\partial T_l}{\partial y} \right|_{y=Y} \tag{2.2.4}$$

In the solidification experiment, liquid is initially at a uniform temperature T_0. Suddenly, the temperature at $y=0$ is lowered to a value T_w below the fusion temperature T_f, while the temperature at large distances ($y \to \infty$) is maintained at T_0. Then, the temperature distribution in both phases may be given by

$$\Theta_s = \frac{T_s - T_f}{T_w - T_f} = 1 - \frac{\mathrm{erf}\,\dfrac{y}{2\sqrt{\alpha_s t}}}{\mathrm{erf}\,\dfrac{\lambda\gamma}{2\sqrt{\alpha_s}}} \tag{2.2.5}$$

$$\Theta_l = \frac{T_l - T_f}{T_w - T_f} = \left[1 - \frac{\mathrm{erfc}\,\dfrac{y}{2\sqrt{\alpha_l t}}}{\mathrm{erfc}\,\dfrac{\lambda}{2\sqrt{\alpha_l}}}\right]\left[\frac{T_0 - T_f}{T_w - T_f}\right] \tag{2.2.6}$$

where λ is a constant to be determined from the equation

$$\left(\frac{1}{Ste_s}\right)\left(\frac{\lambda\gamma}{2}\right) = \sqrt{\frac{\alpha_s}{\pi}}\,\frac{\exp\left(\dfrac{-\lambda^2\gamma^2}{4\alpha_s}\right)}{\mathrm{erf}\,\dfrac{\lambda\gamma}{2\sqrt{\alpha_s}}} + \sqrt{\frac{\alpha_l}{\pi}}\left(\frac{\rho_l c_l}{\rho_s C_s}\right)\left(\frac{T_0 - T_f}{T_w - T_f}\right)\frac{\exp\left(\dfrac{-\lambda^2}{4\alpha_l}\right)}{\mathrm{erfc}\,\dfrac{\lambda}{2\sqrt{\alpha_l}}} \tag{2.2.7}$$

The solid–liquid interface position is given by

$$Y = \lambda\gamma\sqrt{t} \tag{2.2.8}$$

If convection is neglected during the process of melting from below, which is true only at the early stages of melting then the temperature distribution and interface position are governed by the equations

$$\Theta_s = \frac{T_s - T_f}{T_w - T_f}\left[1 - \frac{\mathrm{erfc}\,\dfrac{y}{2\sqrt{\alpha_s t}}}{\mathrm{erfc}\,\dfrac{\lambda}{2\sqrt{\alpha_s}}}\right]\left[\frac{T_0 - T_f}{T_w - T_f}\right] \tag{2.2.9}$$

$$\Theta_l = \frac{T_l - T_f}{T_w - T_f} = 1 - \frac{\mathrm{erf}\,\dfrac{y}{2\sqrt{\alpha_l t}}}{\mathrm{erf}\,\dfrac{\lambda/\gamma}{2\sqrt{\alpha_l}}} \tag{2.2.10}$$

$$Y = \left(\frac{\lambda}{\gamma}\right)\sqrt{t} \tag{2.2.11}$$

The constant λ may be determined from the equation

$$\left(\frac{1}{Ste_l}\right)\left(\frac{\lambda}{2\gamma}\right) = \sqrt{\frac{\alpha_l}{\pi}}\,\frac{\exp\left[-(\lambda/\gamma)^2/4\alpha_l\right]}{\mathrm{erf}\left[(\lambda/\gamma)/2\sqrt{\alpha_l}\right]} + \sqrt{\frac{\alpha_s}{\pi}}\left(\frac{\rho_s C_s}{\rho_l C_l}\right)\left(\frac{T_0 - T_f}{T_w - T_f}\right)\frac{\exp\left(-\lambda^2/4\alpha_s\right)}{\mathrm{erfc}\left(\lambda/2\sqrt{\alpha_s}\right)} \tag{2.2.12}$$

A computer program was written to solve the aforementioned equations with the stated experimental boundary and initial conditions for both solidification and melting experiments. A desktop computer was sufficient to carry out these calculations.

It must be noted that although the test cell depth used in this work was only 100 mm, the experiments showed no significant change in temperature at locations near the upper end of the cell within the time period of all experiments. These observations provided close approximation to the semi-infinite conditions. Therefore, some agreement between Neumann's solution and the solution given by finite differences was expected.

2.2.3.2 MELTING FROM BELOW WITH NATURAL CONVECTION

It is evident from previous work that convection will develop gradually in the liquid phase during the process of melting from below, and hence the Neumann model cannot be applied.

Analytical solution is not possible so numerical methods must be tried. The explicit method of Murray and Landls[3,5] with variable space network approach was selected. In this method, the melted or solidified region was divided into N space increments of size Y/N which increases with time. The other region was also divided into N equal space increments of size $(E-Y)/N$, which decreases as melting or solidification progress.

Equations (2.2.1), (2.2.2), and (2.2.4) were written in a finite difference form as follows (2.2.3):

In the solid phase:

$$\Theta_{n,m+1} = \frac{n}{2Y_m}\left(\Theta_{n+1,m} - \Theta_{n-1,m}\right)\left(Y_{m+1} - Y_m\right) + \frac{\alpha_s \Delta t N^2}{Y_m^2}\left(\Theta_{n-1,m} - 2\Theta_{n,m} + \Theta_{n+1,m}\right) + \Theta_{n,m}$$

$$(2.2.13)$$

In the liquid phase:

$$\Theta_{n,m+1} = \frac{2N-n}{E-Y_m}\left(\Theta_{n+1,m} - \Theta_{n-1,m}\right)\left(Y_{m+1} - Y_m\right)$$

$$+ \frac{\alpha_l \Delta t N^2}{\left(E-Y_m\right)^2}\left(\Theta_{n-1,m} - 2\Theta_{n,m} + \Theta_{n+1,m}\right) + \Theta_{n,m}$$

$$(2.2.14)$$

and

$$Y_{m+1} = \frac{\Delta t}{\rho L}\left[\frac{K_s N}{2Y_m}\left(\Theta_{N-1,m} - 4\Theta_{N,m}\right) + \frac{K_l N}{2\left(E-Y_m\right)}\left(\Theta_{N+3,m} - 4\Theta_{N+2,m}\right)\right] + Y_m \quad (2.2.15)$$

These finite difference formulations were written for the process of solidification, and they must be reversed for melting. However, for the case of melting from below, the effect of convection must be included. This has been done by replacing liquid thermal conductivity with an effective thermal conductivity which includes the effect of convection. It is a well-known fact that natural convection develops in the fluid confined between two horizontal plates when heated by the lower plate or cooled by the upper plate. Convection was found to follow correlations of the form $K_e/K_l = cR_a^n$, where c and n are constants that depend on the geometry. It is expected that convection in the melted layer can be estimated from similar correlations. Boger and Westwater[5] have applied this idea and reached a correlation similar to those obtained previously with no phase change. However their results cannot be generalized since water has a peculiar density inversion at 4°C.

In this work, the computation was modified in such a way that the values of c and n were changed until the best agreement between the measured and computed interface positions in all experiments has been achieved. A computer program was written to calculate the change in the interface position with time as well as the temperature distribution in both phases. Values of K_e were calculated at the end of each time increment and were used in the stepwise calculation of the temperature distribution. A desktop computer (COMMODORE 64) was sufficient to carry out such calculation using five equal spatial increments in each phase. However, to obtain accurate temperature distribution using this method, it was necessary to use 20 increments in each phase, which necessitate the use of high-speed computer. Time increment was adjusted automatically according to the stability criteria of the explicit methods. $\Delta t < Y^2/3\alpha_e N^2$. Time increment, Δt, would increase as melting or solidification progress; however, the gradual increase in the values of α_e during melting reduced computational speed at later stages to some extent.

It is obvious from the finite difference formulation that the numerical solution cannot be started if only one phase is present. Hence, the solution was initiated by assuming the presence of a melt or solidified layer of very small thickness. The time required to form this layer may be calculated from latent heat and the known temperature gradient in the formed layer. Different initial thicknesses were tried, and the effect was found to be significant on the accuracy of the interface positions, if initial thickness of the formed layer was greater than 1 mm. Decreasing initial thickness much below 1 mm would increase computational time significantly without great improvement in the accuracy of the results.

Initial temperature distribution in the melted or solidified layer was assumed linear varying between the temperature of lower plate and the fusion temperature. In the other region, temperature was assumed unchanged except in the region bounded by the interface and the first nodal point, where it was assumed to change from the fusion to the initial temperatures. This assumption was reasonable since the time required to form a layer of 1 mm was too small to show any significant variation at larger distances. Accordingly, the computation was started at time t_0 in which interface has reached a distance of 1 mm from the lower plate.

Physical properties of the pentacosane–hexacosane mixture used are given below. It is known that although most of the binary mixtures behave in a rather complicated manner leading to phase segregation, but in this case, it is not so because the two compounds used have almost identical physical properties.

Melting point = 55°C
Latent heat of fusion = 224.36 kJ/kg
Solid phase:

density = 785.3 kg/m³
specific heat = 2.905 kJ/kg °C
thermal conductivity = 0.24 W/m °C

Liquid phase:

density = 769.2 kg/m³
specific heat = 2.905 kJ/kg °C
thermal conductivity = 0.179 W/m °C

Viscosity was measured as a function of temperature and found to vary from 5×10^{-3} to 3.6×10^{-3} kg/m s in the temperature range from 60°C to 70°C.

2.2.4 RESULTS AND DISCUSSION

2.2.4.1 SOLIDIFICATION FROM BELOW

In the experiments of solidification, initial temperature of the pentacosane–hexacosane mixture was nearly the same in all the three experiments (58.5°C, 59°C, and 60°C), while the lower plate temperature was varied from 20°C to 40°C. Figure 2.2.2 shows the interface motion during the three experiments. Interface positions were found directly from the measured temperature profile in both phases shown in Figures 2.2. 3–2.2.5, by measuring the time required for the temperature at known positions to reach 55°C. The solid lines in Figures 2.2.2–2.2.5 represent the theoretical prediction based on both, the analytical solution of Neumann and the numerical solution of Murray and Landis. It was not possible to show the differences between the two solutions within the accuracy of the drawing. The agreement between

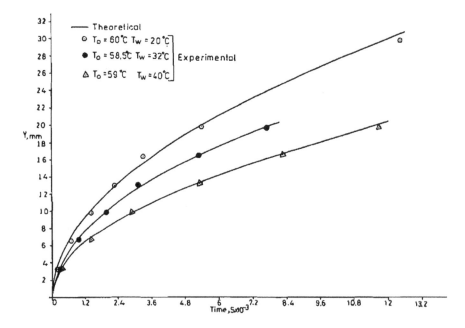

FIGURE 2.2.2 Comparison of experimental and theoretical interface positions during solidification.

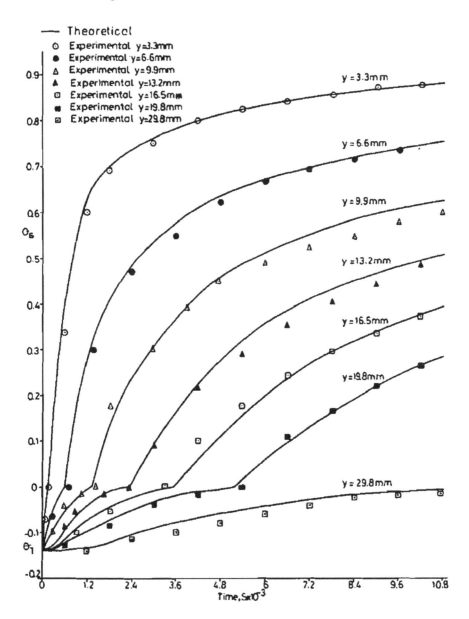

FIGURE 2.2.3 Experimental and theoretical temperature distribution during solidification, $T_0 = 60°C$, $T_w = 20°C$.

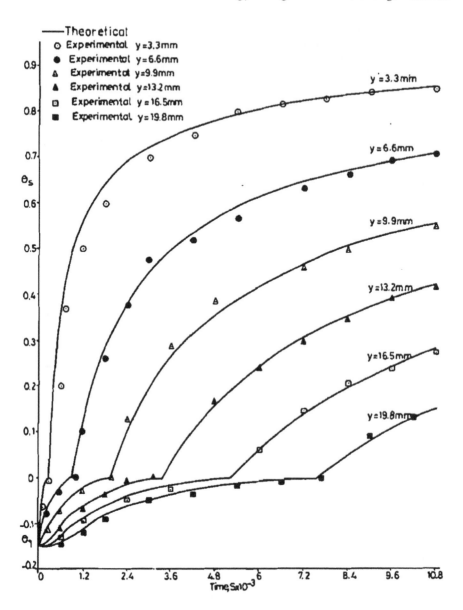

FIGURE 2.2.4   Experimental and theoretical temperature distribution during solidification, $T_0 = 58.5°C$, $T_w = 32°C$.

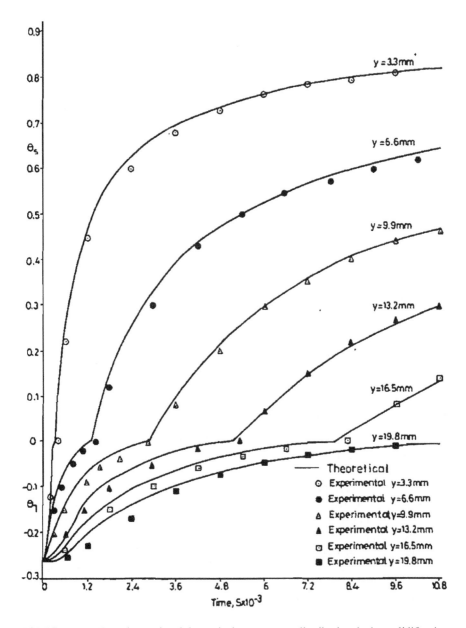

FIGURE 2.2.5 Experimental and theoretical temperature distribution during solidification, $T_0 = 59°C$, $T_w = 40°C$.

experimental and theoretical estimates of interface position and temperature distribution is reasonably good in all experiments. In the experiment of solidification of n-octadecane, Hale and Viskanta[8] have observed some deviation between the experimental and theoretical values of interface positions at the early stages of the experiments. This may have been due to the fact that it was impossible to achieve very rapid changes in the temperature of the lower plate, which may have caused some delayed action during solidification. Yin-Chao Yen[11] suggested that it may take 5 min for the temperature of the plate to reach 97% of the steady-state value. In this work, the lower plate temperature was recorded with time and found to reach the steady-state value within about 2 min. This has been achieved by the proper design of the lower copper box. The deviations in the dimensionless temperature Θ_l and Θ_s, shown in Figures 2.2.3–2.2.5, belong to a very small deviation in the temperature as can be seen from their definitions.

Interface velocities, measured from the slope of the curves in Figure 2.2.2, were found to decrease from about 6×10^{-3} mm/s measured after 5 min, down to about 1×10^{-3} mm/s at the end of the experiment (after nearly three hours). These values naturally depend on both initial and wall temperatures. Boger and Westwater[5] showed similar trends during freezing of water from below; however, interface velocities were higher due to the higher thermal conductivity of water which is about 2.5 times that of pentacosane–hexacosane mixture. Thomas and Westwater[5] measured the interface velocity during the solidification of n-octadecane. The fusion front velocity never exceeded 2.3×10^{-3} mm/s; however, the deviation between the experiments and theoretical prediction was large.

2.2.4.2 Melting with Convection

The good agreement shown in Figures 2.2.2–2.2.5 between experimental measurements and theoretical predictions based on the conduction model gave strong evidence that the process was not affected by convection. However, during melting from below, convection currents develop as the melted layer thickness increases and hence conduction models will fail to predict temperature distribution and interface positions correctly. This is shown in Figures 2.2.6–2.2.8 in which the calculated interface positions using both conduction and convection models were plotted. It is clear that the conduction model describes the process only at the early stages of the experiments. At later stages, the measured interface positions reached as much as four times higher than those predicted by the conduction model. This strong effect of convection has been reported by others.[5,8,11]

The good agreement between the experimental and theoretical predictions of the interface positions shown in these figures is not surprising since convection was included by optimizing the values of c and n in the correlation $K_e/K_l = cR_a^n$, during the computation. Hence, the validity of the convection model may be asserted from comparison of the temperature distribution, as well as the interface motion.

Hale and Viskanta[8] applied correlation originally developed for steady-state natural convection in a fluid confined between two parallel plates to account for the effect of convection during melting from below. The agreement between their experimental

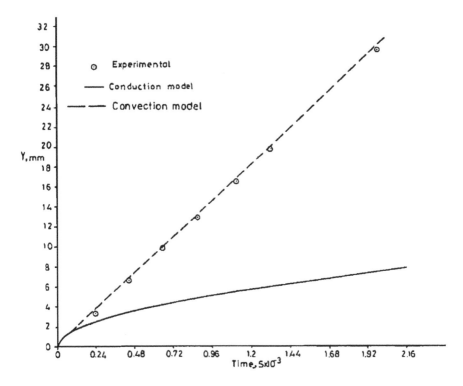

FIGURE 2.2.6 Comparison of experimental and theoretical interface positions during melting, $T_0 = 46.5°C$, $T_w = 73.5°C$.

result and the calculated interface position was only fair. In this work, the optimized values of c and n for the three experiments of melting were 0.159 and 0.34, respectively. These values provided higher estimates of the effective thermal conductivity when compared with correlations based on no phase change such as that suggested by O'Toole and Silveston[12] in which values of c and n were 0.229 and 0.252, respectively. The correlation found in this work covers a good range of Rayleigh number lying between few hundreds and 3×10^6.

Except at the beginning of the melting process, interface velocities remained constant unlike in solidification. Values of the interface velocities were 15×10^{-3}, 7×10^{-3}, and 2.2×10^{-3} mm/s for the experiments in which the lower plate temperatures were 73.5°C, 65.5°C, and 59°C, respectively. The initial temperature was varied between 42.5°C and 50°C.

Temperature change with time at different locations in both phases is plotted in Figures 2.2.9–2.2.15 for only one experiment. Theoretical estimates based on both convection and conduction models were also plotted on the same figures. Good agreement between the experiment and the theoretical prediction based on convection model, while some agreement with conduction model may be seen only for short time from the start of melting. The aforementioned figures show the presence of

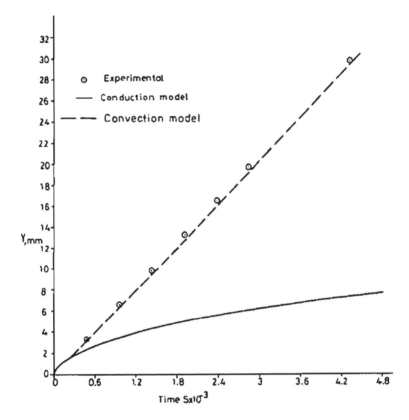

FIGURE 2.2.7 Comparison of experimental and theoretical interface positions during melting, $T_0 = 42.5°C$, $T_w = 65.5°C$.

distinct temperature distribution in the melted layer at any time, unlike what has been reported previously. This is even more evident from Figure 2.2.16 which shows the temperature distribution in both phases during melting with the maximum effect of convection ($T_w = 73.5°C$). The results of Hale and Viskanta[8] for melting of n-octadecane showed that the melted layer remained practically isothermal except at positions near the heating surface and the interface. In studying the melting of ice from below, Yin-Chao Yen[11] stated that temperature distribution in the melted layer was linear at first, but it tends to become flat rapidly as convection developed. Convection is expected to develop in water–ice systems faster than that in paraffin wax systems due to the high viscosity of the latter.

Heitz and Westwater[14] have determined the conditions of onset of convection in water–ice systems. Their results showed that the critical Rayleigh number defining these conditions was a strong function of the ratio of the depth to width of the tested cell (L/D). Values of critical Rayleigh number were of the order of 3000 for L/D equal unity, which represent the dimension of the test unit used in this work. This was not far from the values found for the wax mixture used as they were in the range of 2000–2500.

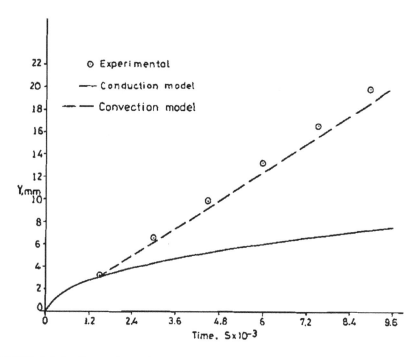

FIGURE 2.2.8 Comparison of experimental and theoretical interface positions during melting, $T_0 = 50°C$, $T_w = 59°C$.

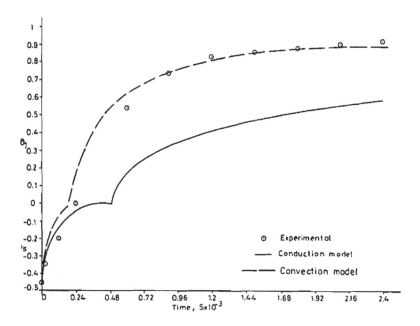

FIGURE 2.2.9 Experimental and theoretical temperature distribution during melting, $T_0 = 46.5°C$, $T_w = 73.5°C$, $y = 3$.

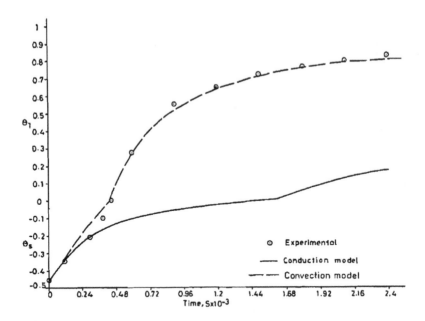

FIGURE 2.2.10 Experimental and theoretical temperature distribution during melting, $T_0 = 46.5°C$, $T_w = 73.5°C$, $y = 6.6$ mm.

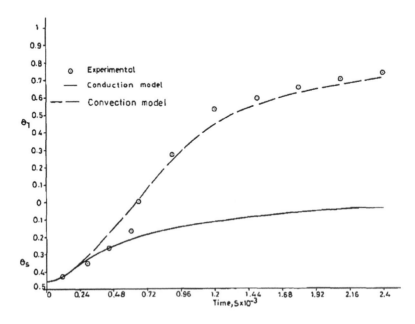

FIGURE 2.2.11 Experimental and theoretical temperature distribution during melting, $T_0 = 46.5°C$, $T_w = 73.5°C$, $y = 9.9$ mm.

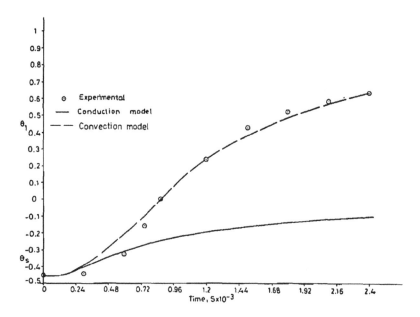

FIGURE 2.2.12 Experimental and theoretical temperature distribution during melting, $T_0 = 46.5°C$, $T_w = 73.5°C$, $y = 13.2$ mm.

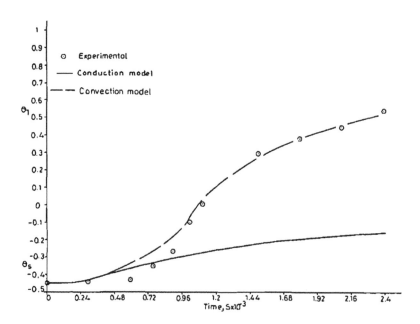

FIGURE 2.2.13 Experimental and theoretical temperature distribution during melting, $T_0 = 46.5°C$, $T_w = 73.5°C$, $y = 16.5$ mm.

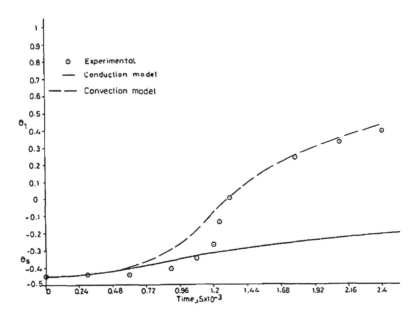

FIGURE 2.2.14 Experimental and theoretical temperature distribution during melting, $T_0 = 46.5°C$, $T_w = 73.5°C$, $y = 19.8$ mm.

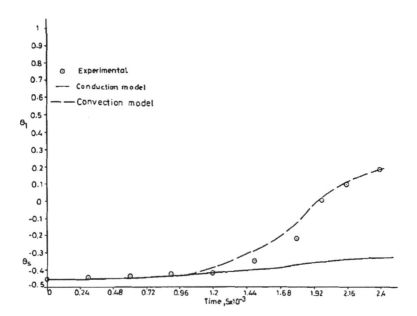

FIGURE 2.2.15 Experimental and theoretical temperature distribution during melting, $T_0 = 46.5°C$, $T_w = 73.5°C$, $y = 29.8$ mm.

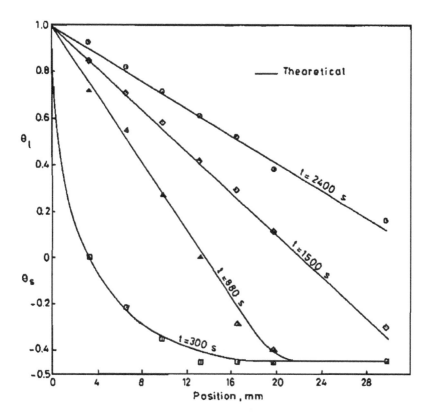

FIGURE 2.2.16   Temperature distribution during melting $T_0 = 46.5°C$, $T_w = 73.5°C$.

2.2.5   CONCLUSION

It was found that if a stable condition exists during heat transfer with phase change as in solidification from below, then the temperature distribution in the two phases and the interface motion may be found by solving the conduction equation in both phases. The variable space network method of Murray and Landis was found to describe the experimental results well when compared with the exact analytical solution of Neumann.

During melting from below, natural convection was found to develop gradually as melt layer thickness increases. The experimental results of this work show the existence of clear temperature distribution in both phases. This is not in agreement with previous work showing a fast development of a uniform temperature throughout the melted phase due to mixing. Accordingly, the conduction equation is expected to describe the process of melting if an effective thermal conductivity in the liquid phase was used to account for convection. This is in agreement with the results discussed in this article.

The correlation obtained for the effective thermal conductivity was different from those usually obtained from steady-state natural convection with no phase change.

ACKNOWLEDGEMENTS

This work was performed at the Mechanical Engineering Department, College of Engineering, University of Basrah, Iraq. The authors thank Dr. A.Z. Ijam for helpful discussions.

NOMENCLATURE

C Specific heat of the material, kJ/kg °C

E Cell depth, m

Gr   Grashof number: $\dfrac{\rho_l^2 g \beta Y^3 \left(T_w - T_f\right)}{\mu^2}$, dimensionless

K Thermal conductivity, W/m °C

L Latent heat of fusion, kJ/kg

n increment of space, dimensionless

N number of spatial increment in each phase

Pr   Prandtl number: $\dfrac{\mu C}{k}$ dimensionless

Ra   Rayleigh number: $\text{Gr} \cdot \text{Pr}$, dimensionless

Ste   Stefan number: $\dfrac{C\left(T_f - T_w\right)}{L}$, dimensionless

t time, s

t_0 initial time necessary to form 1 mm of melted or solidified layer, s

T temperature, °C

y position in the direction of heat flow, m

Y interface position, m

α Thermal diffusivity of the material, m²/s

β thermal expansion coefficient, dimensionless

γ ratio of the densities of the two phases $\left(\rho_l / \rho_s\right)$, dimensionless

Θ dimensionless temperature, $\dfrac{T - T_f}{T_w - T_f}$

Δ increment

μ dynamic viscosity, kg/m s

ρ density, kg/m³

λ Constant defined by Equations (2.2.7) and (2.2.12)

SUBSCRIPTS

e effective value

f referring to fusion

l liquid phase

m evaluated at time $= m\Delta t$

n evaluated at position $= n\Delta y$

N evaluated at $y = Y$

O initial condition

S	solid phase
W	evaluated at the lower plate

REFERENCES

1. Carlslaw, H.S., and Jaeger, L.C., *Conduction of Heat in Solids*, Second edition, pp. 283–296. Oxford University Press, Oxford (1959).
2. Eckert, E.R.F., and Drake, R.M., *Analysis of Heat and Mass Transfer*, pp. 224–228. McGraw-Hill, New York (1972).
3. Murray, W.D., and Landis, F., "Numerical and machine solutions of transient heat conduction problems involving phase change", *Trans. ASME, Ser. C, J. Heat Transfer*, 81, 106–112 (1959).
4. Thomas, J.C., and Westwater, J.W., "Microscopic study of solid-liquid interfaces during melting and freezing", *AIChE Chem. Eng. Prog. Symp. Ser.*, 59(41), 155–164 (1963).
5. Boger, D.V., and Westwater, J.W., "Effect of buoyancy on the melting and freezing processes", *Trans. ASME, Ser, C, J. Heat Transfer*, 89, 81–89 (1967).
6. Pujudo, P.R., Stermole, J.F., and Golden, J.O., "Melting of a finite paraffin slab as applied to phase change thermal control", *J. Spacecraft Rockers*, 6, 280–284 (1969).
7. Ukanaw, A.D., Stermole, F.J., and Golden, J.O., "Phase change solidification dynamics", *J. Spacecraft Rockers*, 8, 193–196 (1971).
8. Hale, N.W. Jr., and Viskanta, R., "Solid-liquid phase change heat transfer and-interface motion in materials cooled or heated from above or below", *Int. J. Heal and Mass Transfer*, 23, 283–292 (1980).
9. Tien, C., and Yen, Y.C., "Approximate solution of a melting problem with natural convection", *AIChE Chem. Eng. Prog. Symp. Ser.*, 62(64), 166–172 (1966).
10. Bathelt, A.G., Viskanta, R., and Leidenfrost, W., "An experimental investigation of natural convection in the melted region around a heated horizontal cylinder", *J. Fluid Mech.*, 90, 227–239 (1979).
11. Yen, Y.-C., "Onset of convection in a layer of water formed by melting ice from below", *Phys. Fluids*, 11(6), 1263–1269 (1968).
12. O'Toole, J.L., and Silveston, P.L., "Correlations of convective heat transfer confined in horozontallayers", *AIChE Chem. Eng. Prog. Symp, Ser.*, 32(57), 81–86 (1961).
13. Goodman, T.R., "The heat-balance integral and its application to problems including a change of phase", *Trans. Am. Soc. Mech. Eng.*, 80, 335 (1958).
14. Heitz, W.L., and Westwater, J.W., "Critical Rayleigh number for natural convection of water confined in square cells with L/D from 0.5 to 8", *Trans. ASME, J. Heat Transfer*, 93, 188–196 (1971).

2.3 The Role of Natural Convection during Melting and Solidification of PCM in a Vertical Cylinder

Mohammed Farid
University of Basrah

Yongsik Kim, Takuya Honda, and Atsushi Kanzawa
Tokyo Institute of Technology

2.3.1 INTRODUCTION

The high storage density of most of the phase change materials (PCMs) makes them attractive for storing all intermittent-type energy, such as solar energy or industrial waste heat. The major drawback of these materials is their low thermal conductivity that limits the rate of heat transfer in the storage units employing them. Although the thermal conductivity of the melt PCM is of the order of half or less than that of the solid, heat transfer in the melt during heat charge is usually enhanced by the effect of natural convection. This has encouraged investigators to study the effect of natural convection during melting of PCM encapsulated in different types of containers.

Cylindrical capsule is considered to be important due to its easy handling and large surface area per unit volume. It may be applied directly in a cross-flow or shell and tube heat exchangers. Sparrow and Broadbent [1,2] have provided a good description for the pattern of melting and solidification of n-eicosane encapsulated in a vertical cylinder. However, their experiments were limited to a single tube length and diameter, and hence, the description of the effect of natural convection cannot be considered complete. Also, they made no attempt to evaluate the effect of convection theoretically.

Melting of PCM in vertical cylinders is complicated by the presence of three effects: (a) natural convection in the melt phase, (b) the downward melting at the upper section of the cylinder, and (c) the effects associated with change of density of the PCM as it melts. This suggests that heat transfer during melting in a vertical cylinder is multidimensional in nature. Theoretical analysis would then require the evaluation of temperature and velocity profiles in the melted layer by solving the momentum, mass, and energy equations numerically, as has been done by Pannu and Rice [3] for melting inside vertical and horizontal cylinders, and Sparrow et al. [4] for melting outside vertical cylinders. Although such analysis provides more realistic two-dimensional interface profiles, the computation time required is usually very significant. This is very serious if the computation has to be done for practical storage unit consisting of multicylinders with a fluid losing its heat as it passes across them. This would require successive calculations for each tube's row, which makes the use of such a two-dimensional solution not practical.

Bareiss and Beer [5] have reported some measurements of melting of different PCMs in vertical cylinders. Their objective was to develop an empirical correlation that can be used for evaluating natural convection heat transfer coefficient as a function of vertical position. The practical use of local value of Nu, which may be obtained from their correlation, is limited since its application requires the use of the two-dimensional model. If the two-dimensional model is to be applied, it would be better to adopt the more appropriate solution based on the simultaneous solution of mass, momentum, and energy equations.

The above discussion suggests that a more simplified approach is required to analyze the effect of natural convection during heat transfer with phase change. In recent years, some efforts have been carried out to quantify the effect of natural convection in rectangular enclosures [6,7]. More recently, we have been successful in evaluating the effect of natural convection in horizontal [8] and vertical [9] rectangular enclosures. In the horizontal enclosure, a one-dimensional heat transfer was found to exist in all the experimental conditions; however, in the case of vertical enclosure, the deviation from one-dimensional heat transfer was observed. Moreover, it was possible to represent the two-dimensional heat transfer by a one-dimensional model, with an effective thermal conductivity (k_e) in the melt that takes into account not only natural convection but also the two-dimensional effect. The effective thermal conductivity may be thought as an equivalent to the axial dispersion coefficient, usually used in a one-dimensional model to describe a two-dimensional molecular and Eddy diffusion.

In this article, a similar approach was followed to account for convection. The unsteady heat conduction equation in the cylindrical coordinates was written in both phases and solved numerically using explicit finite differences. The effective thermal conductivity, used to account for convection, was strongly dependent on the melt layer thickness and the difference between the wall and melting temperatures. The effect of cylinder length was included since Rayleigh number was based on it. The model is expected to predict the average condition of heat transfer along the whole height of the cylinder; since k; provides the true average effect of natural convection. The same model was used to predict the experimental conditions of heat discharge, in which convection may be important only for the first few minutes of solidification. The effect of cylinder length, diameter, melting temperature of the PCM, wall temperature, and the degree of subcooling or superheating were all investigated experimentally and were used to check thoroughly the validity of the model. To show the effect of all these parameters clearly, the results were plotted with real time rather than dimensionless time.

2.3.2 EXPERIMENTAL

The details of the experimental apparatus and procedure are described elsewhere [10]. Figure 2.3.1 shows a sketch of the experimental apparatus. The test section (capsule) comprised coaxial double cylindrical tubes. The inner copper tube, with an internal diameter and wall thickness of 37 mm and 3 mm, respectively, was unusually filled with n-eicosane ($T_m = 36°C$) or n-hexadecane ($T_m = 17°C$). The outer polyvinylchloride tube, with inside diameter of 50 mm, was used to form an annulus for the flow of water used for heating or cooling. Two water baths were used to supply water at a constant temperature. The water bath used during solidification experiments had a refrigeration unit to cool the water to the low temperature required. During both melting and solidification, the wall of the cylinder containing the PCM, initially at a uniform temperature, was subjected to a step change in its temperature. Temperatures were then measured at radial positions: 0, 3, 7, 9,11,13,15,16, and 17 mm from the center of the cylinder having inside diameter of 37 mm, and at positions: 0, 4, 7, 9, 11, 13, 15, 17, 19, and 21 mm from the center of the cylinder having inside diameter of 46.5 mm. Two cylinder lengths of 200 and 280 mm were selected, and the radial temperature distribution was measured at locations: 20, 60, 100, 140, 180 mm, as well as 220 and 260 mm for the 280 mm long cylinder, as measured from the bottom.

Temperature measurements were done using a Cu-Co thermocouple probe as shown in Figure 2.3.1. An electronic sampling system consisted of an A-D converter

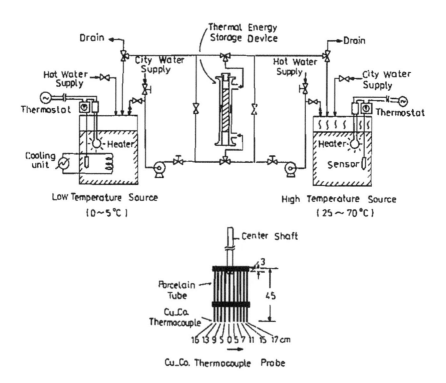

FIGURE 2.3.1   Experimental apparatus.

and a microcomputer was used to store and process the data. Interface locations were found from the recorded temperature-time history at the different radial positions. Heat charged/discharged at all the different vertical positions were calculated from the summation of the latent heat evaluated from the interface movement, and the sensible heat estimated from the integration of the temperature distribution, as described in [10]. A microcomputer was used to process the experimental data and to carry out these calculations.

The quantity, Q, represented in most of the figures shown in this article was estimated from the summation of the heat charged or discharged calculated at different heights. The contribution of the uppermost position was omitted in these calculations, as will be discussed later.

2.3.3 GENERAL PATTERN OF MELTING AND SOLIDIFICATION

2.3.3.1 MELTING

Prior to melting, the upper surface of the solid PCM is usually not flat, but rather concave downward due to volume contraction that occurs during solidification. At the early stages of melting, natural convection in the thin annular melted layer is not yet developed and heat transfer is mainly controlled by conduction. However, the actual situation is complicated by two factors: (a) the rapid melting of the uppermost solid PCM thin layer which progressively flows to form a layer of molten PCM above the solid phase and (b) the upward motion of the molten PCM in the annular space, formed due volume increase associated with melting, that eventually enhance the effect of the first factor. The presence of molten PCM at the top creates downward melting, which leads to a two-dimensional melting even before convection become important. As time passes, natural convection builds up and accelerates the rate of melting in both directions. Figure 2.3.2 shows the interface profiles as measured in the 200 and 280 mm length cylinders during melting. The general shape of these profiles is not different from those reported in the literature [1,5], showing large deviation from the one-dimensional heat transfer at the top of the cylinder. This effect progresses with time to the lower sections of the cylinder. Figure 2.3.2 shows a rapid increase in the quantity of heat charged at about 200 s, a clear evidence for the effect of a non-flat upper solid surface. Ignoring the uppermost measurement gave much smoother increase in the quantity of heat charged. The uppermost location recorded almost complete charge in a very short time, a false indication since the center part of the tube was empty at that location.

2.3.3.2 SOLIDIFICATION

Figure 2.3.2 depicts the interface profiles measured in a solidification experiment which looks very different from those found in melting. Heat transfer is controlled by conduction through the solidified layer leading to a uniform buildup of solid layers over the whole length of the cylinder. However, two minor deviations from the one-dimensional heat transfer may be observed: (a) at short time, the heat discharged is lower at the upper section of the cylinder, opposite to what has been observed in

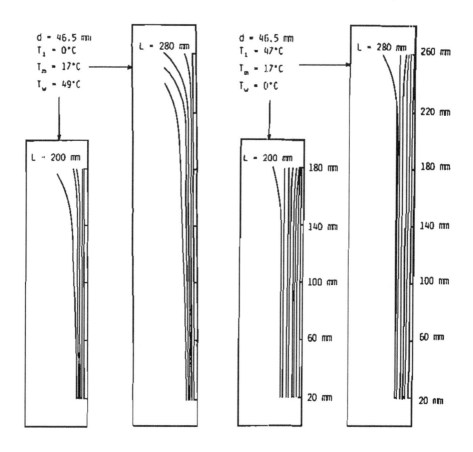

measurement time : 100,300,480,600,900,1600,2100,2700 seconds

FIGURE 2.3.2   Melting pattern.

melting. This is evident from the delayed movement of the interface at the upper positions in the cylinders as shown in Figure 2.3.2, and also from the increase of the rate of heat discharge shown in Figure 2.3.3 when the measurements of the upper section of the cylinder was omitted. This is even more supported by the results of Figure 2.3.5, showing a higher rate of discharge at the lower location of the cylinder. The melt in these lower locations loses heat not only by radial conduction but also by the upward movement of the melt due to buoyancy. This leads to faster inter-face velocity and higher rate of discharge at these locations. (b) With time, the melt temperature drops close to the melting temperature and convection disappears lead-ing to a more uniform heat transfer. The rapid increase of heat discharged, shown in Figure 2.3.3, is mainly due to the non-uniform solidification at the uppermost position.

To include the effect of all the parameters discussed except for the effect of non-flat end that occurs during solidification, the uppermost position has not been included in the summation of the total heat charged and discharged presented in this article.

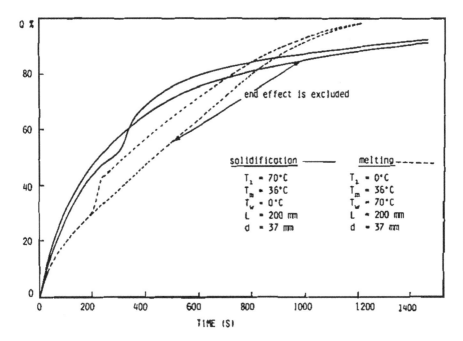

FIGURE 2.3.3   End effect on the measured rate of heal charge and discharge.

Sparrow and Broadbent [1] have eliminated this problem by maintaining a flat upper surface using a special leveling device.

2.3.4   THE ANALYSIS

Heat transfer during phase change of PCM contained in a vertical cylinder may be defined by the unsteady heat conduction equation:

$$\partial^2 T/\partial r^2 + (1/r)\partial t/\partial r = (1/\alpha)\partial T/\partial t \qquad (2.3.1)$$

written for both phases, with an effective thermal diffusivity α_e, in the melt that includes the effect of not only natural convection but also the effect of downward melting usually observed in melting in vertical cylinders. Including these factors, the one-dimensional model, that is developed here, is expected to predict heat transfer rate averaged over the whole length of the cylinder.

The method of solution that was previously used [8] for melting and solidification in rectangular enclosure was followed. Each phase was divided into equal space increment using $N1 + N2 + 1$ *total number of nodes, with* $n = 1$, representing the wall and $n = N1 + N2 + 1$, and representing the center of the cylinder. The boundary equations were as follows:

1. $T = T_0 + (T_i - T_0)\exp(-at)$ at $r = r_w$ or $n = 1$ (2.3.2)

2. $\partial T_s/\partial r = 0$ at $r = 0$ or $n = N1 + N2 + 1$ (2.3.3)

3. $T_s = T_l$ at $r = r_w - R$ or $n = N1 + 1$ (2.3.4)

4. $\rho_s \lambda dR/dt = k_l \partial T_l/\partial r - k_s \partial T_s/\partial r$ at $r = r_w - R$ (2.3.5)

Equation (2.3.2) shows that the wall of the cylinder was not assumed to undergo a step change in its temperature. An exponential change was assumed with a evaluated from the experimental measurements. Equations (2.3.1, 2.3.3, 2.3.5) may be written as follows using explicit finite differences:

In the melt:

$$T_{n,m+1} = T_n + \left(\alpha_e N1^2 \Delta t/R_m^2\right)\left(T_{n-1,m} - 2T_{n,m} - T_{n+1,m}\right)$$
$$+ \left(n/2R_m\right)\left(T_{n+1,m} - T_{n-1,m}\right)\left(R_{m+1} - R_m\right)$$
$$+ \alpha_e N1 \Delta t \left(T_{n-1,m} - T_{n+1,m}\right)/\left\{2R_m\left[r_w - (n-1)R_m/N1\right]\right\} \quad (2.3.6)$$

In the solid:

$$T_{n,m+1} = T_n + \left[\alpha_s N2^2 \Delta t / (r_w - R_m)^2\right]\left(T_{n-1,m} - 2T_{n,m} - T_{n+1,m}\right)$$
$$+ \left[n/2(r_w - R_m)\right]\left(R_{m+1} - R_m\right)\left(T_{n+1,m} - T_{n-1,m}\right)$$
$$+ \left(\alpha_s N2^2 \Delta t\right)\left(T_{n-1,m} - T_{n+1,m}\right)/\left[2(N1 + N2 + 1 - n)(r_w - R_m)^2\right] \quad (2.3.7)$$

while the interface location is given by

$$R_{m+1} = R_m + (\Delta t/\rho\lambda)\left[\left(k_s N1/2\right)\left(T_{N1-1,m} - 4T_{N1,m}\right)/R_m\right.$$
$$\left. + \left(k_e N2/2\right)\left(T_{N1+3,m} - 4T_{N1+2,m}\right)/(r_w - R_m)\right] \quad (2.3.8)$$

and the temperature at the center of the cylinder is given by

$$T_{N1+N2+1,m+1} = T_{N1+N2+1,m} + \left[1 - \alpha_s \Delta t N2^2/(r_w - R_m)^2\right]T_{N1+N2+1,m}$$
$$+ \left[\alpha_s \Delta t N2^2/(r_w - R_m)^2\right]T_{N1+N2-1,m} \quad (2.3.9)$$

A computer program was written to carry out all the computations starting from the known initial temperature distribution. The numerical scheme requires an approximate estimation for the temperature distribution and time to form an initial layer of melting having a thickness usually less than 1 mm, and also the time for melting the remaining solid PCM (less than 1% of the total PCM) at the end of the computation. After total melting, the computation continues by solving Equation (2.3.1) in the melt phase, until the melt reaches nearly uniform temperature close to the cylinder wall temperature.

In addition to the temperature distribution and interface location, the computer program provides an estimate of the heat flux at the cylinder wall and the percentage heat charged or discharged. Although the wall does not add any important resistance to heat transfer; its heat capacity was included in the estimation of the amount of heat transferred as it represents a significant quantity, particularly at the beginning of the experiments. In all the computation, the percentage heat charged/discharged was approaching 100% within 2% due to some error in the numerically estimated temperature derivative at the wall. Using $N1 = N2 = 6$ instead of 4, was found to improve the results on the expense of computation time. It is to be mentioned that there were no major differences in the method of solution for the solidification case.

2.3.4.1 EVALUATION OF NATURAL CONVECTION IN THE MELT

In melting, convection increases as the width of the annular melt region increases while it decreases, in solidification, as the radius of the melt cylindrical region decreases. Nusselt number was correlated with Rayleigh number based on cylinder length and the difference between the wall and melting temperatures during melting or between the liquid bulk and melting temperature during solidification. Accordingly, the effective thermal conductivity that accounts for the effect of natural convection may be written as follows:

$$k_e/k_l = cRa_L^n(\delta/L) \tag{2.3.10}$$

where

$$Ra_L = g\beta C_l \Delta T L^3/\mu k_l,$$

and δ represents the melt layer thickness and the radius of the unsolidified melt during solidification. Equation (2.3.10) may be rewritten for Rayleigh number based on melt layer thickness, rather than cylinder length, as follows:

$$k_e/k_l = cRa_\delta^n(\delta/L)^{1-3n} \tag{2.3.11}$$

or

$$\left(k_e/k_l\right)_{max} = c'Ra_{r_w}^n \tag{2.3.12}$$

where $c' = c(r_w/L)^{1-3n}$, is equivalent to the constants of the correlations used in our previous work [8,9], in which the effect of cell geometry has not been tested. A common fact in all these correlations is that the effective thermal conductivity is always strong function of δ, and only a weak function of L, or even independent. In melting, k_e increases gradually since $T_w - T_m$ remain constant with continuous increase of δ, while in solidification both δ and $T_l - T_m$ decrease leading to a very fast termination of the effect of natural convection.

2.3.5 RESULTS AND DISCUSSION

2.3.5.1 EVALUATION OF THE EFFECTIVE THERMAL CONDUCTIVITY

Melting experiments were chosen for evaluating the constant c and n in Equation (2.3.10) since the control mechanism of heat transfer is only conduction during solidification. In the melting experiments, $T_w - T_m$ was varied from 7.5°C to 60°C, while cylinders with only two lengths of 200 and 280 mm were used in the tests. Since the variation of L was limited in our experiments and because k_e, is a weak function of L, it was decided that the evaluation of c and n would be done solely based on $T_w - T_m$. It may be important to note here that the physical properties of the PCM used in Rayleigh number, particularly melt viscosity, were defined at the melting temperature and not at the average film temperature as is usually done. The agreements between the model and the experiments were always poor when the average film temperature was used due to the exaggerated effect of the viscosity on the measured Rayleigh number. We have found that Sparrow and Broadbent [1] have reached the same conclusion when correlating their experimental results of melting rate as a function of Fo, St, and Gr. This was not surprising since most of the thermal resistance lies on the melting surface, which has the lowest temperature and a continuously decreasing surface area as melting progresses.

Comparison between the theoretical and experimental heat charge showed that the values of c and n, in Equation (2.3.10), were essentially constant even though $T_w - T_m$ was varied between 7.5°C and 60°C. Values of c and n were found to be 0.28 and 0.25, respectively. Moreover, the model was also checked against the experimental measurements of solidification, and some of the measurements of others [1], as will be discussed.

In our experiments, the maximum variation in the value of $(r_w/L)^{0.25}$ was from 0.507 to 0.58. Taking the average value, c' in Equation (2.3.12) was found to be 0.15. As Table 2.3.1 shows, maximum convection in melting occurs when the PCM is melted from below in a rectangular enclosure, as expected. Natural convection

---

TABLE 2.3.1
Some Empirical Correlations Developed for Evaluating Natural Convection during Melting or PCM

Correlation	Practical Range	Condition	Ref.
$Nu = 0.159 Ra_\delta^{0.34}$	$3 \times 10^2 < Ra_\delta < 3 \times 10^6$	Melting from below in horizontal enclosure	[8]
$Nu = 0.25 Ra_\delta^{0.25}$ or: $Nu = 0.072 Ra_\delta^{0.333}$	$10^3 < Ra_\delta < 4 \times 10^6$	Melting in two-dimensional rectangular enclosure	[7]
$Nu = 0.10 Ra_\delta^{0.25}$	$10^4 < Ra_\delta < 10^7$	Melting in vertical plane wall	[9]
$Nu = 0.15 Ra_\delta^{0.25}$ or $Nu_L = 0.28 Ra_L^{0.25}$	$2 \times 10^3 < Ra_\delta < 10^7$	Melting in vertical cylinder	this work

intensity during melting in two-dimensional enclosure is lying between those found in vertical and horizontal enclosures, which is a realistic finding. The correlation developed in this work predicts 50% higher convection than that found in a vertical plane wall which is attributed to the observed strong effect of downward melting in vertical cylinders.

2.3.5.2 Experimental Measurements and Model Predictions

Figure 2.3.4 shows the large variation in the experimentally measured heat charge to the cylinder at different heights. The higher rate of heat transfer in the upper-most position is attributed to the factors discussed earlier. Convection model predicts the average condition of heat transfer well as shown in the same graph, which also includes the results of computations based on pure conduction that predicts complete melting (not shown) after 2274 s, compared to that of only 727 s by the convection model.

The condition shown in Figure 2.3.5 for an experiment of solidification is different. The variation in the measured heat charged with time at different heights was not large except at the beginning, before conduction dominates fully the heat transfer. The maximum heat discharge rate occurs at the lowest location in the cylinder, as represented by the uppermost curve. This is in agreement with the observation found both in Figures 2.3.2 and 2.3.3, as discussed earlier. As shown in Figure 2.3.4, the convection model predicts the average condition of the measurements better than the conduction model, but with only small differences; since heat transfer is mainly

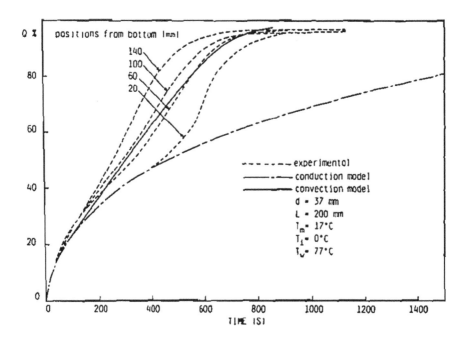

FIGURE 2.3.4 Measured heat charge at different heights with model prediction.

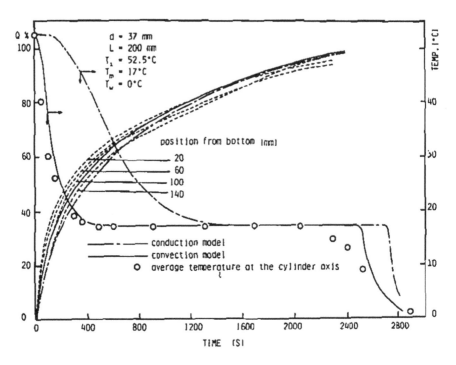

FIGURE 2.3.5   Measured and predicted heat discharged with average centre temperature.

controlled by conduction in the solid. To observe the effect of natural convection during solidification, the average temperature of the PCM at the center axis of the cylinder was plotted versus time in the same figure, together with the theoretical predictions of both conduction and convection models. The excellent agreement of the convection model with the measured average temperature at the axis is a good support to the correlation of the effective thermal conductivity, developed totally from melting experiments.

Parameters that have been investigated during both experiments of melting and solidification were: cylinder length, diameter, melting temperature, wall and initial temperature as shown in Figures 2.3.6–2.3.8. Good agreement between model and experiments may be observed in Figure 2.3.6, which shows the effect of the most important parameter, $T_w - T_m$. Although curves numbered 2 and 6 belong to two PCMs with different melting temperatures of 17°C and 36°C, they lie close to each other because the corresponding values $T_w - T_m$ in these two experiments were 38°C and 33°C, respectively, with a difference of only 8°C. The model was also used to describe the experiments of melting carried out in a larger diameter cylinder as shown in Figure 2.3.7 with some deviation in some of the experiments. Comparison between curves numbered 4 and 5 in the same figure shows some increase in the measured rate of heat charged using shorter cylinder, which was in agreement with the model prediction. The change in the heat charge rate was small since cylinder length was varied only from 200 to 280 mm. The scale of the scatter in the measurements requires further investigation in which cylinder length should be varied over much wider range.

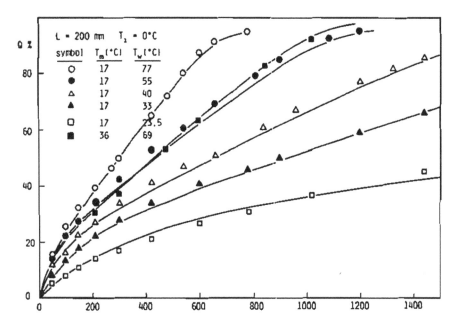

FIGURE 2.3.6   Model prediction for melting experiments in the 37 mm diameter cylinder.

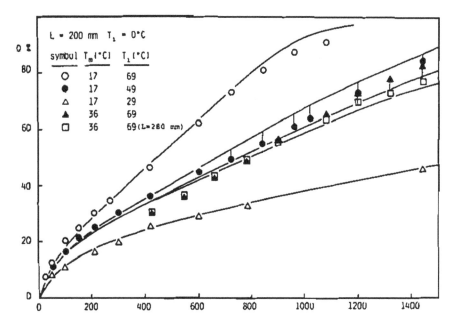

FIGURE 2.3.7   Model prediction for melting experiments in the 46.5 mm diameter cylinder.

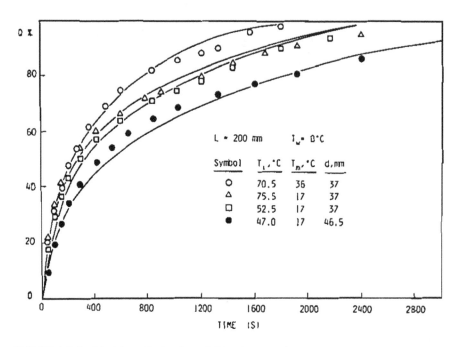

FIGURE 2.3.8   Model prediction for solidification experiments.

Effect of the diameter of the cylinder, melting temperature of the PCM, and the initial temperature of the melt wax were studied during heat discharged and the results were represented in Figure 2.3.8. The deviation between the model prediction and the experiments might have been due to some errors in the solid thermal conductivity used. As may be seen from both experimental and theoretical results presented by curves 2 and 3 in the same figure, the effect of superheating is important at the beginning of the experiments and diminishes gradually with time. Comparing curve 1 in Figure 2.3.8 with curve 2 in Figure 2.3.6 which belong to close values of ΔT, it may be observed that the rate of heat discharge at the first few minutes is much higher than the rate of heat charge. This is due to the high thermal conductivity of the solid compared to that of the melt with a ratio of about 2.7 for n-eicosane. After some time, convection is developed that increases the rate of heat charge beyond that of heat discharged, as may be seen from the same two curves. Both experiment and model (not presented in this paper) showed no effect of cylinder length on the rate of heat discharge within a variation in length between 200 and 280 mm.

The computer model was also checked against the experimental measurements of Sparrow and Broadbent [1], for melting in a vertical cylinder which was presented as the variation of the fraction of the PCM melted with time, as shown in Figure 2.3.9. The agreement between the model and the experiments of melting with 8.3°C initial subcooling was excellent. However, other experiments carried out with the solid PCM initially at the melting temperature showed that nearly 10% of the wax was already melted at the start of the experiments. This is not surprising for such a condition since 10% belongs to only 1.3 m layer of wax melted. If the initial 10% is

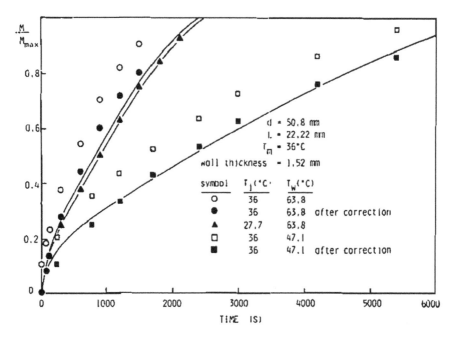

FIGURE 2.3.9 Model prediction for the experimental measurements of Sparrow and Broadbent [1] for melting.

subtracted, then the agreement between the model and experiments (as presented by the solid symbols) is reasonable. Similar observation was found when the model was used to predict the experimental measurements of Chaboki and Sparrow [11] for melting of n-eicosane in a vertical cylinder without rotation.

The model was also checked against the experimental measurements of Sparrow and Broadbent [2] and Larson and Sparrow [12] for the solidification of n-eicosane in vertical cylinders for two different geometries. Theoretical interface positions, as predicted by the model, were always below the experimental values as shown in Figures 2.3.10 and 2.3.11. However, the model used in this work shows a better agreement with the experimental results of Larson and Sparrow [12] than the model employed by them, as shown in Figure 2.3.11. This is expected since the model used by Larson and Sparrow is a conduction model that does not take into account the effect of natural convection in the melt. Larson and Sparrow [12] have concluded that the most likely cause of the deviation between the computed and experimental results is the uncertainty in the values of the solid thermal conductivity of the n-eicosane used to evaluate FoSte$_s$ for the experimental data. Using higher values of solid thermal conductivity would certainly bring the experimental and theoretical results closer.

During solidification, liquid superheat may be defined by the ratio: $\left(T_{b,L} - T_m\right)/\left(T_i - T_m\right)$, in which the bulk liquid temperature, $T_{b,L}$, is defined as follows:

$$T_{b,L} = \frac{2}{R^2} \int_0^R rT\,dr \qquad (2.3.13)$$

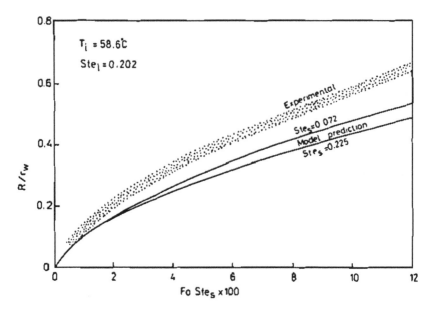

FIGURE 2.3.10   Model prediction for the measurements of Sparrow and Broadbent [2] for solidification.

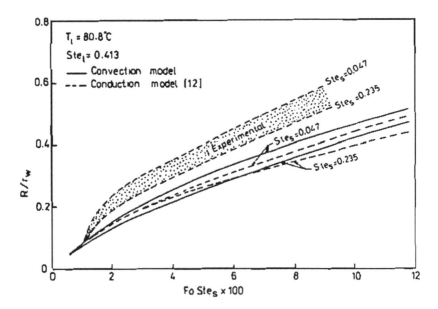

FIGURE 2.3.11   Model prediction for the measurements of Larson and Sparrow [12] for solidification.

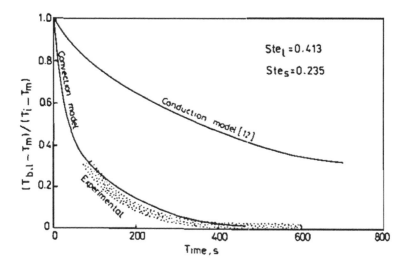

FIGURE 2.3.12   Comparison of the measurements of Larson and Sparrow [12] for the liquid superheat during solidification.

Figure 2.3.12 shows that some measurements [12] for the liquid superheat together with the theoretical predictions of the conduction model [12] and the model used in this work. The convection model, with parameters evaluated from the experimental measurements of this work, shows excellent agreement with the values of liquid superheat measured by Larson and Sparrow as may be seen in Figure 2.3.12.

Figure 2.3.13 shows model prediction for the effect of cylinder wall thickness, which is significant at the beginning of the heat charge/discharge. The copper wall loses (or gains) its heat content rapidly due to the high thermal conductivity of the copper.

2.3.6   CONCLUSION

Although natural convection in the melt PCM may be large during both melting and solidification in vertical cylinders, its effect of heat transfer is important only in melting, as heat transfer is controlled mainly by conduction during solidification. The effect of natural convection during solidification may be verified from the rapid decrease of the melt temperature at the center of the cylinder. Solidification progresses towards the center of the cylinder uniformly, while melting rate varies along the height of the cylinder significantly, being maximum at the top. Downward melting that makes heat transfer two-dimensional, is influenced by three factors: natural convection in the melt, the upward melt flow due to volume increase, and the initial rapid melting of the non-flat upper surface of the solid PCM.

A one-dimensional model was developed to predict heat discharge rates during phase change in a vertical cylinder. However, the model includes the effects of both convection and downward melting using an effective thermal conductivity (k_e),

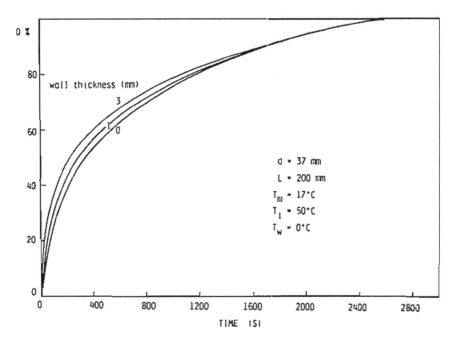

FIGURE 2.3.13 Model prediction for the effect of the cylinder wall thickness (melting).

evaluated under such conditions. The model was found to predict well the rate of heat charged and discharged, averaged over the whole length of the cylinder. Values of c and n in Equation (2.3.10) were found essentially constant and equal to 0.28 and 0.25, respectively. The model was successfully used to describe the experiments carried out at different conditions of tube length, diameter, melting, and wall temperatures. Moreover, the model also describes the experimental measurements of others. The correlation developed from melting experiments to account for convection was also used to describe convection during solidification. This correlation predicts natural convection during melting in a vertical cylinder almost 50% higher than that in a vertical rectangular enclosure due to the strong effect of downward melting. The effect of cylinder height on heat transfer has not been fully investigated in this work since only two cylinder lengths of 200 and 280 mm were tested. It is interesting to note that the effect of natural convection was better correlated when the physical properties of the melt PCM were evaluated at the melting temperature rather than the average film temperature as is usually done in the study of natural convection in a single phase.

ACKNOWLEDGEMENTS

The first author wishes to express his gratitude to the Matsumae International Foundation, Japan, for the award of the fellowship that enabled him to carry this work.

NOMENCLATURE

C	specific heat of the PCM
d	cylinder diameter
Fo	Fourier number, $\alpha_s t / r_w^2$
g	acceleration due to gravity
h	natural convection heat transfer coefficient in the melt
k	thermal conductivity of the PCM
L	cylinder length
M	time nodal position
M/M_{max}	fraction of PCM melted or solidified
n	space nodal position
$N1$	number of division in the melt during melting
$N2$	number of division in the solid during solidification
Nu_L	Nusselt number based on cylinder length, hL/k_l
Nu_δ	Nusselt number based on melt layer thickness, $h\delta/k_l$
Q	percent heat charged or discharged
r	radial coordinate
R	solidified or melted thickness
Ra_L	Rayleigh number, based on cylinder length
Ra_δ	Rayleigh number based on melt layer thickness
Ste_l	initial liquid-phase Stefan number, $(T_i - T_m)C_l/\lambda$
Ste_s	solid-phase Stefan number, $(T_m - T_w)C_s/\lambda$
t	time
T	temperature in excess of the melting temperature
ΔT	$T_w - T_m$ in melting
	$T_l - T_m$ in solidification
α	thermal diffusivity of the PCM
β	volumetric thermal expansion coefficient
δ	melt layer thickness during melting, or: radius of the unsolidified region during solidification
ρ	density of the PCM
λ	latent heat of fusion of the PCM
μ	viscosity of the melt

Subscript

b	bulk
e	effective condition
i	initial condition
l	liquid
m	time increment
n	space increment
o	steady-state condition
s	solid
w	wall

REFERENCES

1. Sparrow, E.M., and Broadbent, J.A., "Inward Melting in a Vertical Tube Which Allows Free Expansion of the Phase-Change Medium", *ASME Journal of Heat Transfer*, 104, 309–315 (1982).

2. Sparrow, E.M., and Broadbent, J.A., "Freezing in a Vertical Tube", *ASME Journal of Heat Transfer*, 105, 217–225 (1982).

3. Pannu, J., Joglekar, G., and Rice, P.A., "Natural Convection Heat Transfer to Cylinders of Phase Change Material Used for Thermal Storage", *AIChE Symposium Series*, 76, 47–55 (1980).

4. Sparrow, E.M., Patankar, S.V., and Ramadhyani, S., "Analysis of Melting in the Presence of Natural Convection in the Melt Region", *ASME Journal of Heat Transfer*, 99, 520–526 (1977).

5. Bareiss, M., and Beer, H., "Influence of Natural Convection on the Melting Process in a Vertical Cylindrical Encloser", *Letters in Heat and Mass Transfer*, 7, 329–338 (1980).

6. Hale, N.W., and Viskanta, R., "Solid-Liquid Phase Change Heat Transfer and Interface Motion in Materials Cooled or Heated from Above or Below", *International Journal of Heat and Mass Transfer*, 23, 283–292 (1980).

7. Marshall, R.H., "Natural Convection Effect in Rectangular Enclosers Containing a Phase Change Material", F. Kreith, R. Boehm, J. Mitchell, and R. Bannerot (Eds.), *Thermal Storage and Heat Transfer in Solar Energy System*, ASME, New York, pp. 61–69 (1970).

8. Farid, M.M., and Mohamed, A.K., "Effect of Natural Convection on the Process of Melting and Solidification of Paraffin Wax", *Chemical Engineering Communications*, 57, 297–316 (1987).

9. Farid, M.M., and Husian, R.H., "AN Electrical Storage Heater Using Phase-Change Method of Heat Storage", under publication.

10. Honda, T., Kim, Y., Kishigami, H., and Kanzawa, A., "Charge and Discharge Heat Transfer Rates in the Latent Heat Thermal Energy Storage", *World Congress III of Chemical Engineering, Tokyo*, 629–632 (1986).

11. Chaboki, A., and Sparrow, E.M., "Melting in a Vertical Tube Rotating About a Vertical, Colinear Axis", *International Journal of Heat and Mass Transfer*, 30, 613–622 (1987).

12. Larson, E.D., and Sparrow, E.M., "Effect of Inclination on Freezing in a Sealed Cylindrical Capsule", *ASME Journal of Heat Transfer*, 106,394–401 (1984).

2.4 Thermal Performance of a Heat Storage Module Using PCMs with Different Melting Temperatures
Mathematical Modeling[1]

Mohammed Farid
University of Basrah

Atsushi Kanzawa
Tokyo Institute of Technology

2.4.1 INTRODUCTION

Latent heat methods of heat storage have attracted many investigators in different fields such as solar energy and nuclear energy, since the available heat may be stored at the temperatures required in these applications. Organic and inorganic materials are available for thermal storage at low, medium, and high temperatures. Most of the phase change materials (PCMs) have high storage density with a small temperature swing that makes them attractive for thermal storage. The major drawback of the PCMs is their low thermal conductivity which limits heat transfer rates in the storage units employing them. Many investigations have been carried out to improve the heat transfer rates in the phase change storage units.

Fundamental studies are available in the literature on heat transfer during melting and solidification of PCM encapsulated in cylindrical [1–5] and flat [6,7] geometries. The objective in most of these studies was to evaluate the temperature distribution in both phases and to locate the interface position with time during melting and solidification. Significant efforts have been made to evaluate the effect of natural convection in the melt layer, which was found significant only in melting and not in solidification for both geometries. Most of the reported works are on cylindrical

[1] Contributed by the Solar Energy Division of the AMERICAN SOCIETY OF MECHANICAL ENGINEERS for publication in the JOURNAL OF SOLAR ENERGY ENGINEERING. Manuscript received by the Solar Energy Division, July 14, 1988.

123

geometries because they may be used to construct a shell and tube or cross-flow heat storage unit.

The unsteady heat conduction equation was solved in both phases to evaluate the temperature distribution and interface positions during phase change. Analytical solutions were limited to simple boundary conditions, whereas the numerical solutions, available for the two-dimensional model, provide detailed information not only about temperature distribution in the radial and axial positions in the cylindrical capsule but also about the convection current in the melt layer [3], Unfortunately, such solutions require large computational time with high-speed computers. Farid et al. [5] have shown that heat transfer with phase change in a vertical cylinder may be simplified using a one-dimensional model and employing an effective thermal conductivity that takes into account the effect of natural convection in the melt layer.

In thermal storage units employing any type of encapsulation, the fluid temperature varies in the flow direction while the PCM temperature varies in the direction perpendicular to the flow as well as in the flow direction. If a detailed analysis, usually employed for single capsules, is used for the analysis of the storage unit, the computational time will be very large.

This has encouraged previous investigators to follow simplified approaches, which may be classified into two types:

1. Quasi-steady-state analysis based on the assumption that the PCM has negligible heat capacity. This is the simplest analysis which is correct only if the storage unit operates near the melting temperature. This is favorable but unlikely to occur in practice because phase-change storage units usually operate with a significant temperature swing which makes this type of analysis incorrect. The quasi-steady-state analysis has been widely used for different types of encapsulation [8,9].

2. Analysis based on the assumption that the system has a very low Biot number so that the temperature variation normal to the flow direction can be ignored. This approach was first applied by Morrison et al. [10] and Marshal [11] for modeling of a storage unit consisting of a number of thin rectangular containers with either air or water flowing between them. More recently, Abe et al. [12] and Kamimoto et al. [13,14] have followed a similar approach for a storage unit consisting of bundles of thin polyethylene rods as a PCM using ethylene glycol as a heat transfer fluid. The solid-solid transition occurred within a wide range of temperature, and, hence, latent heat was accounted for using as effective heat capacity that included its effect. The good agreement found between the experimental results and theoretical prediction was due to the use of small diameter rods which may permit the assumption of no radial temperature distribution within the rods. However, PCMs such as organic and inorganic compounds are usually contained in larger diameter capsules and may exhibit a large temperature variation in the direction perpendicular to the flow, making the analysis less accurate.

Dietz [15] has applied the NTU-effectiveness method to analyze the performance of a storage unit that consisted of a bundle of vertical cylindrical capsules containing a PCM with air flowing across them for heat exchange. The analysis was limited to the condition in which the heat transfer resistance in the wax is small as compared to the external resistance in the air, which may be true only if the capsule diameter is sufficiently small.

The first objective in the present work was to develop a mathematical model, which may be used to predict the transient behavior of a phase-change thermal storage unit consisting of bundles of tubes filled with PCM with air flowing across (or along) them. Unlike the previous models, the radial temperature distribution in the PCM was included.

The second objective of this work was to increase the heat transfer rate using cylindrical capsules containing PCM's of different melting temperatures in the same storage unit. During heat charge, air must flow across the capsules containing the PCM's in the order of decreasing melting temperature.

This arrangement was expected to provide a nearly constant heat flux to the capsules even though the air temperature decreases as it flows across the capsules. The decrease in the air temperature was compensated by the use of lower melting temperature PCMs. Airflow direction should be reversed during heat discharge. The idea of using PCMs of different melting temperatures was previously suggested by Farid [16] for a storage unit with rectangular geometry.

2.4.2 ANALYSIS

Theoretical simulation has been done for heat storage unit consisting of vertical cylindrical capsules containing the phase change materials (PCMs) with hot or cold air flowing across them. The capsules are fixed in a duct using an on-line arrangement with a specified number of rows, as shown in Figure 2.4.1.

During heat charge, the air temperature drops as it flows across the capsules while the PCM undergoes a phase change as well as a change in sensible heat. Accordingly, the air temperature is assumed to vary with time and position in the direction of airflow. The heat flow causes a radial temperature distribution in the capsules and melting starts in the first row and then gradually in other rows. The same may be said for heat discharge.

To reduce the computation time, the storage unit was divided, in the direction of flow, into a few sections each containing a limited number of rows of capsules.

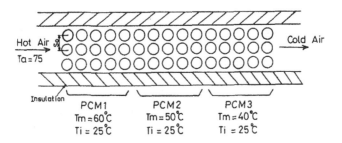

FIGURE 2.4.1 Capsules arrangement during heat charge.

The decrease (or increase) in the temperature of the air leaving each section was calculated by a simple heat balance using values of the heat flux estimated at the surface of the capsules lying in the center row of each section. The computer program, used to carry out these calculations, automatically stops all the computation if the selected number of rows per section was too big to achieve accurate results.

To evaluate the heat flux at the surface of the cylindrical capsules, it was necessary to calculate the radial temperature distribution in both melt and solid layers of the PCM. This has been achieved by solving the heat conduction equation in the cylindrical coordinates,

$$\partial^2 T / \partial r^2 + (1/r)\partial T / \partial r = (1/\alpha)\partial T / \partial t \qquad (2.4.1)$$

in both phases, with the following boundary conditions:

$$\partial T / \partial r = 0 \qquad (2.4.2)$$

$$\partial T / \partial r = (h/k)(T_w - T_a) \quad \text{at} \quad r = r_w \qquad (2.4.3)$$

together with the interface boundary conditions:

$$T_s = T_l \quad \text{at} \quad r = R \qquad (2.4.4)$$

$$\rho \lambda dR / dt = k_l \, \partial T_l / \partial r - k_s \, \partial T_s / \partial r \quad \text{at} \quad r = R. \qquad (2.4.5)$$

Equations (2.4.1)–(2.4.5) were written in a finite difference form and were solved using the explicit method of solution as described elsewhere [5]. The only variations to the solution described in the mentioned reference are as follows:

1. A convective boundary condition due to airflow across the capsules, as given by Equation (2.4.3), has been included. The heat-transfer coefficient, h, was evaluated using the following known empirical correlation, developed for forced convection of fluid flowing across bundles of tubes:

$$hD/k_a = c(\mathrm{Re})_a^n \qquad (2.4.6)$$

   Reynolds number, Re, for the air was based on the capsule diameter. Values of c and n were selected according to the space gaps between the capsules in the airflow direction and in the direction perpendicular to it [17].
2. Since melting and solidification occur only after some time has passed, the first stage in the computation must cover a sensible heating or cooling without phase change, which required the solution of Equation (2.4.1) in one phase with the boundary conditions (2.4.2) and (2.4.3) and the known initial condition.

Sensible heating or cooling must also be considered after complete phase change. In practical storage units employing latent heat method of storage, sensible heat represents a significant part of the total heat stored and, hence, it must be included.

Effect of natural convection in the melt layer was included employing an effective thermal conductivity for the liquid using the following empirical correlation developed by Farid et al. [5]:

$$k_e/k_l = 0.28(\mathrm{Ra})_{\mathrm{wax}}^{0.25}(\delta\,/\,L) \tag{2.4.7}$$

This equation shows that the effective thermal conductivity increases as melting progresses during heat charge.

In the computation, an initial radial temperature distribution in all the capsules was assumed. The air outlet temperature from the first section was calculated from the known inlet temperature and was fitted by the computer program as a function of time using linear regression. The computation for the first section continued until the difference between the air inlet and outlet temperatures became very small. The calculated air outlet temperature of the first section during the whole period was then used as inlet temperature to the second section and procedure was continued till the last section in the storage unit.

The amount of heat charged and discharged is estimated from the surface heat flux in each section and the total transfer area. Summation of these quantities during the whole period provided estimates for the total heat charged/discharged, which were also calculated from the known initial and final temperatures of the PCM, taking into account both sensible and latent heats. The deviation between the two estimates, which was of the order of $\pm2\%$, was a measure of the error involved in the numerical differentiation of the temperature with position during the evaluation of the surface heat flux.

The mathematical model developed in this work was used to study the effects of system parameters such as airflow rate, and there was space gap between the capsules. However, the main objective was to compare the performance of the storage unit with the capsules filled with one-type PCM or with PCMs having different melting temperatures.

2.4.3 RESULTS OF SIMULATION

In the simulation, the physical properties of the three PCMs used in the thermal storage unit were assumed identical to those of *n*-octadecane except their melting temperatures, which were assumed to be 40°C, 50°C, and 60°C. This was done deliberately to see the effects of testing PCMs with different melting temperatures. The simulation was carried out for the storage unit with the capsules either filled with a single PCM having a melting temperature of 50°C or with the three PCMs. Each type of PCM was contained in five rows of cylindrical capsules with three capsules in every row, arranged in the order of increasing melting temperature during heat charge. The capsules were 0.3 m long copper tubes having an internal diameter and wall thickness of 15.0 and 1 mm, respectively. Distance from center to center of the capsules was 47 mm in the flow direction and 35.8 mm in the direction *perpendicular to the flow. This was to achieve a high convec*tion heat-transfer coefficient for the flowing air, however, the effect of the space gap between the capsules was also studied throughout the simulation. For the purpose of comparison, initial wax temperature and air inlet temperature were assumed to be 25°C and 75°C, respectively, during heat charge and 75°C and 25°C, respectively, during heat discharge. This provided a similar driving temperature, $T_m - T_a$, during heat charge and discharge.

Figures 2.4.2–2.4.6 present the results of the simulation. Air outlet temperature from the storage unit, as obtained from the simulation, is plotted with time during heat charge and discharge. The difference between this temperature and the air inlet temperature, as shown by the horizontal line in the figure, represents the power output or input to the system (after multiplication by MC_p for the air). Percentage heat charged or discharged is also plotted in Figures 2.4.4 and 2.4.6 for easier comparison of the thermal performance of the storage unit under different conditions.

In any air-based system, a significant heat transfer resistance is expected to be present in the air itself. This is due to the moderate heat-transfer coefficient of the air flowing through the storage unit. Figure 2.4.2 shows (in the legend) that the air heat-transfer coefficient increases with decreasing flow area available for airflow (i.e., increasing air velocity). Energy charged to the system increases during the first period as shown from the decrease in the air outlet temperature.

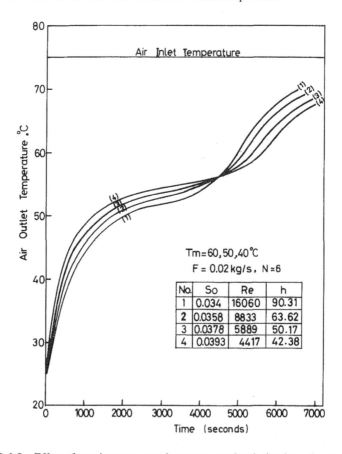

FIGURE 2.4.2 Effect of varying space gap between capsules during heat charge.

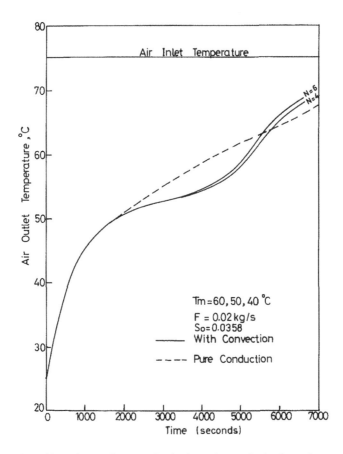

FIGURE 2.4.3   Effect of natural convection in the melt wax during heat charge.

The explicit finite difference solution required the melt and solid phases to be divided into N space divisions. Figure 2.4.3 shows very little variation in the air outlet temperature when the number of divisions is increased from 4 to 6. Using finer mesh increases computational time significantly (a requirement for numerical stability). The effect of neglecting natural convection in the melt is also shown in the same figure. The effect is significant soon after melting was started, however, such effect was not observed during heat discharged because solidification was solely con-trolled by conduction in both phases. These findings were similar to the observations *reported* [1,2,5] *for single capsule.*

The results presented in Figures 2.4.2 and 2.4.3 were obtained from the simula-tion carried out for a storage unit with capsules filled with PCM's having melting temperatures of 60°C, 50°C, and 40°C. Figures 2.4.4 and 2.4.5 show a comparison in the performance of the storage unit, during heat charge, when the capsules were filled either with a single type of PCM ($T_m = 50$°C) or with *three types of PCMs. Air outlet temperature was significantly* lower when three types were used. The effect was more pronounced at lower airflow rate of 0.01 kg/s and usually appear as soon as melting was started. In Figure 2.4.4, the percentage heat charged to the unit was

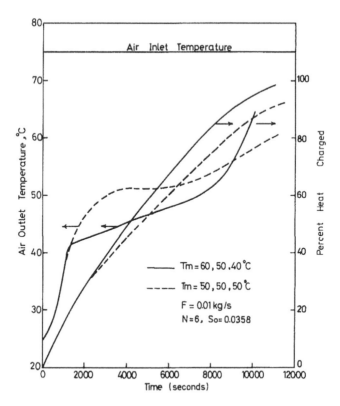

FIGURE 2.4.4 Effect of using PCs with different melting temperatures during heat charge.

also plotted with time. An improvement of about 10% was obtained in the rate of heat charge by using PCs with melting temperatures of 60°C, 50°C, and 40°C. Similar findings may be seen in Figure 2.4.6 for heat discharge. Air outlet temperature was significantly higher when three types of PCs were used, except at the end of the discharge period in which a rapid decrease was observed. The same effect may be seen for heat charge in Figures 2.4.4 and 2.4.5.

During both heat charge and discharge, the variation in the air temperature leaving a single capsule goes through three distinct regions: a rapid change due to sensible heating or cooling, a slow change due to latent heat storage, and followed by a rapid change due to sensible heating or cooling *after complete melting or solidification as may be seen in* Figures 2.4.4 and 2.4.6. In the storage unit, melting in the capsules starts at different times and, hence, these three regions are less distinct when one type of PCM was used. However, melting or solidification, in the three sections containing PCs having different melting temperatures, starts almost at the same time. This causes a sharp decrease in the rate of change of air outlet temperature as shown in Figures 2.4.4 and 2.4.6.

Increasing airflow rate, during heat charge, produces higher air outlet temperature as shown in Figure 2.4.5. Due to this, the last rows containing the PCM with the lowest melting temperature will melt first. This causes a more smooth change in the

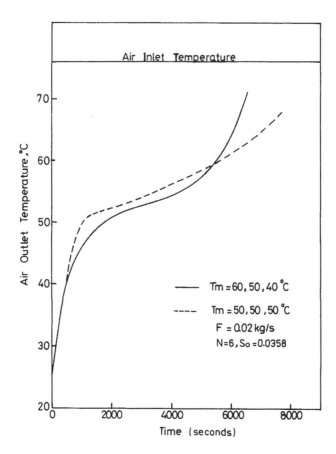

FIGURE 2.4.5   Effect of using PCs with different melting temperatures during heat charge.

air outlet temperature even when PCs with different melting temperatures are used. The air outlet temperature was constant for longer periods when a single PCM was used as compared to that obtained using PCs with different melting temperatures. Although a constant air outlet temperature is favorable, the use of three PCs is preferred because of the increase in the air outlet temperature during heat discharge accompanied by an increase in the heat transfer rates.

As described earlier, the storage unit consists of three sections, each containing 15 capsules filled either with a single type of PCM or with different types of PCM. The percentage of heat charged to the three sections with different PCs is plotted in Figure 2.4.7 and with one PCM in Figure 2.4.8. The summation of the heat charged to the three sections gives the total heat charged to the storage unit. Comparison of the results presented in the two figures shows that similar rates of heat charge to the three sections may be achieved by using PCs with different melting temperatures.

The lower heat transfer rate obtained, when the capsules in the storage unit were filled with one PCM, is due to the decrease in the driving force $(T_a - T_m)$, in the

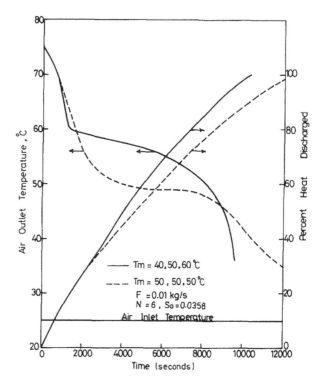

FIGURE 2.4.6 Effect of using PCs with different melting temperatures during heat discharge.

direction of airflow. However, during heat charge, PCs with decreasing melting temperatures in the direction of airflow compensates for the decrease in the airflow temperature resulting in a more uniform heat transfer rate throughout the entire storage unit. The same may be said for heat discharge. This speaks well for using PCs with different melting temperatures in the same storage unit.

2.4.4   CONCLUSION

A theoretical model was developed to predict the transient behavior of a thermal storage unit that consists of vertical cylindrical capsules containing PCs with air flowing across them for heat exchange. Previous models were based either on quasi-steady-state analysis or on the assumption of negligible temperature variations in the radial direction in the capsules. The model presented here does not carry any such assumption, and also takes into account the effect of natural convection in the melt layer of the PCM.

Simulation based on the model developed shows that some improvement in the performance of the storage unit during both heat charge and discharge may be achieved by using PCs with different melting temperatures in the same unit. Higher rates of heat transfer are achieved during both heat charge and discharge.

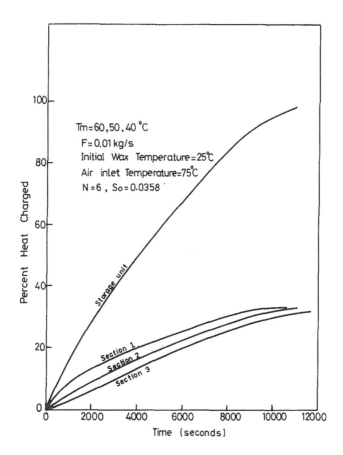

FIGURE 2.4.7   Percentage heat charged to the three sections of the storage unit, using PCs with different melting temperature.

ACKNOWLEDGEMENT

The first author wishes to express his gratitude to the Matsumae International Foundation, Japan, for the award of the fellowship that enabled him to conduct this work.

NOMENCLATURE

b = number of capsules in every row
C_p = specific heat capacity, J/kg °C
c = constant in Equation (2.4.6)
d = ordinary derivative
D = capsule diameter, m
F = air mass flow rate, kg/s
h = air heat-transfer coefficient, J/m²°C s
k = thermal conductivity, J/m °C s

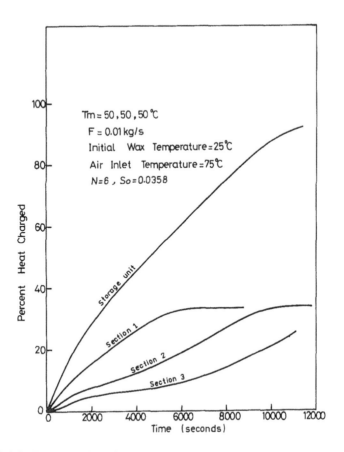

FIGURE 2.4.8   Percentage heat charged to the three sections of the storage unit, using single PCM.

L = capsule height, m
n = constant in Equation (2.4.6)
N = number of space division in each phase
r = radial coordinate
R = location of the wax solid-liquid interface, m
$(Ra)_{wax}$ = Rayleigh number for wax $(\rho C_p \beta / \mu k)_i g L (T_w - T_m)$
$(Re)_a$ = Reynolds number for air, based on capsule's diameter, $FD/[\mu L(b + V)(So - D)]$
So = capsule's center-to-center distance in the direction perpendicular to flow, m
t = time, s
T = temperature, °C

GREEK LETTERS

α = thermal conductivity, m²/s
σ = melt layer thickness, m
λ = latent heat of fusion for the wax, J/kg

ρ = density, kg/m^3
μ = viscosity, kg/m s
β = thermal expansion, K^{-1}

SUBSCRIPT

a = air
e = effective value
i = initial
l = liquid wax
m = melting
s = solid wax
w = wall

REFERENCES

1. Sparrow, E. M., and Broadbent, J. A., "Inward Melting in a Vertical Tube Which Allows Free Expansion of the Phase-Change Medium," *ASME Journal of Heat Transfer*, Vol. 104, 1982, pp. 309–315.
2. Sparrow, E. M., and Broadbent, J. A., "Freezing in a Vertical Tube," *ASME Journal of Heat Transfer*, Vol. 105, 1983, pp. 217–225.
3. Pannu, J., "Natural Convection Heat Transfer to Cylinders of Phase Change Materials Used for Thermal Storage," *AIChE Symposium Series*, Vol. 198, 1980, pp. 47–55.
4. Honda, T., et al., "Charge and Discharge Heat Transfer Rates in the Latent Heat Thermal Energy Storage," *World Congress III of Chemical Engineering*, Tokyo, 1986, pp. 629–632.
5. Farid, M. M., et al., "The Role of Natural Convection during Melting Solidification of PCM in a Vertical Cylinder," to be published in the Chemical Engineering Communication.
6. Hale, N. W., and Viskanta, R., "Solid-Liquid Phase Change Heat Transfer and Interface Motion in Materials Cooled or Heated from Above or Below," *International Journal of Heat and Mass Transfer*, Vol. 23, 1980, pp. 283–292.
7. Farid, M. M., and Mohammad, A. K., "Effect of Natural Convection on the Process of Melting and Solidification of Parrafin Wax," *Chemical Engineering Communications*, Vol. 57, 1987, pp. 297–316.
8. Smith, R. N., et al., "Heat Exchanger Performance in Latent Heat Thermal Energy Storage," *ASME Journal of Solar Energy Engineering*, Vol. 102, 1980, pp. 112–118.
9. Saitoh, T., and Hirose, K., "High Performance Phase Change Thermal Storage Using Spherical Capsules," *Chemical Engineering Communications*, Vol. 41, 1986, pp. 39–58.
10. Morrison, D. J., and Abdul-Khalik, S. I., "Effect of Phase Change Energy Storage on the Performance of Air-Based and Liquid-Based Solar Heating Systems," *Solar Energy*, Vol. 20, 1978, pp. 57–67.
11. Marshall, R., "Comparisons of Parrafin Wax Storage Subsystem Models Using Liquid Heat Transfer Media," *Solar Energy*, Vol. 29, 1982, pp. 503–511.
12. Abe, Y., et al., "Charge and Discharge Characteristics of a Direct Contact Latent Thermal Energy Storage Unit Using Form-Stable High Density Polyethylene," *ASME Journal of Solar Energy Engineering*, Vol. 106, 1984, pp. 465–474.
13. Kamimoto, M., et al., "Latent Thermal Energy Unit Using Form-Stable High Density Polyethylene; Part II: Numerical Analysis of Heat Transfer," *Transaction of the ASME Journal of Solar Energy Engineering*, Vol. 108, 1986, pp. 290–297.

14. Kamimoto, M., et al., "Latent Thermal Energy Storage Using Pentaerythritol Slurry," *21st Inter-Society Energy Conversion Engineering Conference*, San Diego, California, 1986, pp. 730–736.

15. Dietz, D., "Thermal Performance of a Heat Storage Module Using Calcium Chloride Hexahydrate," *ASME Journal of Solar Energy Engineering*, Vol. 106, 1984, pp. 106–111.

16. Farid, M., "Solar Energy Storage with Phase Change," *Journal of Solar Energy Research*, Vol. 4, 1986, pp. 11–29.

17. Holman, J. M., *Heat Transfer*, 4th ed., McGraw-Hill, New York, 1976, pp. 221–223.

2.5 Performance of Direct Contact Latent Heat Storage Units with Two Hydrated Salts

Mohammed Farid
Jordan University of Science and Technology

Ali Nasser Khalaf
University of Basrah

2.5.1 INTRODUCTION

Hydrated salts are good candidates for storing heat at intermediate temperatures due to their high storage density and low cost compared to other phase change materials. However, all hydrated salts melt incongruently, which causes phase segregation during the repeated cycles of heat charge and discharge. Sub-cooling is another problem that makes heat withdrawal rather difficult. These two problems have been well discussed by Telkes [1], Etherington [2], Biswas [3], and Farid and Yacoub [4].

Etherington [2] was the first to introduce the idea of using direct contact heat transfer between an immiscible heat transfer fluid and a solution of hydrated salt. Light mineral oil as a heat transfer fluid was bubbled through a solution of sodium phosphate dodecahydrate. Heat is exchanged between the bubbles and the solution efficiently without the requirement of expensive metallic heat transfer surface. Different hydrated salts such as $Na_2SO_4 \cdot 10H_2O$, $Na_2HPO_4 \cdot 10H_2O$, $CaCl_2 \cdot 6H_2O$, and sodium acetate have been tested by Etherington [2], Edie and Melsheimer [5], Costello et al. [6], Hallet and Keyser [7], Fouda et al. [8], and Fouda et al. [9], in units of different sizes. Heat is exchanged between the bubbles and the solution efficiently without the requirement of expensive metallic heat transfer surface.

More recently, three commercial-grade salts, $Na_2CO_3 \cdot 10H_2O$, $Na_2SO_4 \cdot 10H_2O$, and Na_2HPO_4 $12H_2O$, have been thoroughly tested by Farid and Yacoub [4] with the same apparatus (as used in the present work).

If heat is available at varying temperature, such as in the application of solar energy, and if a hydrated salt is used to store the heat, then the storage unit may operate for long periods as a sensible heat storage unit due to the difficulty of dissolving the salt when the energy source temperature drops below or close to the

137

crystallization temperature. This will reduce the heat storage capacity significantly since the specific heat of the salt solution is lower than that of water.

To overcome this problem and, hence, allowing the storage unit to operate with larger temperature swings, the idea of using two separate columns containing two hydrated salts with different crystallization temperatures will be tested in this work.

The use of phase-change materials with different melting temperatures was first suggested for heat storage using paraffin waxes encapsulated in different geometries by Farid [10], Farid and Kanzawa [11], and Farid et al. [12].

In the present work, the idea was tested on a system employing direct contact latent heat storage method. Two hydrated salts $Na_2CO_3 \cdot 10H_2O$ and $Na_2S_2O_3 \cdots 5H_2O$ were selected due to their different crystallization temperatures. During heat charge, the hot fluid was first introduced to the column containing $Na_2S_2O_3 \cdot 5H_2O$ (having higher crystallization temperature). The heat transfer fluid leaving the first column was pumped to the second column containing $Na_2SO_4 \cdots 10H_2O$. During heat discharge, the direction was reversed.

This method of operation would maintain reasonable differences between the kerosene and the continuous phase temperature in both columns during heat charge and discharge ensuring high heat transfer rates.

2.5.2   MEASUREMENTS

Figure 2.5.1 shows a sketch of the apparatus used in this study. The two storage columns were fabricated from 1 mm thick galvanized steel plate to form 0.184 m inside diameter cylindrical container of 1 m length. The two columns were fitted

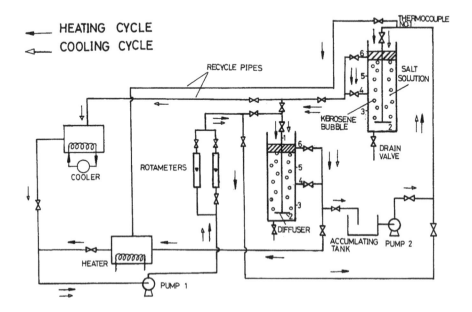

FIGURE 2.5.1   Schematic diagram of the storage system.

with a drain valve at the bottom for solution withdrawal. They were insulated with 50 mm thick glass wool to reduce heat losses. The two columns were fixed at two different levels to allow for gravity flow of the kerosene (the heat transfer fluid) from the higher column to the lower column with sufficient head to overcome the pressure drop in the system.

Two thermostatically controlled units were used to heat or cool the kerosene to a predetermined temperature prior to its feed to the first column. The heating capacity of the unit was 3.1 kW, while the cooling capacity was only 1.2 kW. Hence, it was necessary to add supplementary cooling using a water-cooled heat exchanger.

The distributor used to disperse the kerosene into small bubbles is similar to that used by Farid and Yacoub [4]. The diffuser, which was fixed at the bottom of the column, was made out of eight glass tubes of 3 mm diameter connected radially to a single tube of 16 mm diameter fixed at the axis of the column. The only variation to the distributor system used earlier was the introduction of the kerosene downward through the 16 mm tube fixed along the axis of the column.

Preliminary experiments showed severe plugging problem when the kerosene was introduced upward, particularly in the column containing $Na_2S_2O_3 \cdot 5H_2O$.

The two hydrated salts were prepared in an electrically heated tank to the required compositions of 30% and 55% anhydrous $Na_2CO_3 \cdot 10H_2O$ and $Na_2S_2O_3 \cdot 5H_2O$, respectively. These compositions were selected to avoid the formation of anhydrous salts during the heating–cooling cycles as may be seen from the phase diagram of both salts.

Before starting the cooling cycle, the lower column was filled with the hot $Na_2CO_3 \cdot 10H_2O$ to a height of 0.65 or 1.3 m, whereas the higher column was filled to the same height with $Na_2S_2O_3 \cdot 5H_2O$ solution. The cooling unit usually started prior to each run until the kerosene temperature reached the preset value. The cold kerosene was then pumped through the piping bypassing the storage columns to cool the piping system to ensure almost constant low temperature of kerosene as it enters the column.

The kerosene was then allowed to flow to the lower column through the distributor. The kerosene bubbles coalesced at the top of the column was discharged from it by gravity into an insulated accumulating tank where it is pumped continuously to the higher column with the same flow rate. The coalesced kerosene collected at the top of the higher column was discharged to the cooling unit by gravity. Although supplementary cooling was used, it was necessary to send the hot kerosene to a temperature-damping tank before entering the cooling unit (not shown in Figure 2.5.1). This was 0.3 m³ tank filled with kerosene at room temperature.

The large volume of kerosene compared to that circulated through the system was sufficient to reduce the required cooling capacity.

After most of the hydrated salts have been crystallized, the heating cycle was started by pumping the hot kerosene to the upper column the kerosene layer formed at the top of this column was discharged by gravity to the diffuser of the lower column. Hence, only one pump was required during heating run as may be seen in Figure 2.5.1.

At the end of the two runs, the hot solutions were drained from both columns and were kept in the preparation tanks for the next day. At the end of the cooling–heating cycle, the concentration might have changed due to salt deposition in the system.

Hence, it was necessary to measure the composition before starting the next cooling–heating cycle.

Kerosene flow rate was measured with rotameters, while the temperatures of the salt solution, kerosene inlet, and outlet, were continuously measured using Nickel-Alumel (type K) thermocouples.

2.5.2.1 MEASUREMENTS OF HEAT LOSSES AND BUBBLE SIZE

The overall loss coefficient U_w was evaluated using the well-known procedure used by Ward et al. [13]. Values of U_w for both heating and cooling for both columns were found between 1.64 and 2.31 W/m² °K, depending on the operating and ambient conditions.

Bubble size, required in the theoretical prediction of the performance of the storage system, was not measured. It was determined from the following empirical correlation developed by Farid and Yacoub [14] for a diffuser identical to the one used in this work:

$$R = 3.8 - 1.3 \quad (F/A) \tag{2.5.1}$$

2.5.3 THEORETICAL ANALYSIS

The theoretical treatment of this problem is similar to that described by Farid and Yacoub [4]. Hence, the subject will be discussed very briefly, and for more details, the reader is referred to the full description presented in the reference.

A heat balance between the bubble and continuous phases during crystallization gave the following equation:

$$\left[\bar{M}_d C_d (T_o - T_i) + \pi D_t H U_W (T_c - T_r) \right]_{av} \Delta t$$
$$= (M_l C_l + M_s C_s)_{av} \Delta T_c + \Delta M_s H_f + M_{col} C_g \Delta T_c + M_d C_d \Delta T_o \tag{2.5.2}$$

Values of M_s and M_d were evaluated from the phase diagram of the saltwater system using the lever rule. Latent heat of crystallization H_f was not assumed constant since it included the effect of latent heat of dilution.

The transfer efficiency defining the heat transfer between the two phases was defined by the equation

$$E = (T_o - T_i)/(T_c - T_i)$$
$$= \left[1 - \exp(-F_d \pi^2 \alpha \theta / R^2) \right]^{1/2} \tag{2.5.3}$$

This equation was used to evaluate the outlet kerosene temperature T_o at any time and when used with Equation (2.5.2), the continuous phase temperature may be determined at the end of some time increment.

Bubble radius R was determined from Equation (2.5.1), while bubble rise time θ was evaluated using the known relationship between buoyancy and drag force described by Fouda et al. [8,15].

The effective diffusivity factor F_d, which defines the effect of internal circulation in the drop, was evaluated from the following empirical equation developed for kerosene–water systems by Farid and Yacoub [14]:

$$F_d = 1.1 \times 10^{-8} (2uR / \alpha)^2 \qquad (2.5.4)$$

Successive application of Equations (2.5.1)–(2.5.4) would give the variation of kerosene outlet temperature and the continuous phase temperature with time during cooling and heating cycles. The outlet kerosene temperature from the first column was used as inlet temperature to the second column to evaluate the continuous phase temperature in the second column and its kerosene outlet temperature.

The volumetric heat transfer coefficient based on solution volume was defined by

$$\bar{M}_d C_d (T_o - T_i) - U_v V (\Delta T)_m = \rho_d v' C_d \frac{dT_o}{dt} \qquad (2.5.5)$$

where

$$(\Delta T)_m = ((T_i - T_c) - (T_o - T_c)) / Ln((T_i - T_c)/(T_o - T_c))$$

The overall volumetric heat transfer coefficient of the storage unit consisting of two columns was assumed equal to the average of the coefficients of the two columns.

Thermal efficiency (η) was defined as the ratio of the actual energy removed (or stored) from both columns to the theoretically available energy in the two columns

$$\eta = Q_u / Q_{th} \qquad (2.5.6)$$

where

$$Q_u = \sum_{j=1}^{n} \bar{M}_d C_d |(T_0 - T_i)| \Delta t_j + \sum_{j=1}^{n} \pi D_t H U_w (T_c - T_r) \Delta t_j \qquad (2.5.7)$$

The summation given by Equation (2.5.7) was carried out on both columns. The theoretically available energy Q_{th} may be evaluated from the summation of the terms on the right–hand side of Equation (2.5.2), integrated over the whole period of heat charge or discharge.

2.5.4 RESULTS

2.5.4.1 WATER SYSTEM

Measurements with the storage unit filled with water were used to confirm the accuracy of the experimental measurements of temperature and flow rate. These

TABLE 2.5.1

Effect of Using Two Columns on the Volumetric Heat Transfer Coefficient and Transfer Efficiency for Water System

Kerosene Superficial Velocity (cm/s)	$H=0.65$ m (One Column)				$H=0.65$ m (Two Columns)			
	Cooling		Heating		Cooling		Heating	
	U_v	η	U_v	η	U_v	η	U_v	η
0.06	1592	0.99	2697	0.89	2267	0.92	3405	0.94
0.12	3580	1.00	3971	1.06	4223	0.97	5717	0.96
0.18	5793	0.99	7562	0.98	6800	0.99	10,000	0.96

measurements were also used to demonstrate the improvement that can be achieved in the heat transfer rates when two columns connected in series were used.

The listed values of thermal efficiency in Table 2.5.1 with different kerosene flow rate and column height for both cooling and heating show little deviation from 100%. The little deviation in these values may belong to inaccuracies in the measurements of temperature and kerosene flow rate and also in the determination of heat loss coefficient.

Kerosene bubble size measurement has not been carried out in this work. The distributor used was similar to that used by Farid and Yacoub [4], and hence, the variation of bubble size with kerosene flow rate is expected to be similar. To confirm this, the transfer efficiency was measured for these column heights as shown in Figure 2.5.2. The theoretical transfer efficiency calculated from Equations (2.5.1), (2.5.3), and (2.5.4) is in good agreement with the measured values as may be seen from Figure 2.5.2.

Figure 2.5.3 shows the effect of kerosene superficial velocity (and, hence, the effect of bubble size) on the estimated volumetric heat transfer coefficient. The large dependence of the volumetric heat transfer coefficient on kerosene superficial velocity (i.e., bubble size) is because this coefficient is a product of the area-related heat transfer coefficient and the heat transfer area which is inversely proportional to the bubble diameter. Measured values of U_v were found slightly higher than those previously measured by Farid and Yacoub [4], but they were of the same order of magnitude and follow similar trends.

Table 2.5.1 shows a significant increase in U_v when the same volume of water was contained in two separate columns connected in series. The effect was similar to that found when the column height or volume is reduced by half.

2.5.4.2 PREDICTION OF THE PERFORMANCE OF THE STORAGE UNIT EMPLOYING HYDRATED SALTS

Equations (2.5.1)–(2.5.4) were used to predict the performance of the storage unit. The agreement between the predicted and measured performance is shown in Figures 2.5.4–2.5.6 during heat charge and discharge for the two salts system. The agreement is much better for the cooling run. In the present work, the cooling run

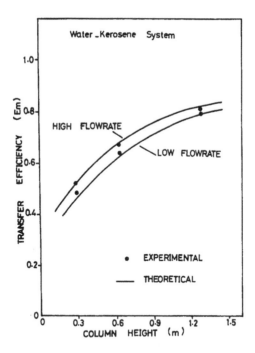

FIGURE 2.5.2   Effect of column height on the transfer efficiency.

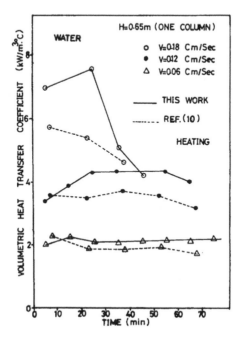

FIGURE 2.5.3   Comparison of the measured volumetric heat transfer coefficient with those measured by Farid and Yacoub [4].

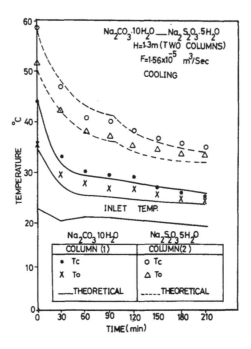

FIGURE 2.5.4   Measured and predicted outlet temperature of the two-salt system.

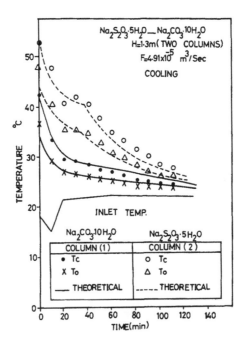

FIGURE 2.5.5   Measured and predicted outlet temperature of the two-salt system.

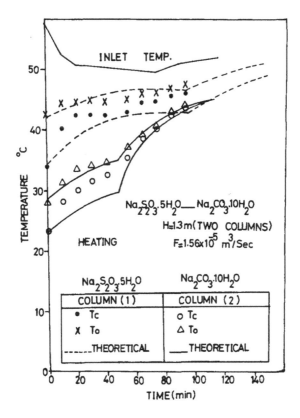

FIGURE 2.5.6   Measured and predicted outlet temperature of the two-salt system.

was usually followed by the heating run. At the end of cooling, most of the thermo-couples were surrounded by the crystals of the salt introducing significant error in the measured temperatures. This did not represent a serious problem in the cooling run since it occurs at the end of the run when the kerosene temperature approached the continuous phase temperature. However, the effect on measurements was signifi-cant for the heating run, as the problem occurred at the beginning of the run where there is a great difference between the continuous phase temperature and kerosene temperature.

Some of the deviation in the heating run may belong to the assumption that bubble size and shape are not affected by the presence of crystals. During heating where a large amount of crystals are available in the system initially, the bubble size is expected to be smaller in size, which may describe the more rapid temperature changes observed in Figures 2.5.6 and 2.5.7.

Figures 2.5.4 and 2.5.5 do not show any significant sub-cooling during heat dis-charge for both $Na_2S_2O_3 \cdot 5H_2O$ and $Na_2CO_3 \cdot 10H_2O$ solutions with crystallization temperatures of approximately 42°C and 30°C for the two salts, respectively. This is in agreement with the crystallization temperature obtained from phase diagrams for the used composition of 30% of anhydrous sodium carbonate and 55% of anhy-drous sodium thiosulphate. The sharp change in the slope of the cooling and heating

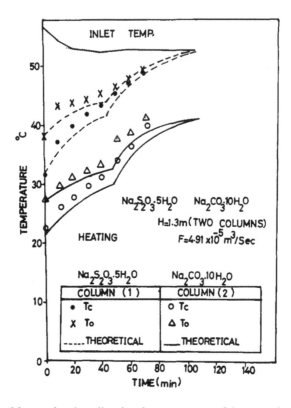

FIGURE 2.5.7   Measured and predicted outlet temperature of the two-salt system.

curves shown in Figures 2.5.4 and 2.5.7 is an indication of the start of crystallization or melting of the salt. The behavior of $Na_2CO_3 \cdot 10H_2O$ solution is in agreement with that reported by Farid and Yacoub [4] for a single column. However, the effect of cycling has not been studied in this work, and it can be expected that the crystallization temperature would gradually be lowered due to the effect of solution dilution as a result of the increased amounts of undissolved salt.

2.5.4.3 VOLUMETRIC HEAT TRANSFER COEFFICIENT

Table 2.5.2 for $Na_2CO_3 \cdot 10H_2O$ and Table 2.5.3 for $Na_2S_2O_3 \cdot 5H_2O$ show that the average heat transfer coefficient was almost doubled by containing the storage solutions into two columns having the same total volume of the single column. This is similar to that observed for the water–kerosene system and follows similar reasoning. Farid and Yacoub [4] have reported similar effect when the solution height was reduced by half. This provided an attractive method of improving heat transfer rate in such storage systems. A similar conclusion may be reached from Figure 2.5.4 for the effect of solution volume on U_v for $Na_2CO_3 \cdot 10H_2O$. The continuous decrease of U_v with time was similar to that previously observed, which was due to the poor contact between the bubble and the solution after crystallization started.

TABLE 2.5.2

Effect of Using Two Separate Columns on the Volumetric Heat Transfer Coefficient of $Na_2CO_3 \cdot 10H_2O$

Kerosene Superficial Velocity (cm/s)	$H=0.65$ m One Column		$H=0.65$ m Two Columns	
	Cooling	Heating	Cooling	Heating
0.06	1845	2870	4810	6311
0.12	5472	5738	11,650	10,940
0.18	6710	7646	12,428	9887

TABLE 2.5.3

Effect of Using Two Separate Columns on the Volumetric Heat Transfer Coefficient of $Na_2S_2O_3 \cdot 5H_2O$

Kerosene Superficial Velocity (cm/s)	$H=0.65$ m (One Column)		$H=0.65$ m (Two Columns)	
	Cooling	Heating	Cooling	Heating
0.06	1526	1808	3160	3990
0.12	5113	4180	6296	7047
0.18	6936	4538	10,856	8471

Figures 2.5.8 and 2.5.9 show a comparison of the volumetric heat transfer coefficient for two columns connected in series containing $Na_2CO_3 \cdot 10H_2O$ or $Na_2S_2O_3 \cdot 5H_2O$, or both. The heat transfer was most efficient when most of the storage medium was still in the liquid form and diminish later as crystallization proceeded as may be seen in Figure 2.5.8. However, values of U_v were found to remain almost constant during melting as shown in Figure 2.5.9. These two figures show that the values of U_v for the salt's systems are of the same order of magnitude as those for the water system, a good indication that most of the heat transfer resistance lies in the bubble, and the bubble size is not affected significantly by the presence of the crystals except at a large crystal concentration.

2.5.4.4 THERMAL EFFICIENCY

Table 2.5.4 shows a comparison of the storage thermal efficiency during heat charge and discharge for different salt systems. The thermal efficiency of the water system which was close to 100% was an indication that all the heat that was stored could be extracted. With hydrated salts, it will not be possible to extract all the theoretically available heat due to some subcooling and salt cake formation with repeated cycling. Thermal efficiency was not affected by kerosene flow rate as reported by Farid and Yacoub [4], Edie and Melsheimer [5], and Hallet and Keyser [7] except at

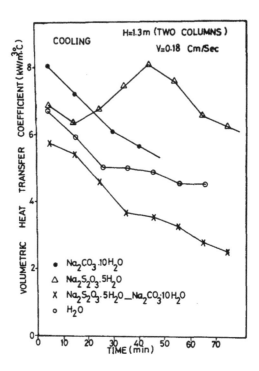

FIGURE 2.5.8   Volumetric heat transfer coefficients obtained from cooling runs of different salts systems.

low kerosene velocity in which subcooling, poor mixing, and crystal sedimentation is expected. The exceptionally high and low thermal efficiency of $Na_2S_2O_3 \cdot 5H_2O$ during cooling and heating, respectively, was due to the high crystallization temperature of this salt compared to other salts studied.

Table 2.5.5 compiles some information available on the average thermal efficiency of some storage units used by different investigators for different hydrated salts. Although different salts exhibit different thermal efficiency, Table 2.5.5 shows a clear dependence of η for all salt systems on storage volumes as has been observed by Farid and Yacoub [4] for $Na_2SO_4 \cdot 10H_2O$. This was due to the better condition of crystallization as the storage volume increased.

2.5.4.5   POWER INPUT AND OUTPUT FROM THE SYSTEM

Power output/input was calculated from the measured differences between inlet and outlet kerosene temperature. Power input/output from the storage system consisting of two units was calculated from the summation of the power input/output from each unit. It is rather difficult to compare power output/input for different salt systems, as this requires similar experimental conditions that are difficult to achieve. This limits the number of runs that can be used for comparison. In addition, the different salts have not only different melting temperatures but also have different latent heat of crystallization. This would cause different variations of power output even if the salts have the melting temperature.

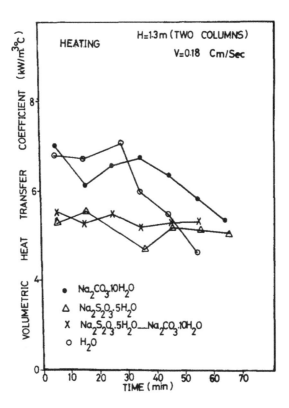

FIGURE 2.5.9   Volumetric heat transfer coefficients obtained from heating runs of different salts systems.

TABLE 2.5.4

Effect of Flow Rate on the Thermal Efficiency ($H = 0.65$ m, One Column)

Kerosene Superficial Velocity (cm/s)	$Na_2CO_3 \cdot 10H_2O$		$Na_2S_2O_3 \cdot 5H_2O$		$Na_2CO_3 \cdot 10H_2O$ $Na_2S_2O_3 \cdot 5H_2O$		Water	
	Cooling	Heating	Cooling	Heating	Cooling	Heating	Cooling	Heating
0.06	62.7	72.3	85.1	60.0	80.0	76.1	99.0	89.0
0.12	88.0	79.3	97.7	72.8	89.2	74.4	100	106
0.18	81.4	82.0	92.8	77.8	91.3	80.5	99.0	98.0
Average Efficiency	77.3	77.8	91.8	70.1	86.8	77.0	99.0	97.0

TABLE 2.5.5

Effect of Column Size on the Thermal Efficiency for Different Salts

| Volume of the Column (m³) | Thermal Efficiency | | Salt | Ref. |
	Cooling	Heating		
0.0086	56.4	50.3	$Na_2SO_4 \cdot 10H_2O$	*
0.0086	55.8	58.6	$Na_2CO_3 \cdot 10H_2O$	*
0.0094	59.7	62.0	$CaCl_2 \cdot 6H_2O$	[4]
0.0094	51.0	45.3	$Na_2SO_4 \cdot 10H_2O$	[4]
0.016	75.0		$Na_2SO_4 \cdot 10H_2O$	[11]
0.0172	77.3	77.8	$Na_2CO_3 \cdot 10H_2O$	*
0.0204	78.2	82.1	$Na_2CO_3 \cdot 10H_2O$	[6]
0.0204	78.0	71.8	$Na_2SO_4 \cdot 10H_2O$	[6]
0.0204	72.2	63.7	$Na_2HPO_4 \cdot 12H_2O$	[6]
0.0344	92.3	75.8	$Na_2S_2O_3 \cdot 5H_2O$	*
0.0408	79.6	82.6	$Na_2CO_3 \cdot 10H_2O$	[6]
0.0408	85.0		$Na_2SO_4 \cdot 10H2O$	[6]
0.129	91.0		Na2SO4 $\cdot 10H_2O$	[11]

* This work.

The effect of using two storage units connected in series and containing the same type of salt is shown in Figures 2.5.10–2.5.13 for both heat charge and discharge. Power output from the system and power input to the system were always high when two storage units were used, as may be seen from these figures. The effects of column height and kerosene flow rate are also shown in Figures 2.5.10 and 2.5.12, respectively, which is in agreement with the observations of Farid and Yacoub [4]

Figures 2.5.14 and 2.5.15 show some comparison of the power input/output for the two storage units when filled with either identical or different salts. Sodium thiosulphate provided the highest power output during heat discharge, while sodium carbonate can absorb energy at rates higher than that of thiosulphate. This was due to the crystallization temperature of the two salts. Using the combination of these two salts, each contained in one storage unit provided intermediate values of power input/output.

Care should be taken in these comparisons as a storage unit that provides a high power output will lose its heat content rapidly and, hence, will provide heat with much lower rates at later periods.

Figure 2.5.16 shows the rate of energy extracted from the storage system during cooling of sodium thiosulphate in two columns. Most of the power obtained was from the first column during the first period of operation. However, at later stages, a significant part of the power output was from the second column. This was not the case when two salts were used, as Figure 2.5.17 shows that both storage units provided a significant share of the total power output during the whole period of operation.

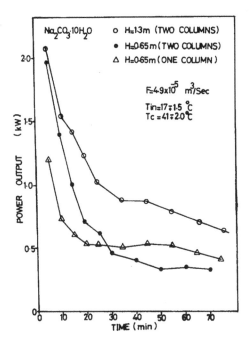

FIGURE 2.5.10   Effect of using two columns on the power output from sodium carbonate system.

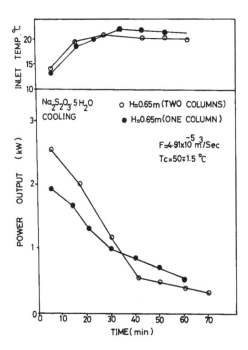

FIGURE 2.5.11   Effect of using two columns on the power output from sodium thiosulphate storage system.

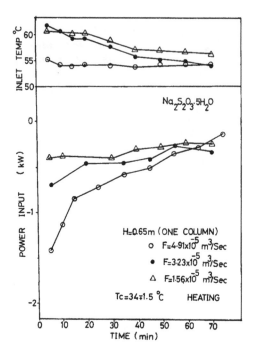

FIGURE 2.5.12   Power input for the storage unit using sodium thiosulphate pentahydrate.

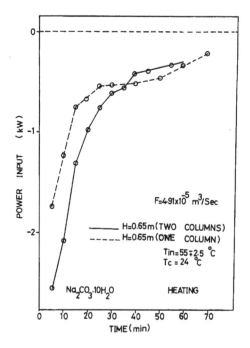

FIGURE 2.5.13   Power input for the storage unit using sodium carbonate decahydrate.

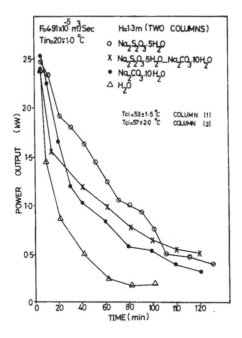

FIGURE 2.5.14 Comparison of the power output of different salts systems with that for water system.

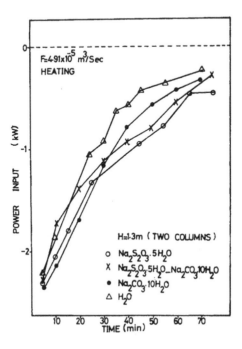

FIGURE 2.5.15 Comparison of the power input of different salts systems with that for water system.

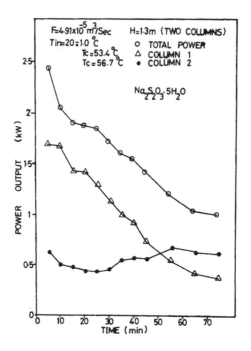

FIGURE 2.5.16   Power output from the two columns using sodium thiosulphate pentahydrate.

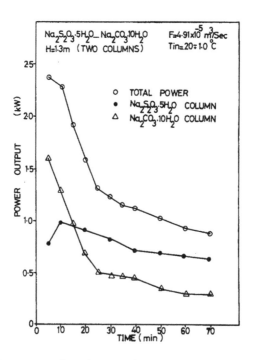

FIGURE 2.5.17   Power output from the two columns using two-salt system.

2.5.5   CONCLUSIONS

The use of two columns connected in series for heat storage using salts hydrates provided higher deriving temperature between the storage medium and the heat transfer
fluid. This gave higher rates of heat transfer during heat charge and discharge when
compared to the rates usually obtained from a single column.

The use of two hydrated salts with different crystallization temperatures would
improve the performance of the latent heat storage system further by allowing the
system to operate as a latent heat storage for longer periods.

Sodium thiosulphate pentahydrate generally showed a behavior similar to that of
other salts. However, the thermal efficiency of a storage unit using this salt was found
lower during heating and higher during cooling due to the higher crystallization temperature of this salt. The difficulty in dissolving all the $Na_2S_2O_3 \cdot 5H_2O$ suggests that
it should be used only in combination with other salts of lower crystallization temperature. The significant increase of thermal efficiency with a volume of storage unit
due to the better conditions of crystallization was in good agreement with previous
observations.

ACKNOWLEDGEMENTS

The facilities provided by the Chemical Engineering Department at Basrah
University are acknowledged. Thanks to Mr. Ahmad Shawaqfeh for his help in typ-
ing the manuscript.

NOMENCLATURE

A	diffuser area (total), m^2
C_d	specific heat capacity of the dispersed phase, J/kg/K
C_g	specific heat capacity of the column glass wall, J/kg/K
C_l	specific heat capacity of the solution, J/kg/K
C_s	specific heat capacity of the solid crystals, J/kg/K
D	diffuser diameter, m
D_t	column diameter, m
E	transfer efficiency, defined by Equation (2.5.3)
F	volumetric flow rate of the dispersed phase, m^3/sec
F_d	effective diffusivity factor defined by Equations (2.5.3) and (2.5.4)
H	continuous phase height, m
H_f	latent heat of crystallization of the salt, J/kg
m	mass fraction of the hydrated salt crystallized from the solution
$\bar{M}_d$	mass flow rate of the dispersed phase, kg/s
M_d	total mass of the dispersed phase in the column, kg
M_l	mass of salt solution, kg
M_s	mass of solid crystals in the column, kg
M_t	total mass of the salt and solution, kg
M_{col}	mass of the glass wall of the column, kg
R	radius of the bubble, m

t	time, s
T_c	continuous phase temperature, °C
T_i	bubble inlet temperature, °C
T_o	bubble outlet temperature, °C
T_r	ambient temperature, °C
u	bubble velocity, m/s
U_v	volumetric heat transfer coefficient, W/m³/K
U_w	heat loss coefficient, W/m²/K
V	volume of the continuous phase, m³
V'	total volume of the dispersed phase in the column, m³
w	weight fraction of the anhydrous salt in the solution

GREEK

α	thermal diffusivity, m²/s
θ	bubble contact time, s
ρ_c	density of the continuous phase, kg m³
Δ	small increment
η	efficiency

APPENDIX 1: PHYSICAL PROPERTIES

The following correlations for the physical properties of kerosene, sodium carbonate, and sodium thiosulphate were developed from data given by Perry and Green [16].

Density and specific heat capacity of the solution of both salts were found concentration dependent rather than temperature. Mass fraction of the salt crystallized (m) due to cooling to different temperatures were calculated from the phase diagrams presented by Farid and Yacoub [4] and Barrett et al. [17], using the lever rule. The temperature in all correlations is in °C. Units of ρ, C_l, C_s, and H_f are defined in nomenclature.

Kerosene
$\rho = 825.87 - 0.69\,(T)$
$C_d = 1890 + 4.60\,(T)$
$\alpha = 9.40 \times 10^{-8} - 1.87 \times 10^{-10}\,(T)$

Sodium carbonate decahydrate
$\rho = 990.45 + 1104.70\,(w)$
$C_1 = 4165 - 418{,}600(7.97 \times 10^{-3} - 2.74 \times 10^{-3}(T))(w)$
$C_s = 1880$
$m = 9.75 \times 10^{-3} + 0.11\,(29.5 - T) - 8.60 \times 10^{-3}\,(29.5 - T)^2 + 2.65 \times 10^{-4}\,(29.5 - T)^3$
For $w = 0.3$, the crystallization temperature was 29.5°C.

Sodium thiosulphate pentahydrate
$\rho = 1017.15 + 978.86(w)$
C_l was calculated from the weighted average of the specific heat of the salt and water.
$C_s = 1465$

$m = 0.022 + 0.019(42 - T) - 1.58 \times 10^{-5} (42 - T)^2 - 5.60 \times 10^{-6} (42 - T)^3$

$H_f = 200,000$ (heat of dilution was not included due to lack of data)

REFERENCES

1. M. Telkes, Nucleation of the supersaturated inorganic salt solution, *Industrial & Engineering Chemistry Research* **44**(6), 1308 (1952).
2. T. L. Etherington, A dynamic heat storage system, *Heating Piping and Airconditioning Journal* **Dec.**, 147–151 (1957).
3. D. R. Biswas, Thermal energy storage using sodium sulphate decahydrate and water, *Solar Energy* **19**, 99–100 (1977).
4. M. M. Farid and K. Yacoub, Performance of direct contact latent heat storage unit, *Solar Energy* **43**, 237–252 (1989).
5. D. D. Edie and S. S. Melsheimer, *An Immiscible Fluid-Heat of Fusion Energy Storage System, Sharing the Sun: Solar Technology in the Seventies*, Vol. 8, pp. 227–237. K. W. Boer (ed.), Published by the American Section of the International Solar Energy Society, Cape Canaveral, Florida and Winnipeg, Canada, (1976).
6. V. A. Costello, S.S. Melsheimer, and D. D. Edie, Heat transfer and calorimatric studies of a direct contact-latent heat storage system. Thermal storage and heat transfer in solar energy system. Presented at *ASME Meeting*, San Francisco, CA, 10–15 Dec. (1978).
7. J. Hallet and G. Keyser, Power characteristics of a continuous crystallization latent heat recovery system, *ASME publication*, 79-SOL-21 (1979).
8. A. E. Fouda, G. J. Despault, J.B. Taylor, and C. E. Capes, Solar storage system using salt hydrate latent heat and direct contact heat exchanger-I, *Solar Energy* **25**, 437–444 (1980).
9. A. E. Fouda, C. E. Capes, and M. Nakamura, Pilot plant study of a heat storage system using salt hydraes and direct contact heat transfer-A summary report, Thermal energy storage, *Word Congress-II of Chemical Engineering*, Tokyo (1986).
10. M. M. Farid, Storage of solar energy with phase change, *Journal of Solar Energy Research* **4**, 11–29 (1986).
11. M. M. Farid and A. Kanzawa, Thermal performance of heat storage module using PCM's with different melting temperatures: Mathmatical modelling: *Transactions of the ASME, Journal of Solar Energy Engineering* **111**, 152–157 (1989).
12. M. M. Farid, Y. Kim, and A. Kanzawa, Thermal performance of heat storage module using PCM's with different melting temperatures: Experimental. *Transaction of the ASME, Journal of Solar Energy Engineering* **112**, 125–131 (1990).
13. J. Ward, W. M. Loss, and G. O. G. Lof, Direct contact liquid-liquid heat exchanger for solar heated and cooled buildings-Pilot plant results, *U.S.D.O.E. Report COO* 12857-2 (1977).
14. M. M. Farid and K. Yacoub, Direct contact heat transfer between two immiscible fluids. Presented at the *Conference Selected Problems in Chemical Engineering*, Sobotka, Poland, 24–27th Sept. (1988).
15. A. E. Fouda, G. L. Despault, J.B. Taylor, and C. E. Capes, Solar storage system using salt hydrate latent heat and direct contact heat exchange-II, characteristics of pilot system operating with sodium sulfate solution, *Solar Energy* **32**, 57–65 (1984).
16. R. H. Perry and D. Green, *Perry's Chemical Engineers Handbook*, 6th ed., McGraw Hill Co., New York, (1984).
17. P. F. Barrett, B. B. Best, and K. B. Oldham, Thermal energy storage in supersaturated salt solutions, *Material, Chemistry and Physics Journal* **10**, 39–49 (1984).

3 Energy Saving, Peak Load Shifting and Price-Based Control Heating
Passive Applications

INTRODUCTION

Phase change materials (PCMs) have been considered for thermal storage in buildings since before 1980. With the advent of PCM implemented in gypsum board, plaster, concrete or other wall covering material, thermal storage can be part of the building structure even for lightweight buildings. Farid et al. [1] summarized the investigation and analysis of thermal energy storage systems incorporating PCMs for use in building applications. The problems associated with the application of PCMs with regard to the selection of materials and the methods used to contain them were also discussed.

Most of the work included in this chapter investigate the importance of thermal energy storage in buildings, with specific objectives of energy saving and improving comfort. Throughout this chapter, experimental work has been based on climatic conditions for Auckland City in New Zealand, but the computer simulation work allowed conclusions to be drawn for other cities with different climate.

An in-situ experimental facility (Figure 3.1) was built, in 2003, at the University of Auckland/Tamaki Campus under Professor Farid's supervision to study the use of PCM for energy saving, thermal comfort and peak load shifting. It is one of a few similar facilities in the world to be designed and built for such a purpose. The objective was to show that PCM impregnated in gypsum wallboards can provide thermal energy storage benefits in a real scale.

Farid et al. [2] examined the impact of using PCMs in building on electricity demand-side management perspective with real case studies, which incorporate wholesale electricity market data of New Zealand. Their results showed that the PCM application leads to peak load shifting, energy conservation and reduction in peak demand for network line companies and potential reduction in electricity consumption and more savings for residential customers. It also would improve the quality and reliability of power while reduces price volatility.

Several tools are available to design the PCM system. Evola et al. [3] developed a new simple methodology to determine the benefits of PCM in building envelopes without using detailed computer simulations. The methodology was based on four indicators that consider both thermal comfort condition and the PCM performance. Later on, Farid et al. [4] validated the methodology using experimental measurements

FIGURE 3.1   Experimental facility of thermal energy storage applications in buildings.

obtained from brick, concrete and timber construction facilities built in Spain and New Zealand, respectively.

An analysis of a price-based control system in conjunction with phase change materials was performed for two applications: space heating in buildings and domestic freezers. Savings of up to 16.5% and 62.64% per day were achieved for the freezer and building applications, respectively, based on New Zealand electricity rates [5]. The importance of weather forecast on the heating energy consumption in buildings was also studied [6].

Despite large experimental works conducted on the application PCMs in buildings, there is little work published on this application in combination with night ventilation. The application of the PCM energy storage in combination with night ventilation for space cooling was investigated experimentally, showing considerable electricity cost saving [7].

Another major application of PCM in buildings is underfloor heating in combination with PCM wallboards for space heating using a price-based control system [8]. It was investigated experimentally and showed significant electricity savings in both consumption and cost through peak load shifting. In a detailed study, two different types of PCMs of different melting points were used experimentally. The use of higher melting point PCM with the underfloor heating system allowed significant peak load shifting, while using lower melting point PCM in the walls and ceiling provided the comfort needed in the building.

In addition to real-time applications presented earlier, simulations codes (Energy Plus and Design Builder) were used to study the thermal behaviour of a typical house construction, including analysis of energy requirements versus comfort levels due to PCM integration into the building materials. An appropriate design enabled significant energy savings, up to 33%, and comfort enhancement of up to 31% [9]. Furthermore, the benefits of PCM when incorporated in walls, ceiling and in combination with PCM underfloor heating system using two different types of PCMs were evaluated. A 10-day period analysis showed a successful morning and evening peak load shifting with energy and cost saving of 32% and 42%, respectively [10].

REFERENCES

1. Khudhair AM, Farid MM. A review on energy conservation in building applications with thermal storage by latent heat using phase change materials. *Energy Convers Manag* 2004;45:263–75. doi:10.1016/S0196-8904(03)00131-6.
2. Qureshi WA, Nair NKC, Farid MM. Impact of energy storage in buildings on electricity demand side management. *Energy Convers Manag* 2011;52:2110–20. doi:10.1016/j. enconman.2010.12.008.
3. Evola G, Marletta L, Sicurella F. A methodology for investigating the effectiveness of PCM wallboards for summer thermal comfort in buildings. *Build Environ* 2013;59:517–27.
4. Castell A, Farid MM. Experimental validation of a methodology to assess PCM effectiveness in cooling building envelopes passively. *Energy Build* 2014;81:59–71.
5. Barzin R, Chen JJJ, Young BR, Farid MM. Peak load shifting with energy storage and price-based control system. *Energy* 2015;92:505–14. doi:10.1016/j.energy.2015.05.144.
6. Barzin R, Chen JJJ, Young BR, Farid MM. Application of weather forecast in conjunction with price-based method for PCM solar passive buildings - An experimental study. *Appl Energy* 2016;163:9–18. doi:10.1016/j.apenergy.2015.11.016.
7. Barzin R, Chen JJJ, Young BR, Farid MM. Application of PCM energy storage in combination with night ventilation for space cooling. *Appl Energy* 2015;158:412–21. doi:10.1016/j.apenergy.2015.08.088.
8. Barzin R, Chen JJJ, Young BR, Farid MM. Application of PCM underfloor heating in combination with PCM wallboards for space heating using price-based control system. *Appl Energy* 2015;148:39–48. doi:10.1016/j.apenergy.2015.03.027.
9. Vautherot M, Maréchal F, Farid MM. Analysis of energy requirements versus comfort levels for the integration of phase change materials in buildings. *J Build Eng* 2015;1:53–62. doi:10.1016/j.jobe.2015.03.003.
10. Devaux P, Farid MM. Benefits of PCM underfloor heating with PCM wallboards for space heating in winter. *Appl Energy* 2017;191:593–602. doi:10.1016/J.APENERGY. 2017.01.060.

3.1 A Review on Energy Conservation in Building Applications with Thermal Storage by Latent Heat Using Phase Change Materials

Amar M. Khudhair and Mohammed Farid
The University of Auckland

3.1.1 INTRODUCTION

Electrical energy consumption varies significantly during the day and night according to the demand by industrial, commercial and residential activities, especially in extremely hot and cold climate countries where the major part of the variation is due to domestic space heating and air conditioning. Such variation leads to a differential pricing system for peak and off-peak periods of energy use. Better power generation management and significant economic benefit can be achieved if some of the peak load could be shifted to the off-peak load period, which can be achieved by thermal storage of heat or coolness.

Solar energy applications require efficient thermal storage. Hence, the successful application of solar energy depends, to a large extent, on the method of energy storage used. The latent heat of melting is the large quantity of energy that needs to be absorbed or released when a material changes phase from solid state to liquid state or vice versa. The magnitude of the energy involved can be demonstrated by comparing the sensible heat capacity of concrete (1.0 kJ/kg K) with the latent heat of a phase change material (PCM), such as calcium chloride hexahydrate $CaCl_2$ $6H_2O$ (193 kJ/kg). It is obvious that any energy storage systems incorporating PCMs will comprise significantly smaller volumes when compared to other materials storing only sensible heat [1]. A further advantage of latent heat storage is that heat storage and delivery normally occur over a fairly narrow temperature range (the transition zone). The operating temperature span of latent heat systems can, therefore, be quite narrow [2].

PCMs have been considered for thermal storage in buildings since before 1980. With the advent of PCM implemented in gypsum board, plaster, concrete or other wall-covering material, thermal storage can be part of the building structure even for lightweight buildings. In the literature, development and testing were conducted for prototypes of PCM wallboard and PCM concrete systems to enhance the thermal energy storage (TES) capacity of standard gypsum wallboard and concrete blocks, with particular interest in peak load shifting and solar energy utilization.

3.1.2 PCM DEVELOPMENTS

Current interest in the PCM technology can probably be traced back to earlier development work on solar heat storage for space heating just after World War II. The search for alternate approaches, more volume efficient than sensible heat storage, naturally led to the investigation of latent heat systems. Most early studies of latent heat storage focused on the fusion-solidification of low cost, readily available salt hydrates, initially showing the greatest promise. Upon phase change, they tend to supercool, and the components do not melt congruently so that segregation results. Therefore, phenomena such as supercooling and phase separation often plague the thermal behavior of these materials and cause random variation or progressive drifting of the transition zone over repeated phase-change cycles. Although significant advances were made, major hurdles remained toward the development of reliable and practical storage systems utilizing salt hydrates and similar inorganic substances [2]. To avoid some of the problems inherent in inorganic PCMs, an interest has turned toward a new class of materials: low volatility, anhydrous organic substances such as polyethylene glycol, fatty acids and their derivatives and paraffins. Those materials were discarded at first because they are more costly than common salt hydrates, and they have somewhat lower heat storage capacity per unit volume. It has now been realized that some of these materials have strong advantages, such as physical and chemical stability, good thermal behavior, and adjustable transition zone [2].

Abhat [3], Lorsch et al. [4] and Farid [5] have reported comprehensive lists of possible candidates for latent heat storage covering a wide range of temperatures. Not all PCMs can be used for thermal storage. An ideal PCM candidate should fulfil a number of criteria such as high heat of fusion and thermal conductivity, high specific heat capacity, small volume change, noncorrosive, nontoxic and exhibit little or no decomposition or supercooling [6]. In building applications, only PCMs that have a phase transition close to human comfort temperature (~20°C) can be used. From the many hydrated salts and organic PCMs investigated, those shown in Table 3.1.1, are possible candidates for applications in buildings.

3.1.3 PHASE-CHANGE THERMAL STORAGE FOR PEAK LOAD SHIFTING

The development of an energy storage system may be one of the solutions to the problem when electricity supply and demand are out of phase. A building integrated with distributed thermal storage materials could shift most of the load coming from residential air conditioners from peak to off-peak time periods [7]. As a result, capital

TABLE 3.1.1

Hydrated Salts and Organic PCMs

PCM	Melting Point (°C)	Heat of Fusion (kJ/kg)
$KF \cdot 4H_2O$	18.5–19	231
Potassium fluoride tetrahydrate		
$CaCl_2 \cdot 6H_2O$	29.7	171
Calcium chloride hexahydrate		
$CH_3(CH_2)_{16}COO(CH_2)_3CH_3$	18–23	140
Butyl stearate		
$CH_3(CH_2)_{11}OH$ Dodecanol	17.5–23.3	188.8
$CH_3(CH_2)_{16}CH_3$	22.5–26.2	205.1
Tech. grade octadecane		
$CH_3(CH_2)_{12}COOC_3H_7$	16–19	186
Propyl palmitate		
45% $CH_3(CH_2)_8COOH$	17–21	143
55% $CH_3(CH_2)_{10}COOH$		
45/55 Capric–lauric acid		

investment in peak power generation equipment could be greatly reduced for power utilities and then could be reflected in less expensive service to customers. Where power utilities are offering time of day rates, building-integrated thermal storage would enable customers to take advantage of lower utility rates during off-peak hours. Stoval and Tomlinson [8] have examined the shifting of heating and cooling loads to off-peak times of the electrical utility but did not reach general conclusions regarding optimal PCM properties. Their analysis looked at potential applications of PCM wallboard as a load management device for passive solar applications and found it saved energy with reasonable payback time periods.

Peippo et al. [9] have shown that a $120\,m^2$ house in Madison, Wisconsin (43°N), could save up to 4 GJ a year (or 15% of the annual energy cost). Also, they have concluded that the optimal diurnal heat storage occurs with a melt temperature of 1°C–3°C above average room temperature. Claims are made that PCM wallboards could save up to 20% of residential house space conditioning cost [10].

3.1.4 PHASE CHANGE MATERIAL ENCAPSULATION IN STRUCTURES

During the last 20 years, several forms of bulk encapsulated PCM were marketed for active and passive solar applications, including direct gain. However, the surface area of most encapsulated commercial products was inadequate to deliver heat to the building after the PCM was melted by direct solar radiation. In contrast, the walls and ceilings of a building offer large areas for passive heat transfer within every zone of the building [11]. Several workers have investigated methods for impregnating gypsum wallboard and other architectural materials with PCM [12–15]. Different types of PCMs and their characteristics are described. The manufacturing techniques, thermal performance and applications of gypsum wallboard and concrete block, which have been impregnated with PCMs are presented and discussed in the following sections.

3.1.4.1 WALLBOARDS IMPREGNATED WITH PCMs

Kedl and Stovall [16] have presented the concept of octadecane wax impregnated wallboard for passive solar application, which was a major thrust of the Oak Ridge National Laboratory (ORNL)-Thermal Energy Storage program. Thus, ORNL has initiated a number of internal efforts in support of this concept. The results of these efforts were that the immersion process for filling wallboard with wax has been successfully scaled up from small samples to full-size sheets. Analysis shows that the immersion process has the potential for achieving higher storage capacity than adding wax filled pellets to wallboard during its manufacture. An analytical model that handles phase-change wallboard for passive solar application has been applied.

Shapiro [17,18] has shown several PCMs to be suitable for introduction into gypsum wallboard with possible thermal storage applications for the Florida climate. These materials were mixtures of methyl esters, methyl palmitate, and methyl stearate and mixtures of short-chain acids and capric and lauric acids. Although these materials had relatively high latent heat capacity, the temperature ranges required in achieving the thermal storage did not fall sufficiently within the range of comfort for buildings in hot climates. Neeper [19] found that the thermal storage provided by the PCM wallboard would be sufficient to enable a large solar heating fraction with direct gain.

Salyer and Sircar [20] defined a suitable low-cost linear alkyl hydrocarbon PCM from petroleum refining and developed methods of containing the PCM in plasterboard to eliminate leakage and problems of expansion in melting and freezing. Processes whereby this PCM could be incorporated into plasterboard either by post manufacturing imbibing of liquid PCM into the pore space of the plasterboard or by an additive that could be incorporated into the wet stage of plasterboard manufacture have been successfully demonstrated.

A laboratory-scale energy storage gypsum wallboard was produced by the direct incorporation of 21%–22% commercial grade butyl stearate (BS) at the mixing stage of conventional gypsum board production. The incorporation of BS was strongly facilitated by the presence and type of small amounts of dispersing agents. The physio-mechanical properties of the laboratory-produced thermal storage wallboard compare quite well with values obtained for standard gypsum board. The energy-storing capability of the board has a tenfold increase in capacity for the storage and discharge of heat when compared with gypsum wallboard alone [21]. The gypsum wallboard matrix makes an ideal supporting medium for the PCM, since approximately 41% of the wallboard volume is air voids. Based on a small-scale initial differential scanning calorimetry test, coconut fatty acid was selected for the room-scale PCM wallboard tests due to its favorable melting and freezing temperature range (24.9°C). Thus far, the PCM wallboard development work has experimentally shown that the concept is workable on a large scale and that PCM can be successfully integrated and distributed within a building with a significant thermal storage effect [22].

Building energy simulations help to estimate the potential applications of PCM storage in buildings. Therefore, Fraunhofer Institute [23] simulates the thermal behavior of building components to compare the dynamic performance of different types of wall constructions incorporating different amounts of PCMs. The basis is an empirically validated model for the phase transition. The model is being experimentally validated with a measurement facility for wall samples with an area of $0.5 \times 0.5\,m^2$. They investigate the effect as a function of the temperature range of the phase transition, the proportion of PCM, and the structure and usage of a building. The thermographs illustrate qualitatively the effect of the PCM in construction materials: four wall samples with differing amounts of PCM, which had been heated in an oven, and were then monitored during cooling. The variation of temperature with time clearly indicates the effect of the PCM. The larger the proportion of PCM, the longer the cooling process lasts. Thus, in a certain temperature range, the thermal mass of a building component can be significantly increased due to the phase-change process, so that the thermal comfort associated with massive buildings can be approached with light construction materials.

Athienitis et al. [24] conducted an extensive experimental and numerical simulation study in a full-scale outdoor test room with the PCM gypsum board as inside wall lining, as sketched in Figure 3.1.1. The PCM gypsum board used contained about 25% by weight proportion of BS. An explicit finite difference model was developed to simulate the transient heat transfer process in the walls. Their one-dimensional nonlinear numerical simulation successfully predicted the measured temperature history of the walls (Figure 3.1.2). It was shown that the utilization of PCM gypsum board may reduce the maximum room temperature by about 4°C during the daytime and can reduce the heating load at night significantly.

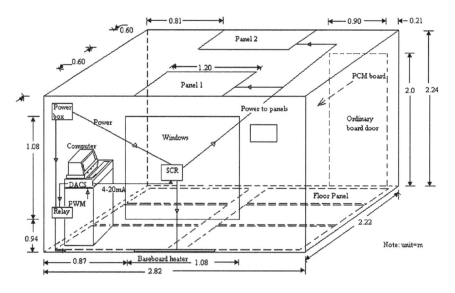

FIGURE 3.1.1   Schematic of outdoor test-room [24].

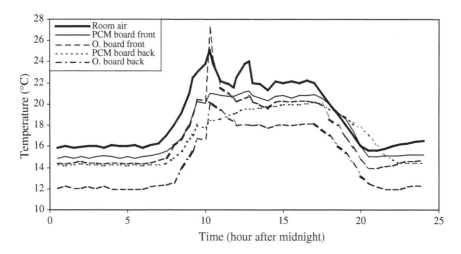

FIGURE 3.1.2 Experimental results from a typical winter sunny day (O. board denotes ordinary board) [24].

Kissock et al. [25] have presented the results of an experimental study on the thermal performance of wallboards imbibed to 30-wt% with commercial paraffinic PCM (K18) in simple structures. In the simulations conducted by them, solar radiation, ambient temperature, and interior temperatures in the test cells were continuously monitored for 14 days. The results indicated that peak temperature in the phase-change test cell was up to 10°C less than in the control test cell during sunny days. A modified finite difference simulation was able to predict the interior the wall temperature in the test cells with reasonable accuracy (average error < 1.7°C) based on measured property and environmental data.

An economic analysis of the energy savings was presented by Tomlinson and Heberle [26] to determine the thermal and economic performance of a PCM wallboard. Stetiu and Feustel [27] have presented a thermal building simulation program based on a finite difference approach to evaluate numerically the latent heat storage performance of treated PCM wallboard. Thermal mass can be utilized to reduce the peak power demand and downsize the cooling or heating systems. PCMs have two important advantages as storage media: they can offer an order of magnitude increase in thermal storage capacity, and their discharge is almost isothermal. This allows storing large amounts of energy without significantly changing the temperature of the sheathing (room envelope). A computer-based numerical model for optimizing the phase-change temperature with the composition of the PCM mixture was developed by Ma and Jin [28].

Neeper [11] has examined the thermal dynamics of a gypsum wallboard impregnated by fatty acids and paraffin waxes as PCMs that are subjected to the diurnal variation of room temperature but are not directly illuminated by the Sun. The melt temperatures of these PCMs are adjusted by using a mixture of ingredients. He has examined three parameters of PCM wallboard that may influence the energy that can be passively absorbed and released during a daily cycle: (a) the melt temperature of the PCM, (b) the temperature range over which melt occurs and (c) the latent capacity per unit area of wallboard. The following findings have been presented. The maximum diurnal energy storage occurs at a value of the encapsulated PCM melt temperature that is close to the average comfort room temperature in most circumstances. Diurnal energy storage decreases if the phase-change transition occurs over a range of temperature. Adapted from Kedl [29], the properties of the wallboard containing 10-wt% and 20-wt% octadecane were used in the numerical simulation of Neeper [11] to optimize the design of such type of walls and to investigate the energy absorbed and released daily by an idealized PCM wall for sinusoidal and square wave daily cycles of room temperature.

3.1.4.2 CONCRETE BLOCKS IMPREGNATING WITH PCMs

Chahroudi [30,31] has discussed potential application areas for thermocrete materials in general terms. Thermocrete is a heat storage medium combining an appropriate PCM with the concrete matrix, resulting in almost isothermal storage. Thermocrete blocks have been specifically analyzed in more detail concerning modes of operation, performance specifications and energy savings in various applications. Thermocrete energy storage materials were made by combining PCMs with open cell cements to produce low-cost energy storage materials with structural and thermostatic properties.

Hawes et al. [32] and Hawes and Feldman [33] have studied the thermal performance of PCMs (BS, dodecanol, paraffin, tetradecanol) in different types of concrete blocks. These studies have covered the effects of concrete alkalinity, temperature, immersion time and PCM dilution on PCM absorption during the impregnation process. They examined the mechanisms of absorption and established a means of developing and using absorption constants for PCM in concrete to achieve diffusion of the desired amount of PCM and, hence, the required thermal storage capacity. The effects of concrete modification and improved PCM incorporation techniques were shown to have increased thermal storage up to about 300%.

The most promising PCM containment methods have been studied by Salyer et al. [34] and applied to solite hollow core building blocks were as follows:

1. Imbibing the PCM into porous materials.
2. Absorbing the PCM into finely divided special silicas.
3. Permeating the PCM into polymeric carriers.

They have noted that the PCM in the form of PCM melt-mix, PCM/silica dry powder, or the PCM/HDPE (cross-linked pellets of high-density polyethylene) when inserted into the hollow-core space of concrete blocks can accommodate large quantities of PCM and, correspondingly, very large amounts of thermal storage.

Lee et al. [35] have studied and presented the results of macro-scale tests that compare the thermal storage performance of ordinary concrete blocks with those that have been impregnated with two types of PCMs, BS and commercial paraffin (P). Two types of blocks were tested and compared, regular block (R) that is made of Portland cement and autoclaved block (A) that consists of Portland cement plus

TABLE 3.1.2

Calculated Heat Storage Values for Various PCM-Block Combinations[a] [35]

Heat Storage Values and Temperature Range	Block and PCM (%PCM in Block)		
	A–BS (5.6%)	A–P (8.4%)	R–P (3.9%)
Temperature range (°C)	15–25	22–60	22–60
Storable sensible heat in blocks (kJ)	1428	5337	7451
Storable sensible heat in PCM (kJ)	233	1136	705
Storable latent heat in PCM (kJ)	977	2771	1718
Total storable heat (kJ)	2638	9244	9874
Total storable heat/storable sensible heat in blocks	1.9	1.7	1.3

[a] Notes: specific heat of concrete block: 0.88 kJ/kg °C; specific heat of BS: 2.41 kJ/kg °C; specific heat of P: 2.04 kJ/kg °C; latent heat of BS: 101 kJ/kg (in 15°C–25°C range); latent heat of P: 189 kJ/kg (in the 22°C–60°C range).

silica. The impregnation was achieved by immersing the heated concrete block in a bath filled with molten PCM until the required amount of PCM was absorbed (3.9%–8.6%). The results showed that the concrete blocks are capable of storing the latent and sensible heats of the incorporated PCM, as well as the sensible heat of concrete as shown in Table 3.1.2. This permits heating and cooling supplied by public utilities to be shifted away from the demand times.

Hadjieva et al. [36] have applied the same impregnation technique for concrete but with sodium thiosulfate pentahydrate ($Na_2S_2O_3 \cdot 5H_2O$) as a PCM. They used porous concrete and managed to fill most of the pore and capillary spaces of the concrete up to as high as 60%. They concluded that the large absorption area of autoclaved porous concrete serves as a good supporting matrix of an incongruently melting $Na_2S_2O_3 \cdot 5H_2O$ and improves its structure stability during thermal cycling. Such microencapsulation may prove very efficient in containing the PCM in an inexpensive way and also in eliminating problems associated with the use of hydrated salts, such as supercooling and phase segregation.

3.1.4.3 UNDERFLOOR HEATING WITH LATENT HEAT STORAGE

One of the common ways of utilizing building mass for thermal storage is floor warming or underfloor heating. This form of heating provides a large heating surface, which makes it possible to heat the room uniformly with low heat fluxes in the order of 100 W/m² or less. The development of a third-order explicit nonlinear finite difference network model to study the performance of a floor heating system with high solar gain has been presented by Athienitis [37]. Simplified analysis of radiant floor heating can be found in the literature [38].

Farid and Chen [39] have developed a computer model to simulate underfloor heating with and without the presence of a layer of PCM. The floor was assumed to be electrically heated with a constant heat flux for 8 h using off-peak electricity. The simulation for the concrete floor showed undesirably large variations in the floor surface temperature, and the concrete lost most of its stored heat within a short period after the heating was switched off. On the contrary, the simulation for the modified system incorporating PCM showed a much lower surface temperature fluctuation and the ability of the floor to provide the necessary warmth during the remaining period of the day. A similar investigation has been conducted by Amir et al. [40] who have studied the performance of two heating floor panels containing either water or *n*-octadecane paraffin in a concrete structure. Laouadi and Lacroix [41] have conducted a theoretical study on similar panels. The simulation results showed that the panels might be charged and discharged twice a day, making such panels useful for common cases in which the peak load occurs in the morning and evening.

Farid and Kong [42] have constructed two concrete slabs, one of them containing encapsulated PCM. The PCM used was commercial $CaCl_2 \cdot 6H_2O$ encapsulated in spherical plastic nodules as manufactured by Cristopia Ltd. The plastic spheres contained about 10% empty space to accommodate volume expansion of the PCM during melting. Rigid plastic was used in the manufacturing of the spheres to prevent any harmful stresses that it may cause on the surrounding structure. However, it is necessary to evaluate the mechanical strength of the concrete floor with the presence of the PCM spherical nodules before such practice can be applied. The performances of the two concrete slabs were tested experimentally and compared with each other. The thermal mass of the concrete was increased significantly by embedding the PCM nodules. The PCM concrete slab maintained sufficient heat storage for the whole day from a heating of only 8 h, while the plain concrete lost most of the stored heat in a few hours, as shown in Figure 3.1.3.

3.1.5   MICRO- AND MACROENCAPSULATION METHODS

Containment costs and attendant problems have been major problems with many of the PCM systems developed in the past. Hawes and Feldman [33] have considered the

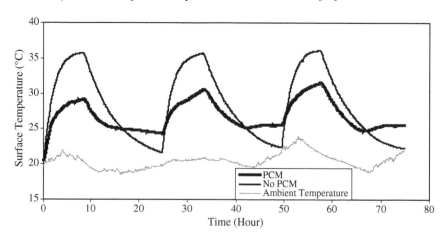

FIGURE 3.1.3 Comparison between the performance of the under floor heating system with and without phase-change storage [42].

means of PCM incorporation: direct incorporation, immersion and encapsulation. The third one can be defined as the containment of PCM within a capsule of various materials, forms and sizes prior to incorporation so that it may be introduced to the mix in a convenient manner. There are two principal means of encapsulation. The first is microencapsulation, whereby small, spherical or rod-shaped particles are enclosed in a thin, high molecular weight polymeric film. The coated particles can then be incorporated in any matrix that is compatible with the encapsulating film. It follows that the film must be compatible with both the PCM and the matrix. The second containment method is macroencapsulation, which comprises the inclusion of PCM in some form of package such as tubes, pouches, spheres, panels or other receptacle. These containers can serve directly as heat exchangers or they can be incorporated in building products.

The PCM must be encapsulated so that it does not adversely affect the function of the construction material. Previous experiments with large volume containment or macroencapsulation failed due to the poor conductivity of the PCM. When it was time to regain the heat from the liquid phase, the PCM solidified around the edges and prevented effective heat transfer. With microencapsulation, the dimensions are so small that this effect does not occur. Microencapsulation also allows the PCM to be incorporated simply and economically into conventional construction materials [23]. Both methods of PCM encapsulation in concrete (micro- and macroencapsulation) may have some drawbacks. Plastic or metallic encapsulation of the PCM is expensive but safe, as the PCM is not in contact with the concrete. Microencapsulation by impregnating the PCM in the concrete is very effective, but it may affect the mechanical strength of the concrete.

Hawlader et al. [43] have conducted experiments and simulations to evaluate the characteristics and performance of encapsulated paraffin in crosslink agent (HCHO) in terms of encapsulation ratio and energy storage capacity. It was found that a higher coating to paraffin ratio led to a higher paraffin encapsulation ratio and then lower product hydrophilicity. Thermal cyclic tests showed that the encapsulated paraffin kept its geometrical profile and energy storage capacity even after 1000 cycles of operation, as shown in Figure 3.1.4. Recently, Hawlader et al. [44] have prepared and compared encapsulated paraffin waxes by complex coacervation and spray drying methods. The microcapsules have high energy storage and release capacity (145–240 kJ/kg), which depends on core to coating ratio as shown in Table 3.1.3.

3.1.6 FIRE RETARDATION OF PCM-TREATED CONSTRUCTION MATERIALS

In recent years, stringent safety codes and flammability requirements have been imposed on building materials to protect these buildings from fire hazards. The following are some approaches that have been investigated and applied successfully in laboratory tests to fire retard PCM imbibed plasterboard by Salyer and Sircar [20]:

- Adding alternate nonflammable surface to the plasterboard (e.g., aluminum foil and rigid polyvinyl chloride film).
- Sequential treatment of plasterboard, first in PCM and then in an insoluble liquid fire retardant (e.g., Fyrol CEF). The insoluble fire retardant

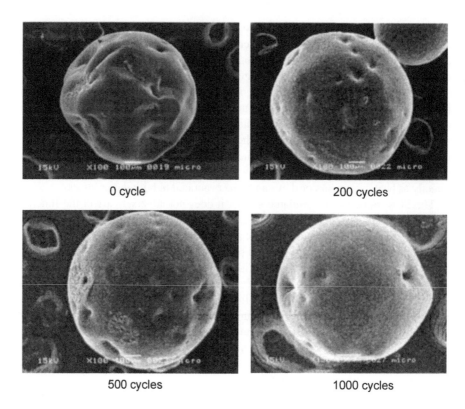

0 cycle

200 cycles

500 cycles

1000 cycles

FIGURE 3.1.4 Microencapsulated paraffin profile evaluated by SEM at different thermal cycles [43].

TABLE 3.1.3

Energy Released and Storage Capacities for Coacervated and Spray-Dried (Core to Coating Ratio) Microencapsulated PCM Samples [44]

Samples	Coacervated Samples			Spray Dried Samples		
Ratio	(2:1)	(1:1)	(1:2)	(2:1)	(1:1)	(1:2)
Energy stored (kJ/kg)	239.78	213.46	193.35	216.44	210.78	145.28
Energy released (kJ/kg)	234.05	224.31	196.38	221.51	211.66	148.32

displaces part of the PCM and some remains on the surface, thus imparting self-extinguishing characterization to the plasterboard.

- Using brominated hexadecane and octadecane as PCMs. It is anticipated that when these halogenated PCM compounds are combined with antimony oxide in plasterboard, the product will be self-extinguishing.
- Fire retardant surface coatings may be used to prevent effectively the wicking action of the plasterboard paper covers.

Banu et al. [15] have conducted flammability tests on energy storing wallboard, ordinary gypsum wallboard impregnated with approximately 24% organic PCM. The flammability tests evaluated the surface burning characteristics. Flames spread and smoke development classifications were determined in a Steiner tunnel, and heat and smoke release rates were determined by cone calorimeter. Comparison of the test results with similar data for other building materials indicates the possibility of reducing the flammability of energy-storing wallboard by incorporation of a flame retardant.

3.1.7 CONCLUSIONS

Limited experimental studies of PCM wallboard have been conducted, but few general rules pertaining to the thermal dynamics of PCM wallboard are available. The main conclusion is that a PCM wall is capable of capturing a large proportion of the solar radiation incident on the walls or roof of a building. Because of the high thermal mass of PCM walls, they are also capable of minimizing the effect of large fluctuations in the ambient temperature on the inside temperature of the building. They can be very effective in shifting the heating and cooling load to off-peak electricity periods. Gypsum wallboard impregnated with PCM could be installed in place of ordinary wallboard during new construction or rehabilitation of a building. It will provide thermal storage that is distributed throughout a building, enabling passive solar design and off-peak cooling with frame construction. Little or no additional cost would be incurred for installation of the PCM wallboard in place of ordinary wallboard. At the University of Auckland, New Zealand, efforts are underway to develop a gypsum wallboard containing a PCM with a melting temperature suitable for New Zealand weather conditions.

REFERENCES

1. Ip KCW, Gates JR. Thermal storage for sustainable dwellings. In: *Proceedings of International Conference Sustainable Building*. Maastricht, The Netherlands, 2000.
2. Paris J, Falardeau M, Villeneuve C. Thermal storage by latent heat: a viable option for energy conservation in buildings. *Energy Sources* 1993;15:85–93.
3. Abhat A. Low temperature latent heat thermal energy storage. Heat storage materials. *Sol Energy* 1983;30:313–32.
4. Lorsch HG, Kauffman KW, Denton JC. Thermal energy storage for heating and air conditioning, future energy production system. *Heat Mass Transfer Process* 1976;1:69–85.
5. Farid MM. A review on energy storage with phase changes. In: *Proceedings of Chicago/Midwest Renewable Energy Workshop*, Chicago, USA, 2001.
6. Ghoneim AA, Klein SA, Duffie JA. Analysis of collector-storage building walls using phase-change materials. *Sol Energy* 1991;47:237–42.
7. Neeper D. Benefits of wallboard impregnated with phase-change materials for residential heating and cooling. Subcontract Report to Solar Energy Research Institute, 1990.
8. Stoval TK, Tomlinson JJ. What are the potential benefits of including latent storage in common wallboard? *Trans ASME* 1995;117:318–25.
9. Peippo K, Kauranen P, Lund PD. A multi-component PCM wall optimized for passive solar heating. *Energy Build* 1991;17:259–70.
10. NAHBRC Web Site. Phase change materials: advanced technology in housing. United States National Association of Home Builders, Available at <www.nahbrc.org>, 2002.

11. Neeper DA. Thermal dynamics of wallboard with latent heat storage. *Sol Energy* 2000; 68:393–403.

12. Salyer IO, Sircar AK, Chartoff RP, Miller DE. Advanced phase-change materials for passive solar storage applications. In: *Proceedings of the 20th Intersociety Energy Conversion Engineering Conference*, Warrendale, Pennsylvania, USA, 1985, pp. 699–709.

13. Shapiro M, Feldman D, Hawes D, Banu D. PCM thermal storage in drywall using organic phase change material. *Passive Sol J* 1987;4:419–38.

14. Babich MW, Benrashid R, Mounts RD. DSC studies of new energy storage materials. Part 3. Thermal and Flammability Studies. *Thermochimica Acta* 1994;243:193–200.

15. Banu D, Feldman D, Haghighat F, Paris J, Hawes D. Energy-storing wallboard: flammability tests. *J Mater Civ Eng* 1998;10:98–105.

16. Kedl RJ, Stovall TK. *Activities in Support of the Wax-Impregnated Wallboard Concept.* U.S. Department of Energy: thermal energy storage researches activity review. New Orleans, LA, 1989.

17. Shapiro M. Development of the enthalpy storage materials, mixture of methyl stearate and methyl palmitate. Subcontract report to Florida Solar Energy Center, 1989.

18. Shapiro M. Development of the enthalpy storage materials, mixture of capric acid and lauric acid with fire retardants. Subcontract report to Florida Solar Energy Center, 1989.

19. Neeper DA. Potential benefits of distributed PCM thermal storage. In: Coleman MJ, editor. *Proceedings of 14th National Passive Solar Conference*, Denver, Colorado, USA: American Solar Energy Society; 1989. pp. 283–8.

20. Salyer IO, Sircar AK. Phase change materials for heating and cooling of residential buildings and other applications. In: *Proceedings of 25th Intersociety Energy Conversion Engineering Conference*, 1990. pp. 236–43.

21. Feldman D, Banu D. Obtaining an energy storing building material by direct incorporation of an organic phase change material in gypsum wallboard. *Sol Energy Mater* 1991; 22:231–42.

22. Rudd AF. Phase change material wallboard for distributed thermal storage in buildings. *ASHRAE Trans: Res* 1993;99:339–46.

23. Fraunhofer ISE. Available at <www.ise.fhg.de/english>, 2002.

24. Athienitis AK, Liu C, Hawes D, Banu D, Feldman D. Investigation of the thermal performance of a passive solar test-room with wall latent heat storage. *Build Environ* 1997;32:405–10.

25. Kissock JK, Hannig JM, Whitney TI, Drake ML. Testing and simulation of phase change wallboard for thermal storage in buildings. In: *Proceedings of 1998 International Solar Energy Conference*, New York, USA, 1998. pp. 45–52.

26. Tomlinson JJ, Heberle DD. Analysis of wallboard containing a phase change material. In: *Proceedings of the 25th Intersociety Energy Conversion Engineering Conference-IECEC'90*, Reno, NV, USA, 1990. pp. 230–5.

27. Stetiu C, Feustel HE. Phase change wallboard as an alternative to compressor cooling in Californian residences. In: *Proceedings of 96 ACEE Summer Study for Energy Efficient Building*, California, USA, 1996.

28. Ma F, Jin L. Improvement on the properties of energy-storing constructional material and optimization of the phase change temperature. *J Huazhong Univ Sci Technol* 1997;25:82–5.

29. Kedl RJ. Wallboard with latent heat storage for passive solar applications. Oak Ridge National Laboratory Report ORNL/TM-11541, National Technical Information Service, 5285 Port Royal Road, Springfield, VA 22161, USA, 1991.

30. Chahroudi D. Thermocrete heat storage materials: applications and performance specifications. In: *Proceedings of Sharing the Sun Solar Technology in the Seventies Conference*, Winnipeg, USA, 1976.

31. Chahroudi D. Development of thermocrete heat storage materials. In: *Proceedings of the International Solar Energy Congress*, vol. 1, New Delhi, India, 1978.
32. Hawes DW, Banu D, Feldman D. Latent heat storage in concrete II. *Sol Energy Mater* 1990;21:61–80.
33. Hawes DW, Feldman D. Absorption of phase change materials in concrete. *Sol Energy Mater Sol Cells* 1992;27:91–101.
34. Salyer IO, Sircar AK, Kumar A. Advanced phase change materials technology: evaluation in lightweight solite hollow-core building blocks. In: *Proceedings of the 30th Intersociety Energy Conversion Engineering Conference*, Orlando, FL, USA, 1995. p. 217–24.
35. Lee T, Hawes DW, Banu D, Feldman D. Control aspects of latent heat storage and recovery in concrete. *Sol Energy Mater Solar Cells* 2000;62:217–37.
36. Hadjieva M, Stoykov R, Filipova T. Composite salt-hydrate concrete system for building energy storage. *Renew Energy* 2000;19:111–5.
37. Athienitis AK. Numerical model of floor heating system. *ASHRAE Trans* 1994;100:1024–30.
38. Chun WG, Jeon MS, Lee YS, Jeon HS, Lee TK. A thermal analysis of a radiant floor heating system using SERIRES. *Int J Energy Res* 1999;23:335–43.
39. Farid MM, Chen XD. Domestic electric space heating with heat storage. *Proc Inst Mech Eng* 1999;213:83–92.
40. Amir M, Lacroix M, Galanis N. Comportement thermique de dalles chauffantes electriques pour le stockage quotidien. *Int J Therm Sci* 1999;38:121–31.
41. Laouadi A, Lacroix M. Thermal performance of latent heat energy storage ventilated panel for electric load management. *Int J Heat Mass Transfer* 1999;42:275–86.
42. Farid MM, Kong WJ. Underfloor heating with latent heat storage. *Proc Inst Mech Eng* 2001;215:601–9.
43. Hawlader MN, Uddin MS, Zhu HJ. Encapsulated phase change materials for thermal energy storage: experiments and simulation. *Int J Energy Res* 2002;26:159–71.
44. Hawlader MN, Uddin MS, Khin MM. Microencapsulated phase change materials. In: *Proceedings of 9th APCChE Congress and CHEMECA 2002*, Christchurch, New Zealand, 2002.

3.2 Impact of Energy Storage in Buildings on Electricity Demand Side Management

Waqar A. Qureshi, Nirmal-Kumar C. Nair, and Mohammed M. Farid
The University of Auckland

3.2.1 INTRODUCTION

Phase change materials (PCMs) are being investigated for a large number of applications as reviewed in Refs. [1,2]. One of them is in buildings because of the capacity of PCM to store thermal energy from the natural solar cycle and utilize it when required. The use of PCM in building applications has been discussed in several publications [3–6]. In [3], experiments performed on a test facility recommended PCM with a large surface area to be used for solar energy storage and also compared the results with simulations. In [4], it has been concluded that RT-20, which is a commercial-grade PCM, can be used as energy storage material to achieve the objective of reduction in electrical energy for heating and enable peak load shifting. In [5], the use of PCM results in a significant saving in cooling by utilizing its property to store heat and coolness. PCM has also been explored for use as a daylighting element along with an energy storage element in [6]. All the above publications have discussed the benefits of PCM with regard to energy storage with a few having mentioned the possibility of having load reduction and load shifting, but none of them have quantified the energy saving and peak load shifting in the context of actual electricity price variations.

PCM properties such as thermal conductivity, density change and stability under extended cycles such as phase segregation and sub-cooling are important and needs to be considered prior to the application of PCM. The material to be used for phase-change thermal storage must have large latent heat of melting. The melting temperature needs to lie within the practical range of operation. Further, they should be non-corrosive, non-toxic and chemically stable with a minimum sub-cooling. Materials having melting range between 15°C and 90°C are used for solar heating application with materials having melting temperature below 15°C to be used for storing coolness and those having temperature above 90°C typically used for air-conditioning using absorption refrigeration cycle. For building application, the suitable PCM should have melting point close to human comfort level, which is around 20°C.

From the available list of phase change materials, commercial paraffin waxes are most suitable with moderate thermal storage densities (~200 kJ/kg). These waxes are available over a wide range of melting temperature, undergoes minimum sub-cooling, chemically inert and stable with no phase segregation. Generally, only technical grade paraffin is used since pure paraffin waxes are expensive.

If solar energy is harnessed efficiently, then it could be utilized to decrease the energy needs for heating. PCMs have high latent heat of greater than 150 kJ/kg compared to concrete (1 kJ/kg) for the small temperature swing associated with the comfort level. This property of phase change material can be utilized for storing solar energy during days for further use at night to keep buildings warm.

Maintaining a sustainable supply of electrical energy and the increasing emphasis towards renewable energy generation are challenges that are being faced worldwide. Supply-side options have mostly been the preferred choice in the past to manage the security of supply, but the focus is increasing toward demand-side alternatives. Demand side management (DSM) attributes like energy efficiency and peak load shifting in electrical power systems are gaining increased attention. In deregulated electricity market environment, like New Zealand, wholesale price for electricity varies with time during a day. If this price varying signal is taken into consideration while consuming, then Thermal Energy Storage (TES) can enable opportunities for reduction of electricity charges paid by consumers.

From New Zealand Household Energy End-Use survey statistics, 34% of the total energy is consumed for space heating of which 24% use electrical means of heating, and on average 8.16% of New Zealand electricity generation is used for space heating [7]. The idea behind TES as a DSM option is to store thermal energy from electricity during off-peak time for later use. This action can benefit the network line companies by reducing overall peak load, which typically occurs during times of peak wholesale prices, and for customers in the form of reduced energy bills if charged on real-time pricing.

The traditional energy management for space heating is done by operating and controlling thermal devices using time-clock instruments, which allows loads to operate during specific times of day or night. The other technique is to estimate the energy price based on weather forecast but this requires a large amount of data collection and processing [8]. The abovementioned methods have some constraints or limitations such as dependence on weather forecasting, large amount of data handling and processing. Electricity is a commodity that cannot be stored economically in large quantities for future usage; moreover, the energy consumption varies temporally and seasonally. Space heating and air conditioning loads are one of the major contributors to this variation as mentioned above. In countries, operating in an electricity market environment, usually wholesale electricity nodal price is strongly correlated with load especially in winter [9,10]. In New Zealand, deregulation in electricity industry has been attributed to effects such as reliability, price volatility and market power [11]. It is generally agreed that the current pricing system could be one of the critical factors affecting electricity generation and consumption in the future. Due to electricity industry deregulation, security of supply and electricity price spikes gets more attention and concern among policymakers, utility companies and consumers [12].

This article is based on the research being conducted at the University of Auckland, New Zealand, investigating the application of commercial-grade PCM "RT-20" [9,13], which has a melting range of 18°C–22°C. This article focuses on detailed analysis of PCM application in the context of DSM measures in deregulated electricity market environment of New Zealand. The results from these studies show significant reductions in the temperature variations between day and night in offices having PCM impregnated in the walls. To relate benefits of PCM application towards energy efficiency and peak load shifting goals, the concept of Electricity DSM in the context of New Zealand electricity market is introduced in this article. Peak load shifting, energy conversation and pricing efficiency is thereafter assessed for estimating potential benefits to distribution line companies and electricity consumers through TES in buildings.

3.2.2 ELECTRICAL DSM AND NEW ZEALAND CONTEXT

DSM is defined as the planning, implementation, and monitoring of distribution network utility activities designed to influence customer use of electricity in ways that will produce desired changes in the load shape, i.e. changes in the time pattern and magnitude of the network load. This definition includes, six broad categories of load-shape DSM objectives: Peak clipping, valley filling, load shifting, strategic conservation, strategic load growth and flexible load shape [9,14] as illustrated in Figure 3.2.1.

There are two important parameters while managing power system distribution network operation i.e. peak usage time pattern and magnitude of demand. Controlling these two parameters using DSM programmes could result in cost-effective use of electricity for utility and customers alike [8].

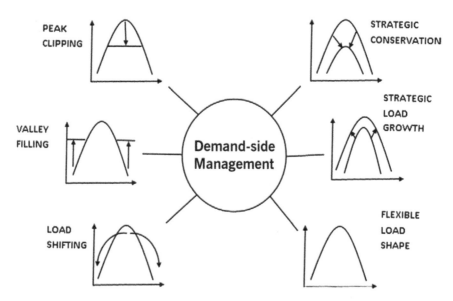

FIGURE 3.2.1 Demand-side management categories [14].

Thermal energy storage (TES) is gaining more attention as the world is becoming more conscious about future energy requirements and shortfall in energy supply with increased dependence on electrical energy. Effective TES in buildings can potentially impact several categories of DSM. These categories include peak load shifting, valley filling and strategic conservation [14].

Real-time pricing (RTP) is an emerging reality in electricity market environment, which is primarily an incentive-driven DSM technique [15]. Though the concept of RTP is not new, it has not been implemented due to the unavailability of the required technology and the lack of economic incentives. The advancements in smart metre technology in recent years have realized more opportunities to implement RTP system especially in power systems operating in an electricity market environment [12]. RTP, which is a type of dynamic pricing mechanism, together with energy conservation programs can motivate the consumer to modify their energy consumption pattern to realize monetary incentives. From network line companies' viewpoint, such an action will introduce demand flexibility into their network operation [16]. With the infrastructure installed for RTP and associated tariffs, efficient TES could be a very useful option for utility companies and consumers. Recent investigations on the effect of RTP on electricity consumption patterns by applying some form of direct control on thermal appliances have been reported [5,7–10]. In [9,13], the authors have focused on peak load shifting as an advantage of PCM building application. Ref. [12] describes the design of the RTP system for residential consumers. Different tariff types and design of the RTP system for few case studies have also been discussed in [16]. Ref. [17] has also discussed RTP as one of the useful tools for DSM programmes. None of the above references has tried investigating contribution of TES in the presence of RTP system. This article is mainly focused on relating RTP and TES while using PCM in building application.

New Zealand Electricity Market (NZEM) has a deregulated electricity market that operates on locational marginal pricing (LMP) wholesale model. Wholesale electricity price is very dynamic in nature and is normally affected by multiple factors such as load, seasons, time of use and offer electricity price from electricity generators. A large rise in demand can lead to capacity limitations which can affect the wholesale prices and result in price volatility. One approach to control this volatility is to manage the retail load on RTP by being responsive to capacity or energy limitation. At present only very large industrial loads with Time-of-Use (ToU) meters have incentives to participate in NZEM. Currently, the New Zealand electricity residential load is not directly participating using wholesale pricing signals. The recently issued Advanced Metering Policy by New Zealand Electricity Commission (NZEC) in May 2008, could facilitate RTP-based DSM for all retail customers like residential loads in future [18].

DSM can also attract distributors and retailers to participate in valley filling or strategic load growth. DSM programs could realize significant savings at the Transmission and Distribution (T&D) level [19]. The DSM programs have the potential for distributors and retailer, in deregulated environment, for cost reduction or revenue raising [17,19,20]. Some of the DSM categories like valley filling, load shifting, and strategic load growth, can be applied in the T&D network to minimize the average losses in the system. Minimizing the losses would ultimately result in increased revenues and reduction in cost and life efficiency of the network resources [19]. Thus, DSM has been recognized as a very important tool for power industry to address various issues mentioned above. In this article, the PCM building application is being investigated along with the RTP system to assess its impact on the various DSM measures which can be used to benefit distribution network companies and consumers.

3.2.3   RESEARCH SETUP

The research and experimental setup consist of three major constituents, i.e. test offices, controlled heating arrangement, and data acquisition system. To assess typical energy efficiency gains for space heating using PCM, a comprehensive facility consisting of two identically designed and shaped test offices, Office-1 and Office-2, and a control centre (Office-3) has been built at Tamaki Campus of the University of Auckland (New Zealand). The interior walls and ceiling of Office-1 are finished with ordinary gypsum wallboards. It is used as a base-case office. The interior walls and ceiling of Office-2 are finished with PCM gypsum wallboards (PCMGW). The two offices are equipped with specialized data acquisition system connected to thermocouples placed in the walls, ceiling and interior of both offices. The top and elevation view of the complete layout of the identical offices is provided in Figures 3.2.2 and 3.2.3. This layout identifies the design of the building, position of heating, monitoring and control equipment and location of the thermocouples (A–E).

Controlled heating arrangement is setup in each office for maintaining the temperature within the required comfort range in winter. Both offices are located in such a way that there is no shade nearby and are provided with a window facing north. This results in the internal conditions being as severe as possible. Both heaters are controlled by external temperature controllers to keep temperature within the desired comfort range, i.e. 18°C–22°C through the thermocouple measuring indoor air temperature. Humidity and other environmental parameters are also being measured and recorded. Specialized software designed for data collection is installed, known as 'colt'. In addition to the interior room temperature, power consumption of the electric heaters is also being continuously measured and archived locally and remotely. The measurements of temperature and power are transmitted and remotely received and recorded. The experimental procedure is explained in Figure 3.2.4. The complete measuring and monitoring system developed for these test offices can be divided into three major modules listed as follows:

1. Device Interface Unit (DIU).
2. Data Logging Unit (DLU).
3. Online Monitoring Unit (OMU).

DIU is the module which is directly connected to the electric heater. DIU has the following functions [9]:

- monitoring both inside and outside room temperatures,
- monitoring atmospheric parameters such as solar radiations, wind speed, etc.,
- detecting switch ON/OFF state of heater,

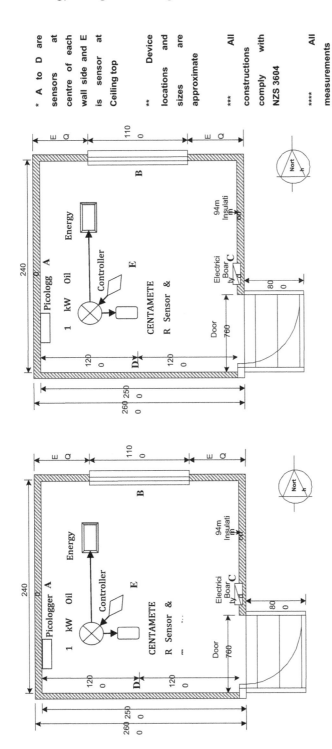

FIGURE 3.2.2 Experimental offices' design (top view).

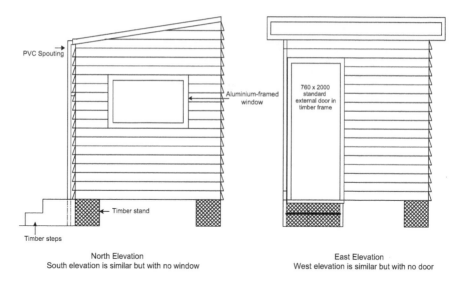

FIGURE 3.2.3   Experimental offices' design (elevation view).

- controlling heater, i.e. switching ON/OFF based on programmed temperature range,
- monitoring of electric load of heaters,
- monitoring and recording of electrical energy consumption of heaters,
- transmitting all parameters through wireless transmitter to control centre.

DLU is a separate control room beside the two offices. This unit has two computers installed in it. A WINDOWS operating system-based computer is dedicated to collect data such as temperatures, and atmospheric parameters, and logging the data into a database archive locally and remotely. While another small LINUX-based computer is dedicated to collect power and energy data and logging the data onto a separate database file. Both databases are managed separately to avoid any data loss. Both machines are connected to the internet at all times [9]. These machines post the instantaneous data to a web server, which can be monitored remotely from any place with Internet access. We have used this facility to access the data of the offices from the City Campus of the University of Auckland and carry out post analysis and real-time visualization (Figure 3.2.5).

OMU is dedicated to ensure the collection of data from offices and making it available remotely. It includes a web server and specially designed software for this application named as colt. Web server collects power and energy data while temperature data are collected through software on local area network [9].

This whole arrangement involves the usage of advanced programming languages such as C# (C-sharp), java and PHP. MySQL is being used to handle power and energy data while temperature and other parameters are archived using MS Excel. All these parameters are stored every half an hour except energy data, which are collected every other second to calculate accurate quantity of energy consumed.

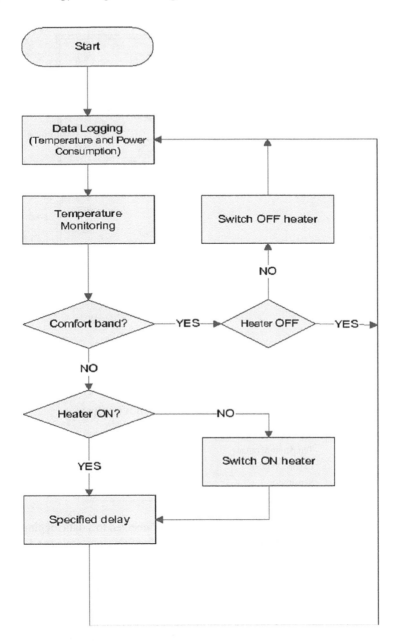

FIGURE 3.2.4 Experimental procedure.

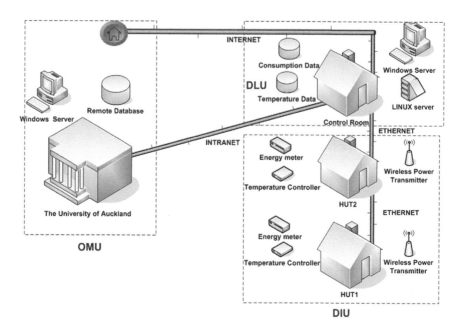

FIGURE 3.2.5   Monitoring and control arrangement for DSM [9].

TABLE 3.2.1
Specification of Type 'K' Thermocouples

Quantity	Value
Temperature range	200°C–1250°C
Extension grade	0°C–200°C
Sensitivity	41 µV/°C

The thermocouples installed in each office are connected to a data logging system "PICO loggers". The specification of the used 'K' type thermocouples and 'TC-08' Pico loggers are as given in Tables 3.2.1 and 3.2.2.

The temperature controller used for controlling heating load, i.e., 1 kW heaters, is an advanced controller. These controllers have the ability to be programmed for different operating scenarios. Each controller has a temperature monitoring probe utilizing thermocouple to sense room temperature and communicate to controller. The specification of the controller is given in Table 3.2.3. Controller logically tests the value of temperature based on a programmed criteria and decides whether to trigger the relay or not. Heating load is connected with this relay in series.

A digital energy meter is also installed to measure energy consumption by the heaters daily. This metre accumulates the total amount of energy consumed by heater.

The specifications of the energy meter is provided in Table 3.2.4. This installation of separate energy meter is done to increase the reliability of the system and enabling comparing the energy consumed measured by online monitoring system.

TABLE 3.2.2
Specification of TC-08 Pico Logger

Quantity	Value
Number of channels	8
Conversion time	100 ms (thermocouple and cold junction compensation)
Temperature accuracy	Sum of ±0.2% of reading and ±0.5°C
Voltage accuracy	Sum of ±0.2% of reading and ±10 μV
Overload protection	±30 V
Maximum common mode	±7.5 V voltage
Input impedance	2 MΩ
Input range (voltage)	±70 mV
Resolution	20 bits
Noise free resolution	16.25 bits
Thermocouple types supported	B, E, J, K, N, R, S, T
Input connectors	Miniature thermocouple

TABLE 3.2.3
Specification of IR33 Temperature Controller

Quantity	Value
Measuring probe	1.5 m
Sensor type	Thermocouple
Transmitter range	30 m
Sensitivity	±0.1°C
Accuracy	±5%
Supply	240 AC, 50 Hz
Relay rating	120/240 V (AC), 12/24 V (DC)

TABLE 3.2.4
Specification of kWh Meter

Quantity	Value
Measuring quantity	kWh
Number phases	1
Sensor type	Series connection
Sensitivity	±0.01 kWh
Accuracy	±5%
Frequency	50
Supply	240 V AC

TABLE 3.2.5

Specification of CentAmeter Device

Quantity	Value
Measuring quantity	Amps
Number phases	3
Sensor type	Clamp
Transmitter range	30 m
Sensitivity	±20 W
Accuracy	±5%
Frequency	50/60 Hz
Batteries	2×AA
Battery life	12 months
Power	<1.0 mW

CentAmeter is the device, which measure the electric load and transmits the load value to the receiver installed in DLU. This device sends the information in form of a data packet in HEX, which is then decoded at the receiver end. Each office is installed with a CentAmeter installed in order monitor load with a unique digital ID. The specification of the CentAmeter sensor and transmitter is shown in Table 3.2.5.

This arrangement has been made to observe the load variations remotely from the City Campus of the University of Auckland and to provide access to others through internet.

3.2.4   RESULTS AND ANALYSIS

Using the setup described in the previous section, actual temperature and electricity consumption data are collected regularly for the two offices. Data are also collected for the actual NZ electricity load and wholesale electricity spot prices at different nodes from the official streaming websites [21,22]. The analysis using these observations is presented in the following two sub-sections. The first sub-section includes the impact analysis of electrical peak load shifting, energy efficiency potential and RTP opportunities, while the second sub-section presents the energy efficiency observed due to PCM for space heating.

3.2.4.1   DSM Opportunity through PCM for NZEM

Winter in New Zealand is moderately cold. Temperatures in early morning and night typically drop below 10°C (Figure 3.2.6) which results the temperature in residential units to go below 15°C or less. Considering the comfort range for homes to be between 17°C and 25°C, controller installed in Offices were programmed to operate the room heaters between 18°C and 22°C.

Figure 3.2.6 presents the data observed from 25th August, 12:00 pm till 26th August, 12:00 pm. In this figure solar radiation, wholesale price (WSP) of electricity and ambient temperature are shown [22]. In New Zealand, the retail loads, like homes, do not pay the WSP rates. During this particular day, solar radiation was maximum

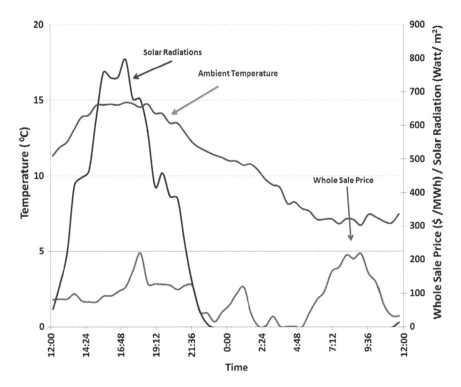

FIGURE 3.2.6 Day profile-25–26 August 2008.

during mid-day while the WSP had two peaks, one between 5:00 pm and 6:00 pm at approximately \$218/MWh and other between 8:00 am and 9:00 am at approximately the same price. The minimum price of electricity was \$0.41/MWh at 5:00 am. This is shown in Table 3.2.6.

To validate the contribution of PCM for DSM, 2 days of measurements are presented for the period from 3rd November 2008, 12:00 pm to 4th November 2008, 12:00 pm (Day 4) and from 5th November 2008, 12:00 pm to 6th November 2008, 12:00 pm (Day 6).

Figures 3.2.7 and 3.2.8 represent the data mentioned, which includes temperatures (internal and external), NZEM data (spot price and electrical load), electric heaters power data based on mentioned controlled scenario and solar radiation data for the same days.

Both offices were set under observation for 12 days (Day 1–Day 12 i.e. 30th October 2008–11th November 2008) under this controlled environment. This section presents data for only 2 days i.e. Day 4 and Day 6. To assess the gains achieved from PCM towards DSM, the control temperature was set between 18°C and 22°C which is the melting range of Paraffin RT-20. This range of temperature also lies within the typical comfort level.

The readings obtained as shown in Figures 3.2.7 and 3.2.8 can be summarized in terms of two aspects of DSM being influenced by PCM, i.e., Load Shifting and Energy Conservation. Summary of results are shown in Tables 3.2.7 and 3.2.8.

TABLE 3.2.6
Peak Temperature and Prices 25–26 August 2008

	$T_{ambient}$ (°C)	Price ($/MWh)
Max	14.85 at 5:30 pm	218.51 at 6:00 pm and 9:00 am
Min	6.87 at 11:30 pm	0.41 at 3:00 am to 5:00 am

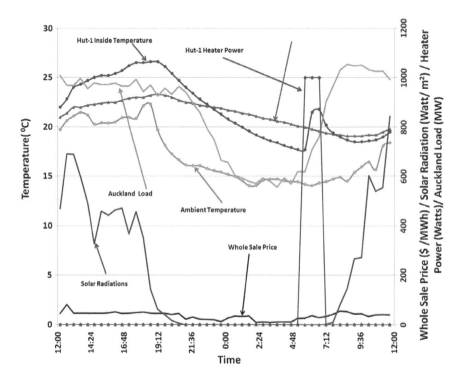

FIGURE 3.2.7   Load shifting and energy conservation on Day 4.

3.2.4.1.1   Load Shifting and Price Efficiency

In New Zealand, it has been found that in winter there are typically two daily peak electricity load periods [21], one typically between 7:00 pm and 10:00 pm and other between 6:30 am and 10:00 am. This winter peak load is partly due to activities such as space heating, hot water and cooking, etc. Household electrical energy consumption is influenced by seasonal changes, and most noticeably because of space heating. According to the Household Energy Efficiency Program (HEEP) report published in 2005, energy consumption in winter increases to approximately 300% than in summer and mostly is due to space heating. The data also reveal that electrical space heating contributes to 8% of the total New Zealand household consumption [7].

From Table 3.2.7 it can be observed that Office-2, which is a PCM office, is capable of shifting heating load by few hours, depending on the amount of energy

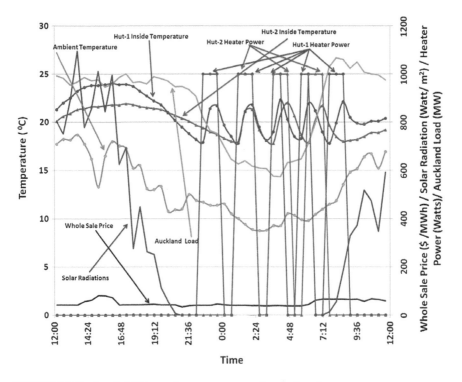

FIGURE 3.2.8   Load shifting and energy conservation on Day 6.

TABLE 3.2.7
Load Shifting

	Office-1 (Non-PCM)		Office-2 (PCM)		
	1st Switch ON Time (hh:mm)	Total ON Period (min)	1st Switch ON Time (hh:mm)	Total ON Period (min)	Load Shift (min)
Day 4	5:00	80	Nil	Nil	Indefinite
Day 6	22:00	435	00:30	270	150

TABLE 3.2.8
Energy Conservation

	Office-1 (Non-PCM) (kWh)	Office-2 (PCM) (kWh)	% Energy Conserved
Day 4	0.41	0	100
Day 6	3.05	2.03	33.44
Total	3.46	0	58.67

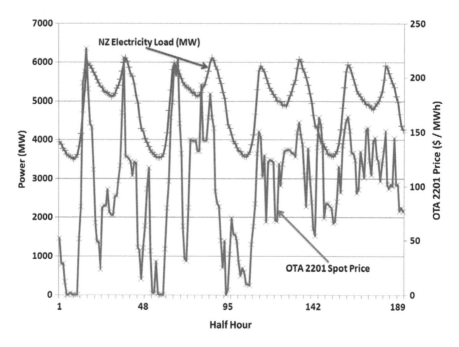

FIGURE 3.2.9   Correlation of demand and price.

stored by PCM during the day. This characteristic of PCM-built environments creates opportunities for peak load shifting during winter season. This shifting will
not only alter the consumption pattern and contribute towards peak shaving but
will also help reducing consumption prices for consumers if paying on the basis of
WSP charges.

From the WSP graph in Figure 3.2.9, it can be seen that prices reach their peaks
during times of early evening and early morning. Price is highly correlated with
load during times of high demand. Especially in winter, if there is shortage of water
in reservoirs due to lesser rain fall, the price of electricity is higher because New
Zealand produces 55%–60% of total generation from hydro sources [7]. Spot price
for one of the major nodes, i.e. OTA 2201 price and NZ electricity load, is given
below to illustrate the relationship between load and price in times of peak season
that is winter. Data from 25th to 28th August are shown in Figure 3.2.9.

The selection of node OTA2201, i.e. Otahuhu 220 kV, GXP was done because
that OTA2201 supplies power to Auckland region and the research site, i.e. Tamaki
campus of University of Auckland, New Zealand. The NZ electricity load was correlated with the OTA2201 price for 25th August 2008. From correlation studies for
the day, it was seen that the load is 77.11% positively correlated with the price which
is high. It has also been observed from previous studies that load and price are highly
correlated at peak times especially in winter season [8,9].

Moreover, if a household operates on a RTP-based tariff, it could potentially
save electricity bill by using power during times of low electricity price. PCM room

acquires some solar energy during the day, stores it in the form of phase-change latent heat and then slowly keeps emitting it as temperature starts to drop down. In this way, PCM-built houses would need heating after a few hours, following temperature drop, as compared to normal houses. Further, heating consumptions can be shifted to times of low wholesale prices using different operating strategies without compromising on the comfort level of the interiors [9].

Thus, it can be concluded that it is possible to shift the electric heating load using PCM-built houses, which if deployed widely can lead to reduced peak demands, reduced network losses and transmission capacity availability for capital deferment.

3.2.4.1.2 Energy Conservation

Table 3.2.8 summarizes the results with regard to energy conservation. It can be observed that it is possible to save energy through the application of PCM in building because of PCM capability to capture and store solar energy during the day. These savings are significant and could result in tens of electricity units saved for each house during a month. Based on our observations, the PCM room conserved 6.41 kWh units of electrical energy in 12 days, which is 31% of the total energy consumed in non-PCM room. Thus, energy conservation is one of the biggest contributions of PCM. At this point of time, these gains have not been statistically extrapolated because energy efficiency is still being investigated in PCM rooms by observing it for a longer-term period.

3.2.4.2 ENERGY CONSERVATION ANALYSIS FOR MULTIPLE DAYS

In the following section, the data and its analysis for 12 days starting from 31/10/200 (Day 1) to 11/12/2008 (Day 12) are presented. Data were collected, archived and analysed for energy efficiency.

Figures 3.2.10 and 3.2.11 represent the complete data for 12 days which include solar radiation, inside office temperature and ambient temperature. The inside office temperature is controlled between 18°C and 22°C as mentioned earlier. These above two illustrations reveal the following facts:

1. While comparing inside temperature for both offices, it can be observed that during hotter days, temperature in Office-2 (PCM) never raised above 25°C (Figure 3.2.11) but temperature above 25°C can be observed in Office-1 (non-PCM) as in Figure 3.2.10.
2. As inside temperatures are controlled heating is provided based on earlier mentioned temperature settings. The frequency of temperature cycles in both offices during night shows heating during particular nights. The more the temperature cycles, the more heating was required.
3. During colder nights such as those 5/11/2008 and 7/11/2008, heating requirement rapidly increased. This is visible from more number of temperature cycles in those particular nights.
4. During some days such as 11/11/2008 and 12/11/2008, when ambient temperature was moderate, no heating was required in PCM office during nights. However, during similar nights, heating in non-PCM office was required for some time.

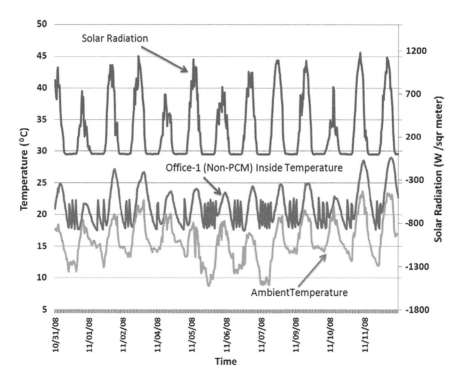

FIGURE 3.2.10 Temperature profile and space heating requirement Office-1 from 10/31/2008 to 11/11/08.

Table 3.2.9 shows, electrical energy (space heating) consumed in both offices, over the period of 12 days. It also shows the difference of energy conserved and the percentage savings gained in the PCM office. The table reveals the following facts.

1. Most of the times, there was some energy saving in terms of kWh. Few times saving was even more than 1 kWh during a single night.
2. In few days, no saving was observed rather PCM office consumed more energy and got charged from internal heating. This is because of not having enough solar radiations during the day but extreme cold nights.
3. In certain days, when ambient temperature was moderate, 100% energy was conserved in PCM office because of lesser heating requirement during night.
4. Day 3, Day 6 and Day 11 were observed to have maximum saving in terms of kWh. These are the days of lower average ambient temperatures and high solar radiation.
5. Day 4, Day 5 and Day 12 were significant in terms of percentage energy savings. These are the days of moderate average ambient temperatures.

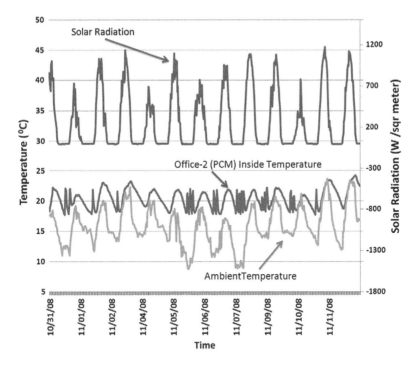

FIGURE 3.2.11 Temperature profile and space heating requirement Office-2 from 10/31/2008 to 11/11/08.

TABLE 3.2.9
Electrical Energy Conservation through PCM

	Office-1 (Non-PCM) (kWh)	Office-2 (PCM) (kWh)	Energy Conserved (kWh)	% Energy Conserved {(Office-1 – Office-2)/ Office-1 } * 100
Day 1	1.88	1.92	−0.04	−2.13
Day 2	1.15	0.67	0.48	41.74
Day 3	1.31	0.5	0.81	61.83
Day 4	0.41	0	0.41	100.00
Day 5	0.58	0	0.58	100.00
Day 6	3.05	2.03	1.02	33.44
Day 7	2.26	1.8	0.46	20.35
Day 8	3.65	3.05	0.6	16.44
Day 9	1.99	2.16	−0.17	−8.54
Day 10	1.82	1.24	0.58	31.87
Day 11	1.88	0.7	1.18	62.77
Day 12	0.5	0	0.5	100.00
Total	20.48	14.07	6.41	31.30

TABLE 3.2.10

Average Temperatures and Electrical Energy Consumption for Space Heating

	Temperature (°C)		
	Mean Temperature Office-1 (T_1, °C)	Mean Temperature Office-2 (T_2, °C)	Mean Temperature Atmosphere (T_0, °C)
Day 1	20.76	20.14	14.50
Day 2	19.75	19.52	15.59
Day 3	21.69	20.70	16.37
Day 4	21.85	21.27	17.11
Day 5	20.49	20.15	16.66
Day 6	20.95	20.1	13.15
Day 7	20.35	19.75	14.33
Day 8	20.92	20.13	12.81
Day 9	20.91	20.19	16.24
Day 10	21.15	20.22	16.59
Day 11	20.47	19.67	16.59
Day 12	23.11	21.51	17.36

Table 3.2.10 shows average offices temperature, ambient temperature and energy consumed over the same period. From the above table, following statements can be drawn:

1. Office-1 average temperature was slightly higher because during the day temperature used to rise above normal comfort level in the non-PCM office which resulted in higher average temperature. This shows that cooling is needed in non-PCM office unlike in PCM office.
2. Energy consumption (space heating) in both offices increased as average ambient temperature decreased, e.g. Day 6, Day 7 and Day 8.
3. During ambient temperatures being above 15°C, the energy consumption was observed to be less than 2 kWh. However, it increased to approximately double due to temperature decreases of only a few degrees.

Table 3.2.11 shows minimum, maximum and average ambient temperature along with energy conserved in PCM office. From the above table, the following facts can be observed:

1. Energy conservation was limited when the minimum temperature is high and the maximum ambient temperature is low. This is because under such condition, the PCM will undergo only partial melting and solidification.
2. Energy conservation was higher when temperature was higher during daytime and temperature was lower in night time as in Day 3 (13.29°C and 20.08°C) and Day 11 (12.66°C and 21.24°C). This wide variation allows the PCM to melt and solidify providing full utilization of its latent heat.

TABLE 3.2.11

Effect of Min and Max Temperature on Electrical Energy Consumption for Space Heating

	Temperatures (°C)			
	Min Ambient Temperature (T_{Min})	Max Ambient Temperature (T_{Max})	Mean Ambient Temperature (T_0)	Energy Conserved (kWh) E1 – E2
Day 1	10.98	18.42	14.50	−0.04
Day 2	11.66	19.15	15.59	0.48
Day 3	13.29	20.08	16.37	0.81
Day 4	14.06	22.3	17.11	0.41
Day 5	13.32	20.13	16.66	0.58
Day 6	8.72	18.71	13.15	1.02
Day 7	10.71	18.96	14.33	0.46
Day 8	8.84	17.1	12.81	0.6
Day 9	13.79	19.82	16.242	−0.17
Day 10	14.15	20.66	16.594	0.58
Day 11	12.66	21.24	16.59	1.18
Day 12	11.92	23.68	17.36	0.5

It can be concluded from statistical analysis presented in this section that energy conservation in the PCM office depends on several variables which include; amount of solar energy absorbed by PCM and the maximum and minimum ambient temperature during the day and night. It can also be observed that non-PCM room required some cooling during certain days which signals the possibility for application of PCM to reduce cooling needs during summer.

3.2.5 CONCLUSIONS

This article assesses the impact following application of using PCM in building material from an electricity demand-side perspective, especially in the wholesale electricity market operating environment of New Zealand. The observed results bring out the fact that using PCM has potential significant advantages from space heating viewpoint such as peak load shifting, energy conversion, reduction in peak demand cost for network line companies and reduction in electricity consumption bills through RTP tariff. It is also observed that energy conservation gains are sensitive to the minimum and maximum temperature during 24 h period. Further, the application of PCM in building material and potential of saving electrical energy for air conditioning during summer has also been identified as a future assessment.

ACKNOWLEDGEMENTS

The author Waqar A. Qureshi would like to acknowledge Higher Education Commission, Pakistan for supporting his ME leading to PhD graduate research at University of Auckland, New Zealand.

REFERENCES

1. Farid MM et al. A review on phase change energy storage: materials and applications. *Energy Convers Manage* 2004;45(9–10):1597–615.
2. Zhu N, Ma Z, Wang S. Dynamic characteristics and energy performance of buildings using phase change materials: a review. *Energy Convers Manage* 2009;50(12):3169–81.
3. Athienitis AK et al. Investigation of the thermal performance of a passive solar test-room with wall latent heat storage. *Build Environ* 1997;32(5):405–10.
4. Khudhair AM, Farid MM. Use of phase change materials for thermal comfort and peak load shifting: theoretical & experimental investigations. In: *ISES Solar World Congress (SWC2007)*, Beijing, China; 2007.
5. Meng Z, Mario AM, Jennifer BK. Development of a thermally enhanced frame wall with phase-change materials for on-peak air conditioning demand reduction and energy savings in residential buildings. *Int J Energy Res* 2005;29(9):795–809.
6. Weinläder H, Beck A, Fricke J. PCM-facade-panel for daylighting and room heating. *Sol Energy* 2005;78(2):177–86.
7. Isaacs N, Camilleri M, Pollard LFA. Energy use in New Zealand households – HEEP year 10 report, Branz; 2006.
8. Gellings CW, Chamberlin JH. *Demand-Side Management: Concepts and Methods*, 2nd ed., vol. 1. The Fairmont Press, Inc.; 1992. p. 451.
9. Qureshi WA, Nair NKC, Farid MM. Demand side management through efficient thermal energy storage using phase change material. In: *Proceedings of Australasian Universities Power Engineering Conference AUPEC 08*, Sydney, Australia; 2008.
10. Nair NKC, Nayagam R, Francis R. New Zealand utility experiences with demand side management. In: *Power and Energy Society General Meeting – Conversion and Delivery of Electrical Energy in the 21st Century*, 2008 IEEE; 2008.
11. Situ RF, Nair N-KC. Application of wavelet transform techniques towards volatility analysis in New Zealand Electricity Market. In: *Australian Universities Power System Conference (AUPEC 2007)*, Perth, WA, Australia; 2007.
12. Wei Z, Feliachi A. Residential load control through real-time pricing signals. In: *Proceedings of the 35th Southeastern Symposium on System Theory*, 2003; 2003.
13. Khudhair AM, Farid MM. A review on energy conservation in building applications with thermal storage by latent heat using phase change materials. *Energy Convers Manag* 2004;45(2):263–75.
14. Gellings CW. The concept of demand-side management for electric utilities. *Proc IEEE* 1985;73(10):1468–70.
15. Wolak FA. Residential customer response to real-time pricing: the Anaheim critical peak pricing experiment 2007. http://repositories.cdlib.org/ucei/csem/CSEMWP-151 [cited November 2008].
16. Moholkar A, Klinkhachorn P, Feliachi A. Effects of dynamic pricing on residential electricity bill. In: *Power Systems Conference and Exposition*, 2004. IEEE PES; 2004.
17. Flolo R, et al. DA/DSM profitability in a deregulated market environment. In: *International Conference on Power System Technology. Proceedings POWERCON'98*; 1998.
18. The Electricity Commission of New Zealand; 2008. <http://www.electricitycommission.govt.nz/> [cited 30.05.08].

19. Yau TS et al. Demand-side management impact on the transmission and distribution system. *Power Syst IEEE Trans* 1990;5(2):506–12.

20. Wallin F, et al. The use of automatic meter readings for a demand-based tariff. In: *Transmission and Distribution Conference and Exhibition: Asia and Pacific, 2005 IEEE/PES*; 2005.

21. System operator New Zealand; 2008. <http://www.systemoperator.co.nz> [cited 30.06.08].

22. Nodal prices of New Zealand Electricity Market; 2008. <http://www.electricityinfo. co.nz> [cited 25–26.08.08].

3.3 Experimental Validation of a Methodology to Assess PCM Effectiveness in Cooling Building Envelopes Passively

Albert Castell
Universitat de Lleida

Mohammed Farid
University of Auckland

3.3.1 INTRODUCTION

Energy consumption has significantly grown in the last decades, especially in the building sector, which has a high contribution to the consumed energy. Thus, reducing building energy demand is a critical issue worldwide. For example, in Europe, new legislations (European Directive 2010/31/EU [1]) state that by 2020 new buildings must consume "nearly zero" energy. Moreover, the Directive also highlights the necessity to reduce global energy consumption and greenhouse gas emissions by 20% before 2020. To achieve these objectives, new technologies will be required for both new and refurbished buildings. One of the best ways to reduce the energy demand of a building is to improve its envelope. Many efforts have been done to reach this objective, by addressing both thermal insulation [2,3]) and thermal inertia [4,5].

Cabeza et al. [2] experimentally measured energy savings in small house-like cubicles, which can be achieved through thermal insulation in a Mediterranean continental climate. Results showed that a good insulation is crucial to reduce energy consumption during both summer and winter periods. Axaopoulos et al. [3] determined the optimum insulation thickness for external walls of different constructions and orientation, considering both heating and cooling period and taking into account wind speed and direction. An economic analysis, based on "life cycle savings method", was performed for each configuration, various thicknesses of insulation material and different orientations.

Gagliano et al. [4] studied the effect of thermal inertia combined with night ventilation. Both experimental and numerical results demonstrated a high potential of such combination to prevent over-heating during summer. Furthermore, Di Perna et al. [5] performed experiments to evaluate the effect of thermal inertia on the indoor temperature with high internal heat loads during summer. They concluded that thermal inertia is a good solution and proposed parameters for assessing thermal inertia that can be adopted in regulations based on semi-stationary assessment methods, which consider the performance of building envelopes with different thermal inertia as analogous.

To combine both thermal insulation and inertia, Aste et al. [6] evaluated the effect of adding thermal inertia to the building envelopes and concluded that thermal inertia could provide some energy saving in buildings, especially when coupled with other energy-saving technologies. Moreover, Bond et al. [7] studied the effect of the distribution of thermal inertia and insulating materials within the wall. Results showed that using distributed insulation and inertia materials is the best solution, allowing the selection of an optimum number of layers.

In the last decades, attention has been drawn to the use of PCM in building envelopes, since it is seen as a potential method to increase thermal inertia with low volume and weight and thus improves energy efficiency of buildings. Several studies have been done to demonstrate the benefits of using PCM by means of experimental work. Cabeza et al. [8] developed and tested an improved concrete having microencapsulated PCM in it. Side-by-side experiments were performed to compare the behavior of the new material with conventional concrete. The results of the experimental measurements showed that the new concrete was successful in smoothing out the indoor temperature oscillations during the day. In a similar approach, Castell et al. [9] evaluated the behavior of macroencapsulated PCM in brick construction for summer passive cooling. The experiments were performed under real weather conditions and results demonstrated the potential of PCM to reduce indoor temperature fluctuations and energy consumption. Kuznik and Virgone [10] and Kuznik et al. [11] also performed experimental measurements under controlled conditions using PCM copolymer composite wallboards. They concluded that the use of PCM reduces the internal temperature up to 4.2°C. Behzadi and Farid [12] determined the benefits of using PCM in wallboards for energy conservation in residential buildings by means of both experimental and numerical work. On a larger scale, Mandilaras et al. [13] studied a two-story typical family house in Greece which incorporated gypsum panels with PCM. Experimental measurements performed with an unoccupied house showed that the thermal mass of the walls was enhanced during late spring, early summer and autumn, leading to a decrease in the decrement factor (which gives an idea of the thermal wave dampening when passing from outside to inside) and an increase in the time lag (which shows the delay between the outer temperature and inner heat flux peaks).

However, most of the experimental work has been focused on demonstrating the potential benefits of using PCM in a passive system, without an attempt to address the issue of design and/or evaluation of PCM benefits using simple and straightforward tools. When aiming to design PCM buildings, it is common to develop numerical models suitable for each application and system. Due to the non-linearity of the PCM behavior, it has been difficult to generalize those solutions. In the last decades, to bridge the gap between research and real applications, many efforts have been made in developing and integrating PCM models into commercial programs. As a result, several tools are nowadays available for the design of PCM systems, although one has to perform detailed building energy simulations with specialized programs, such as Trnsys or EnergyPlus.

As an example of the available tools, Lamberg et al. developed a TRNSYS model (Type 204) to simulate PCM in walls [14,15] using finite difference and effective heat capacity methods. A similar approach was used by Streicher et al. [16] in Type 241, while Ibán̄ez et al. developed Type 232, a PCM module that simulated the effect of adding PCM in walls [17]. In the latest one, the PCM behavior was simulated using active layers with a water flow rate at phase-change temperature. Lately, the TESS library also provided a PCM in wall module (Type 1270) [18]. However, all these models require experimental validation.

Kuznik et al. [19] developed Type 255/260 in TRNSYS following the effective heat capacity approach and using the finite difference analysis to simulate the thermal behavior of an external building wall containing a PCM layer. In this case, the model was experimentally validated and showed a good agreement with the experimental results for both indoor and wall surface temperatures [20].

EnergyPlus includes the capability to simulate PCM in building envelopes [21] since a new implicit finite difference thermal model (CondFD) of building was incorporated in 2007. The model simulated the performance of PCMs using an enthalpy or heat content formulation and was experimentally validated by Tabares-Velasco et al. [22], showing some limitations that need to be taken into account when using the model. Tardieu et al. [23] also validated experimentally an EnergyPlus model for the case of using PCM in building envelopes.

On the other hand, the benefit of using PCM usually relies on the reduction of the daily temperature fluctuation. Up to date, no simple methodology has been successfully used to determine the benefits of PCM in building envelopes without using detailed computer simulations. Evola et al. [24] presented a new methodology which they claimed could be used for both design purposes (in combination with building energy simulations) and also for evaluation of building performance. The methodology was based on four indicators that address both the thermal comfort condition and the PCM performance. In their work, they also presented a demonstration example on how to use these indicators for design purposes, supporting their methodology with numerical simulations performed with EnergyPlus. The methodology used by Evola et al. was simple and effective but was not tested against experimental measurements.

In this article, experimental results obtained from brick, concrete and timber construction facilities build in Spain and New Zealand, respectively, having very different weather conditions are used to test the methodology proposed by Evola et al. Moreover, some modifications of the indicators are proposed and tested in this article, so that the method maybe extended to cover a wider range of situations.

3.3.2   INDICATORS FOR THE PCM EVALUATION

Evola et al. [25] presented a new methodology to determine the effectiveness of using PCM in building envelopes for summer conditions. This methodology is based on four indicators (intensity of thermal discomfort (ITD), frequency of thermal comfort

(FTC), frequency of activation (FA), and PCM storage efficiency) that are described in the following sections.

3.3.2.1 INTENSITY OF THERMAL DISCOMFORT FOR OVERHEATING (ITD~OVER~)

This parameter quantifies the intensity of an uncomfortable thermal sensation due to overheating in a living space. It is based on the difference between room operative temperature [25] and a threshold value, as well as the duration of such overheating.

The ITD for overheating (ITD$_{over}$) is defined as the time integral, over the occupancy period P, of the positive difference between the current operative temperature and the upper threshold for comfort temperature (Figure 3.3.1). This parameter should be evaluated only during the occupancy period:

$$\text{ITD}_{over} = \int_P \Delta T^+(\tau) \times \delta\tau$$

where

$$\Delta T^+(\tau) = \begin{cases} T_{op}(\tau) - T_{lim} & \text{if } T_{op}(\tau) \geq T_{lim} \\ & \text{if } T_{op}(\tau) < T_{lim} \end{cases}$$

3.3.2.2 FREQUENCY OF THERMAL COMFORT (FTC~OVER~)

The frequency of thermal comfort (FTC$_{over}$) [24] is defined as the percentage of time within a given period P, during which the indoor thermal comfort conditions are met. This parameter should be evaluated only during the occupancy period:

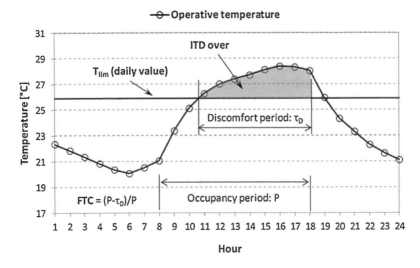

FIGURE 3.3.1 Intensity of thermal discomfort and frequency of thermal comfort: graphic definition [13].

$$\text{FTC}_{\text{over}} = \frac{P - \tau_D}{P}$$

where τ_D is the time under discomfort conditions.

3.3.2.3 FREQUENCY OF ACTIVATION FA

Benefits of using PCM are only obtained when the material undergoes phase change. To determine the real behavior of a PCM, it is necessary to evaluate whether it is utilizing its latent heat or not, and how effectively it is used. To this aim, Evola et al. [24] propose to use the FA.

This parameter determines the percentage of time within a given period during which the PCM is undergoing phase change. Therefore, low values of FA mean that the PCM remains in its liquid or solid phase for a long time, thus not taking advantage of its latent heat capacity. On the other hand, values close to 100% mean that the PCM operates within the phase-change range during most of the period considered.

For this parameter, the period of integration is the whole day and not just the period of occupancy:

$$\text{FA} = \frac{\tau_{\text{PC}}}{P}$$

where τ_{PC} is the time when the PCM is undergoing phase change.

3.3.2.4 PCM STORAGE EFFICIENCY

Finally, it is important to consider that not all the conditions inside the melting range have the same importance from an energy point of view. The specific heat capacity at a temperature close to the peak phase-change temperature is much higher than that close to the phase-change onset temperature. This means that, despite the PCM being activated, its storage capacity can be very different depending on temperature. Since the FA indicator is not capable of accounting for this difference another parameter called PCM storage efficiency (η_{PCM}) was defined by Evola et al. [24]. This parameter is the ratio of the thermal energy actually stored by the PCM to its maximum storage capacity. Since the PCM is subjected to daily temperature cycles, the actual energy storage E_{st} must be evaluated over a 24 h period:

$$\eta_{\text{PCM}} = \frac{E_{\text{st}}}{L} = \frac{E_{\text{st}}}{\int_{T_m}^{T_s} M \times C_{\text{eq}}(T) \times dT}$$

$$E_{\text{st}} \int_P \left(M \times C_{\text{eq}} \times \frac{dT_{\text{PCM}}}{d\tau} \right) \times d\tau \int_P \left(q_L{}^+ + q_R{}^+ \right) \times d\tau$$

From the definition presented by Evola et al., one can understand that low values of the η_{PCM} mean a low utilization of the latent heat of the PCM, while values close to 100% represent a high utilization. However, the definition of this parameter is not clear enough and could lead to a misinterpretation. It is not clear if the energy stored refers only to the energy absorbed by the PCM (not considering the discharge process) or to the energy balance in the PCM during the full day (thus taking into account both charged and discharged energy).

If only the energy absorbed is considered, the parameter is not valid for the PCM evaluation, since the use of phase change materials in building envelopes is based on full cycles (charging–discharging) and therefore both processes need to be accounted for. If the discharging process is neglected, the efficiency may reflect the energy absorbed by the PCM but will fail to predict its availability for the next day, when the PCM needs to be solid again. Therefore, the storage efficiency of the PCM should consider both processes. Evola et al. may have assumed a full solidification of the PCM during nighttime; however, this is unlikely and unrealistic.

On the other hand, if the storage efficiency is defined by considering both the charging and discharging periods, the best possible result for the PCM behavior should be a full daily cycle. This means that the energy charged during the daytime was discharged during night (resulting in a storage efficiency of zero over 24 h period). In such case, the parameter is not well suited to evaluate PCM efficiency, and hence this parameter was not used in the analysis presented in this article.

This article aims to use the mentioned indicators to assess the performance of specific cases and compare them with experimental results to determine the suitability of the methodology. For the sake of clarity and simplicity, the evaluation is performed using indoor temperatures (instead of the operative temperature) with the adaptive comfort theory [26]. Since the objective of the study is not to demonstrate the benefits of using PCM in building envelopes but to validate the methodology, this simplification is valid.

Sicurella et al. [27] stated in their work that this methodology was most suitable for high levels of expectations in terms of comfort (thus accepting a thermal comfort range of 4°C from those presented in [26]), since the indicators may be less effective for lower levels of expectation. This assumption is even more important with the use of PCM, where the main benefit is obtained during the phase-change range. Therefore, the comfort range will be calculated considering high levels of expectations.

3.3.3 PROPOSED MODIFICATIONS TO THE INDICATORS

Evola et al. defined their indicators based on the analysis of an office building incorporating PCM. In that application, the comfort level at night is not important. In this article, we would like to extend this analysis to cover housing in which comfort level at night can also be important. Experimental results have shown that the benefit of the PCM is not limited to daytime, when the indoor temperature can be reduced, but also during nighttime, when the indoor temperature can drop below comfort level [12]. This becomes important in lightweight construction, where night indoor temperatures can drop below comfort levels. In this situation, the PCM may minimize the drop of the indoor temperature by the discharge of its latent energy absorbed during the day. Hence, comfort level was defined in this article as a range of upper and lower temperatures and the indicators proposed by Evola et al. are calculated taking into account the comfort conditions within the built environment during 24 h as discussed in the following section.

3.3.3.1 FULL-PERIOD ITD

To fully evaluate the benefits of using PCM the full period of charging/discharging must be considered, as discussed earlier. The PCM here is intended to prevent overheating during the daytime and minimize the drop in indoor temperatures during the night. Therefore, a comfort range must be defined by means of an upper and a lower comfort temperature limit.

Considering the new comfort range, the ITD evaluates the intensity of uncomfortable thermal sensation due to both overheating and low temperatures. The modified indicator is still based on the difference between the room temperature and the threshold values and the duration of the discomfort. Therefore, it is defined as the time integral of the positive difference between the current temperature and the upper/lower threshold for comfort, when overheating/low temperatures are occurring, respectively. Nevertheless, in this case, the ITD must be evaluated during the whole day ($P=24$ h) (Figure 3.3.2).

In contrast to [27], in this article, both the ITD_{over} and the $\text{ITD}_{\text{under}}$ have been evaluated separately and grouped into one single indicator, ITD.

$$\text{ITD} = \int_P \Delta T^+(\tau) \times \delta\tau$$

where

$$\Delta T^+(\tau) = \begin{cases} \left|T_{\text{op}}(\tau) - T_{\text{upper}}\right| & \text{if} \quad T_{\text{op}}(\tau) > T_{\text{upper}} \\[2mm] \left|T_{\text{op}}(\tau) - T_{\text{lower}}\right| & \text{if} \quad T_{\text{op}}(\tau) < T_{\text{upper}} \end{cases}$$

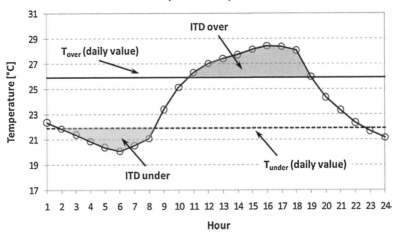

FIGURE 3.3.2   Definition of intensity of thermal discomfort (ITD) as presented by [27].

3.3.3.2 Full-Period Frequency of Thermal Comfort

This indicator is still defined as the percentage of time during which the indoor thermal comfort conditions are met; thus, the indoor temperature is within the comfort range. However, in this case, the evaluation period is the whole day ($P = 24\,h$).

$FTC = 1 - \dfrac{\tau_D}{24}$ where τ_D is the time under discomfort conditions.

In the present work, both the ITD and the FTC will be evaluated for three different occupancy profiles: office, house, and daily. The office profile evaluates the thermal behavior of the building from 9:00 to 17:00, while the house profile does it from 18:00 to 8:00. The daily profile evaluates the whole day behavior. To our knowledge, this approach in the analysis has not been done before.

3.3.4 EXPERIMENTAL SETUP

Two different experimental setups (one located in Auckland, New Zealand, and the other one in Lleida, Spain) were used to provide the data needed for the evaluation of the parameters described earlier. Both experimental setups consist of small house-like constructions where PCM has been introduced in walls. The goal of such constructions is to analyze and characterize the behavior of incorporating PCM in different constructions under real climatic conditions. Side-by-side experiments were performed to allow for direct comparison. Results demonstrated the potential benefits of using PCM in building envelopes [8,9,12].

However, although both experimental setups are similar in concept, there are important differences between them that need to be highlighted as shown in Table 3.3.1.

3.3.4.1 Lightweight Constructions

Two experimental house-like constructions built in New Zealand (called huts from here on) were built with the same construction and orientation to compare the effect of introducing PCM in its envelope [12]. Both huts are 2.6 m long, 2.6 m wide, with a height of 2.6 m on low end/2.8 m on high end (due to a five single slope roof). The huts are 0.4 m off the ground and supported by timber stands. The wall construction is made of a wooden frame with 0.1 m weatherboard exterior, 0.025 m particle

---

TABLE 3.3.1

Main Differences between the Two Experimental Setups Used for the Evaluation

Auckland Huts		Lleida Cubicles	
Constructive system	Lightweight (wood)	Massive (concrete)	Massive (brick)
Openings	A window facing north	Windows facing south, east, and west	No windows
PCM	RT-21	Micronal® PCM	SP-25 A8
PCM incorporation	Embedded	Microencapsulated	Macroencapsulated

FIGURE 3.3.3　Huts used for the experimentation. WOOD hut in the center and WOOD + PCM hut in the right.

board, 0.075 m fiberglass insulation, and 0.013 m gypsum board interior. The floor is made of 0.025 m particle board and carpet. The roof is made of corrugated iron with 0.025 m particle board, 0.075 m fiberglass insulation, and 0.013 m gypsum board interior. There is a window on the north-facing wall and a door on the east-facing wall. All construction complies with the New Zealand Building Code (Figure 3.3.3). The two huts are defined below but for more details, we refer to [12].

1. Wood hut (WOOD): Built with wood, it acts as the control unit.
2. Wood and PCM hut (WOOD + PCM): Built with wood, it incorporates gypsum board impregnated with Rubitherm RT-21 as PCM, at 26.2 wt% in all the walls and the ceiling.

Table 3.3.2 presents some key parameters of RT21 PCM and the 26.2 wt% PCM gypsum board (PCMGB).

To monitor all the required parameters, the huts were fully instrumented.

- Temperatures at north outside wall skin, north wall cavity, north inside wall skin, south inside wall skin, east inside wall skin, and ceiling inside wall skin (Type-K thermocouples).
- Inside air (Type-K thermocouple).
- Inside relative humidity.

Weather conditions were also monitored:

- Outside air (Type-K thermocouple).
- Global radiation (Pyranometer VAEQ08E).
- Wind speed (global wind speed sensor GWWE550).

TABLE 3.3.2
PCM Properties of Pure RT-21 and PCMGB

Property		RT-21	PCMGB
Melting range (°C)		18°C–23°C	18°C–23°C[a]
Congealing range (°C)		22°C–19°C	22°C–19°C[a]
Heat storage capacity (kJ/kg)		134	34[b]
Density (kg/L)	Solid	0.88	0.893[c]
	Liquid	0.77	
Specific heat capacity (kJ/kg K)	Solid	1.8	1.34[d]
	Liquid	2.4	
Heat conductivity (W/m K)		0.2	0.2

[a] Calculations based on 25 wt% PCM.
[b] 25% of pure RT 21 PCM heat storage capacity.
[c] Based on pure gypsum board density of 0.67 kg/L.
[d] Based on averaged PCM specific heat capacity of 2.1 kJ/kg K and gypsum board specific heat capacity of 1.09 kJ/kg K.

FIGURE 3.3.4 Concrete cubicle with microencapsulated PCM.

3.3.4.2 MASSIVE BUILDINGS—CONCRETE

A setup consisting of several house-like constructions (called cubicles from here on), with internal dimensions of 2.4×2.4×2.4 m, simulating real buildings and operated under real conditions was built in Lleida, Spain, to evaluate passive systems for building energy efficiency. Some of those cubicles were used to evaluate the effect of incorporating PCM in building envelopes. Detailed information about this work can be found in [9].

Two of these cubicles are presented here and will be evaluated using the methodology proposed. The cubicles were built with precast concrete panels 0.12 cm thick (Figure 3.3.4). They incorporated one window (1.7×0.6 m) at the east and west walls and four windows (0.75×0.4 m) in the south wall. Summary of the construction used in this study is given below but details can be found in [6].

1. Concrete cubicle (CON): Built with precast concrete panels. No insulation is added to the cubicle.
2. Concrete and PCM cubicle (CON + PCM): Built with precast concrete panels, where the concrete of the South and West wall as well as on the ceiling has been modified to incorporate 5% in weight of microencapsulated PCM (Micronal® PCM). No insulation is added to the cubicle.

The PCM used is Micronal® PCM with a melting point of 26°C, and a phase-change enthalpy of 110 kJ/kg. The experiments were performed under continental Mediterranean weather conditions with high temperatures (25°C to 35°C) and solar radiation (1000–1100 W/m²).

To monitor and analyze the system different data were registered:

- Weather conditions (solar radiation (Middleton Solar pyranometers SK08), ambient temperature and humidity (ELEKTRONIK EE21), wind velocity (DNA 024 anemometer)).
- Internal ambient temperature (ELEKTRONIK EE21).
- Internal surface temperature of the walls, roof and ceiling (Pt-100 DIN B).
- External surface temperature of the south wall (Pt-100 DIN B).
- Energy consumption of the heating/cooling systems with an electrical network analyzer (MK-30-LCD).

3.3.4.3 MASSIVE BUILDINGS—BRICK

Two additional cubicles from the experimental setup in Lleida (Spain) will be also evaluated using the methodology proposed by Evola et al. [23]. These cubicles were built with an alveolar brick construction system, with an internal and external finishing of gypsum and mortar, respectively (Figure 3.3.5). The roof was constructed using concrete precast beams and 5 cm of concrete slab. The insulating material is placed over the concrete, protected with a cement mortar roof with an inclination of 3% and a double asphalt membrane. Extended details can be found in [9] and summarized below.

1. Alveolar cubicle (ALV): Built with alveolar brick, no additional insulation was used.
2. Alveolar and PCM cubicle (ALV + PCM): Built with alveolar brick and an additional layer of macroencapsulated PCM (SP-25 A8) in the internal side of the south and west wall as well as on the ceiling. No additional insulation was used.

Table 3.3.3 presents some key parameters of the PCM used in the constructions.

The experiments were performed under similar weather conditions and with similar instrumentation to those explained earlier.

FIGURE 3.3.5 Alveolar brick cubicle with macroencapsulated PCM.

TABLE 3.3.3

PCM Properties of Pure SP-25 A8 from Manufacturer and DSC Experiments

		SP-25 A8 (Manufacture)	SP-25 A8 (DSC)
Phase-change range (°C)		15°C–30°C	23–29
Heat storage capacity (kJ/kg)		180	137
Density (kg/L)	Solid	1.38	
	Liquid		
Specific heat capacity (kJ/kg K)	Solid	2.5	
	Liquid		
Heat conductivity (W/m K)			0.6

3.3.5 RESULTS AND DISCUSSION

To compare the results of the different parameters between no-PCM and PCM systems, the internal temperature must be monitored for both cases. Figures 3.3.6–3.3.8 present the internal temperature for each constructive system. One can see that for the alveolar brick the temperature fluctuations of both cubicles are small (around 1°C). Although the PCM does reduce these fluctuations, the high thermal inertia of the alveolar brick and its insulating properties result in a stabilized internal temperature even without the use of PCM.

On the other hand, the concrete cubicles present much higher fluctuations (up to 18°C). Although this system also provides high thermal inertia, there is no insulation to reduce heat transfer, and the effect of windows is also important in terms of solar heat gains. Thus, thermal oscillations are higher and the effect of PCM is very clear.

The wood huts also show high thermal fluctuations (up to 14°C), in this case, fostered by the low thermal inertia of the construction and by the effect of the window. The effect of the insulation is also seen for the lower temperature oscillations compared to the concrete.

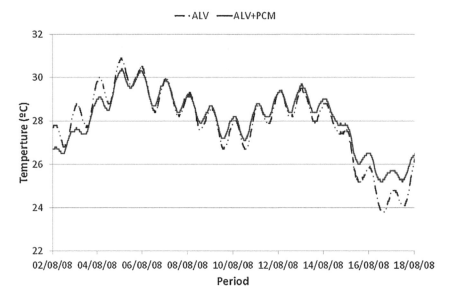

FIGURE 3.3.6   Internal air temperature evolution for the alveolar brick cubicles.

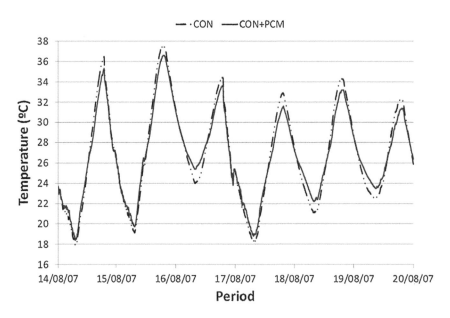

FIGURE 3.3.7   Internal air temperature evolution for the concrete cubicles.

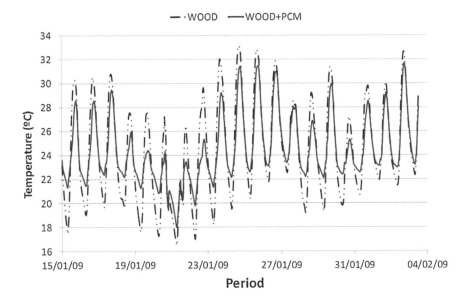

FIGURE 3.3.8   Internal air temperature evolution for the wood huts.

When analyzing the ITD, a direct correlation to the internal temperature is observed. For the alveolar brick cubicles the difference between ALV and ALV + PCM is only reflected in few days for the "office profile" (Figure 3.3.9), while it is more visible for the "house profile" and "daily profile" (Figures 3.3.10 and 3.3.11). These results highlight the importance of the evaluation period when using this methodology, since changes in the occupancy can lead to an underperformance of the system.

Furthermore, during many days the ITD remains zero in both cubicles, although temperature differences are achieved between them (as seen in Figure 3.3.6). These results also suggest that the indicator is not suitable for construction with high thermal inertia and insulation and no significant heat gains, since temperature fluctuations are small.

Regarding the concrete cubicles, their behavior is more clearly reflected by the ITD indicator. In this case, the thermal fluctuation reduction achieved by the PCM (Figure 3.3.7) is represented by lower ITD (Figures 3.3.12–3.3.14). Here the effect of solar radiation and heat losses is very significant and it allows a good ITD representation of the thermal behavior of the cubicles despite their high thermal inertia. Thus, thermal inertia is not the only parameter that affects the ITD, which means the methodology presented by Evola et al. can be used to evaluate other influencing variables and care must be taken when used to determine the efficiency of PCM.

For the wood huts, the ITD also reflects the behavior of the internal temperature in terms of thermal comfort. The stabilizing effect of the PCM on the internal temperature is more important in this lightweight building, as can be seen from the calculated ITD showing a higher difference between the PCM and no-PCM huts (Figures 3.3.15–3.3.17).

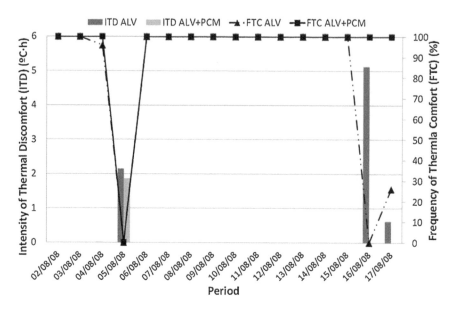

FIGURE 3.3.9   ITD and FTC evaluation for the alveolar brick cubicles and for the "office profile".

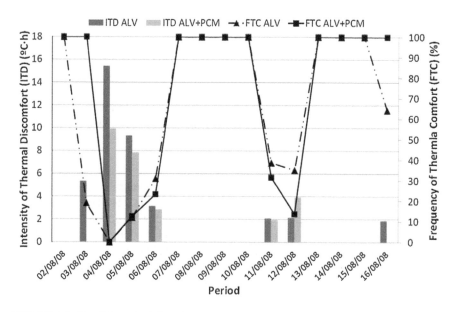

FIGURE 3.3.10   ITD and FTC evaluation for the alveolar brick cubicles and for the "house profile".

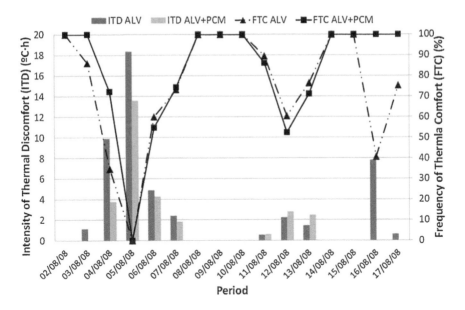

FIGURE 3.3.11 ITD and FTC evaluation for the alveolar brick cubicles and for the "daily profile".

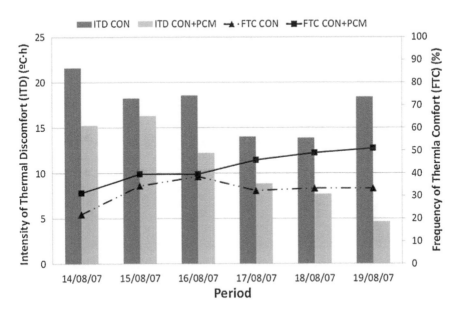

FIGURE 3.3.12 ITD and FTC evaluation for the concrete cubicles and for the "office profile".

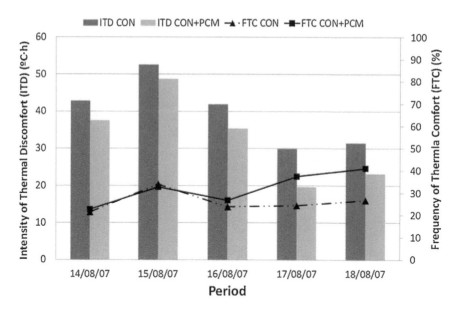

FIGURE 3.3.13 ITD and FTC evaluation for the concrete cubicles and for the "house profile".

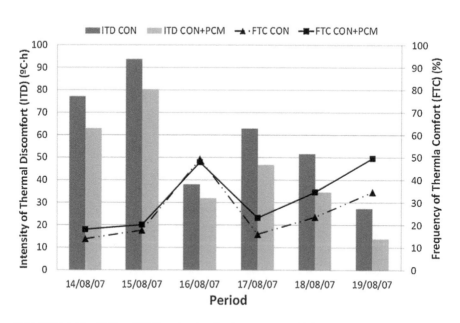

FIGURE 3.3.14 ITD and FTC evaluation for the concrete cubicles and for the "daily profile".

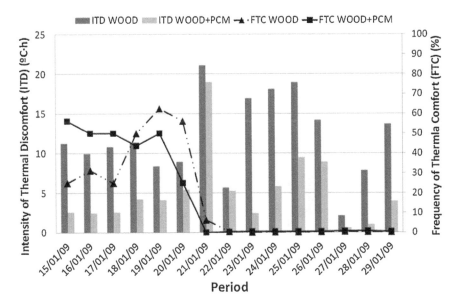

FIGURE 3.3.15   ITD and FTC evaluation for the wood huts and for the "office profile".

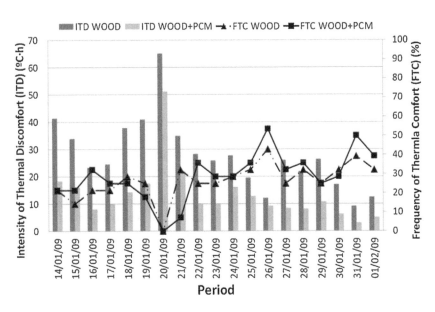

FIGURE 3.3.16   ITD and FTC evaluation for the wood huts and for the "house profile".

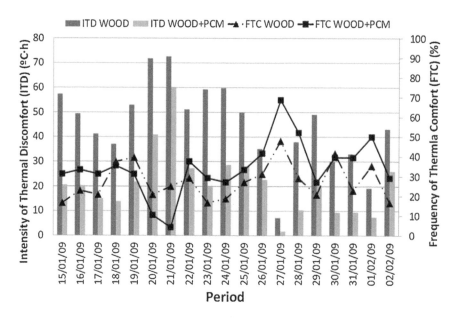

FIGURE 3.3.17   ITD and FTC evaluation for the wood huts and for the "daily profile".

Moreover, the importance of the evaluation period is highlighted again when looking at the results for the day 22/01/09. For the house and the daily profile the hut with PCM behaves clearly better, while for the office profile results show a similar performance. This difference in the performance is because the PCM is not being fully activated (Figure 3.3.8). Therefore, the PCM remains in its melting range, keeping the PCM hut below the thermal comfort range longer than the no-PCM one. Since the office period takes into account most of the time the PCM-hut is under thermal discomfort, the performance of the PCM is penalized while for the house profile or the daily evaluation this underperformance is overtaken by the benefits achieved during the rest of the day.

The second parameter proposed by Evola et al. is the FTC, which by definition is clearly related to the ITD. Higher ITD usually results in low FTC. Therefore, most previous discussion on ITD applies to FTC. However, it is also important to mention that there are cases where the FTC can reflect a different tendency than that of the ITD. While the ITD gives a quantitative value of the thermal discomfort (how far from comfort the building is), the FTC provides a qualitative one (comfort is or is not achieved). Therefore, cases with continued periods of low thermal discomfort (low ITD values) can result in lower ITD and FTC for a specific system. This is shown in all the construction analyzed in this work, but it is more evident in the wood construction. For the daily evaluation of 21/01/09 (Figure 3.3.17), the WOOD + PCM hut shows lower intensity of thermal discomfort, but also lower FTC.

This behavior is in a way related to the fluctuation of thermal discomfort (FD) defined in Sicurella et al. [27]. Therefore, FD could be a good indicator for the design process, defining the cases in which PCM is effective.

Furthermore, Tables 3.3.4–3.3.6 present a detailed evaluation of the ITD for each of the construction, dividing it into ITD_{over} and ITD_{under}, depending if the thermal discomfort is due to high or low temperatures. Notice that only the days when thermal discomfort occurs are presented.

From the experimental results, one can see two different patterns. The alveolar brick cubicles present most of the ITD due to overheating, while for the concrete cubicles and the wood huts it was both overheating and low temperatures. These results demonstrate the importance of evaluating the benefits of PCM in preventing both overheating and low temperatures, as suggested in Section 3.3.3. PCM can be useful in preventing overheating during summer, but also in preventing low temperatures during spring or autumn. The use of a comfort range in the evaluation leads to a more accurate analysis.

To determine the real behavior of the PCM, its FA is presented in Tables 3.3. 7–3.3.9 for the alveolar and concrete cubicles. While for the case of the ALV + PCM cubicle the temperature used for the evaluation is based on direct measurements of the

TABLE 3.3.4

ITD_{over} and ITD_{under} for Alveolar Brick Cubicles and Day Profile

	ALV		ALV + PCM	
	ITD_{over}	ITD_{under}	ITD_{over}	ITD_{under}
03/08/08	1.13	0.00	0.00	0.00
04/08/08	9.88	0.00	3.77	0.00
05/08/08	18.34	0.00	13.57	0.00
06/08/08	4.89	0.00	4.28	0.00
07/08/08	2.40	0.00	1.83	0.00
11/08/08	0.55	0.00	0.63	0.00
12/08/08	2.23	0.00	2.80	0.00
13/08/08	1.46	0.00	2.47	0.00
16/08/08	0.00	7.80	0.00	0.00
17/08/08	0.00	0.62	0.00	0.00

TABLE 3.3.5

ITD_{over} and ITD_{under} for Concrete Cubicles and Day Profile

	CON		CON + PCM	
	ITD_{over}	ITD_{under}	ITD_{over}	ITD_{under}
14/08/07	43.76	33.33	34.86	28.13
15/08/07	71.89	21.83	62.41	17.88
16/08/07	38.07	0.00	31.93	0.00
17/08/07	23.76	39.10	14.86	31.89
18/08/07	33.74	17.82	25.29	9.35
19/08/07	16.30	11.07	10.28	3.59

TABLE 3.3.6

ITD$_{over}$ and ITD$_{under}$ for Wood Huts and Day Profile

	WOOD		WOOD+PCM	
	ITD$_{over}$	ITD$_{under}$	ITD$_{over}$	ITD$_{under}$
15/01/09	11.51	45.86	1.20	19.45
16/01/09	12.28	37.13	2.02	18.36
17/01/09	16.31	24.86	6.63	8.37
18/01/09	0.00	37.07	0.00	13.90
19/01/09	0.00	52.95	0.00	21.83
20/01/09	0.00	71.85	0.00	40.86
21/01/09	0.00	72.63	0.00	59.94
22/01/09	7.32	43.81	0.00	27.30
23/01/09	28.08	31.17	7.49	12.39
24/01/09	39.40	20.42	22.19	6.46
25/01/09	36.43	13.51	24.56	3.55
26/01/09	27.76	7.39	21.04	1.67
27/01/09	3.36	3.84	1.27	0.51
28/01/09	5.09	32.78	0.00	10.32
29/01/09	21.15	27.95	10.27	9.55
30/01/09	0.00	30.01	0.00	9.43
31/01/09	12.54	20.51	2.80	6.70
01/02/09	11.04	8.14	5.64	1.87
02/02/09	30.43	12.73	22.74	3.35

TABLE 3.3.7

Frequency of Activation (FA) for Alveolar Brick Cubicles

	FA ALV+PCM (%) South Façade	ITD Reduction (%) ALV+PCM Cubicle
02/08/08	100.00	0.00
03/08/08	100.00	100.00
04/08/08	90.28	61.88
05/08/08	48.26	26.00
06/08/08	81.60	12.53
07/08/08	89.24	23.70
08/08/08	100.00	0.00
09/08/08	100.00	0.00
10/08/08	100.00	0.00
11/08/08	100.00	−13.64
12/08/08	100.00	−25.24
13/08/08	100.00	−69.59
14/08/08	100.00	0.00
15/08/08	100.00	0.00
16/08/08	100.00	100.00
17/08/08	100.00	100.00

TABLE 3.3.8

Frequency of Activation (FA) for Concrete Cubicles

	FA CON+PCM (%)			ITD Reduction (%) CON+PCM Cubicle
	South Façade	West Façade	Roof	
14/08/07	47.92	48.96	43.40	18.29
15/08/07	51.04	54.51	44.44	14.32
16/08/07	34.38	39.24	33.68	16.12
17/08/07	45.49	49.31	42.36	25.62
18/08/07	56.60	59.72	55.56	32.83
19/08/07	52.78	56.25	52.43	49.32

TABLE 3.3.9

Frequency of Activation (FA) for Wood Cubicles

	FA WOOD+PCM (%)					ITD Reduction (%)- WOOD+PCM Cubicle
	South Façade	West Façade	East Façade	North Façade	Roof	
15/01/09	60.42	39.58	52.08	56.25	56.25	64.01
16/01/09	54.17	47.92	50.00	54.17	50.00	58.77
17/01/09	47.92	31.25	43.75	45.83	45.83	63.56
18/01/09	70.83	43.75	66.67	64.58	64.58	62.51
19/01/09	85.42	66.67	77.08	77.08	70.83	58.78
20/01/09	97.92	97.92	93.75	100.00	87.50	43.13
21/01/09	95.83	100.00	97.92	91.67	83.33	17.47
22/01/09	81.25	72.92	77.08	79.17	72.92	46.61
23/01/09	56.25	52.08	47.92	54.17	50.00	66.46
24/01/09	45.83	33.33	35.42	41.67	41.67	52.11
25/01/09	33.33	20.83	27.08	29.17	31.25	43.71
26/01/09	25.00	16.67	20.83	22.92	27.08	35.39
27/01/09	16.67	6.25	14.58	12.50	16.67	75.36
28/01/09	56.25	39.58	50.00	54.17	54.17	72.74
29/01/09	47.92	43.75	43.75	47.92	47.92	59.62
30/01/09	66.67	52.08	60.42	60.42	64.58	68.58
31/01/09	50.00	43.75	43.75	47.92	47.92	71.25
01/02/09	37.50	20.83	31.25	27.08	35.42	60.88
02/02/09	39.58	29.17	35.42	33.33	37.50	39.55

macroencapsulated PCM, for the CON+PCM cubicle and the WOOD+PCM hut the wall temperatures are used (in a similar approach to that used in Evola et al. [24]), since the microencapsulated/embedded PCM is distributed through the wall and therefore its effect is considered uniform. Phase-change temperature ranges considered were 23°C–30°C for SP-25 A8 (ALV+PCM), 21°C–27°C for Micronal® PCM, and 18°C–23°C for the PCMGB with RT-21.

Experimental results from the ALV+PCM cubicle are contradictory to the expected behavior. When comparing the ITD reduction achieved when using PCM to the FA of the PCM (Table 3.3.7), two unexpected situations are observed. The only days when FA is lower than 100% (from 04/08/08 to 07/08/08) the benefits of using PCM (ITD reduction) are high. On the other hand, from 11/08/08 to 13/08/08, the ITD values show higher discomfort in the ALV+PCM cubicle, while the FA is always 100%.

On the other hand, when analyzing the concrete cubicles one can see that the value of FA is different depending on the orientation of the wall. The lowest FA was for the roof, followed by the south wall and finally west wall being the most active one. This is the result of the solar radiation incident to the surfaces, leading to PCM being inactive due to overheating during longer periods of the day. When comparing the FA to the ITD reduction results are contradictory again. Higher FA does not always result in higher ITD reduction. As an example, FA is at its highest values during the days 15/08/07, 18/08/07 and 19/08/07. However, while ITD reduction is the highest for the last 2 days (33% and 49%, respectively), it is the lowest for the first one (14%).

For the wood hut, similar inconsistencies are observed (Table 3.3.9). Days with high FA (especially the 21/01/09) show low ITD reduction. On the other hand, high ITD reductions are observed in days with low FA (i.e. 27/01/09). In general, there is no clear relation between the benefits of using PCM and the FA indicator.

Based on the experimental results, the FA indicator has proved to poorly present the degree of PCM benefit and may result in misleading evaluations if used.

Furthermore, the FA only determines if the PCM operates within its melting range, but it fails to show if the PCM has performed a complete melting/solidification cycle, which is the ideal situation to take the greatest advantage of the phase-change energy storage.

3.3.6 CONCLUSIONS

In the present article, the methodology presented by Evola et al. [24], which was applied only to results obtained from simulations, has been applied to three very different constructions. The intensity of thermal discomfort is found to be the most relevant indicator since it can give an idea about potential energy savings, which could be achieved through the application of PCM when comfort is guar-anteed by the use of the HVAC system. However, this parameter seems to fail in presenting the benefits of PCM when buildings have high thermal inertia, heavy insulation, and with no solar gains. On the other hand, for construction with significant heat gains, and with limited insulation or low thermal inertia, the ITD reflects the true situation well.

In addition, the evaluation of the ITD indicator for both over-heating and low temperatures discomfort must be considered to fully understand the benefits of using

PCM, since both scenarios can be found for different construction and weather conditions.

The FTC is not representative where HVAC systems are used in the buildings, since comfort will be achieved by those means. However, this indicator presents similar behavior to that of the ITD but may differ depending on the profile of the overheating period (small but continued or large but punctual). This effect is related to another indicator defined by Sicurella et al. [27], the fluctuation of thermal discomfort (FD), which can be a useful parameter for design purposes. The suitability of this new indicator requires further investigation.

The FA failed to show the benefits of PCM since results were contradictory and misleading.

ACKNOWLEDGMENTS

The research leading to these results has received funding from the European Union's Seventh Framework Programme (FP7/2007–2013) under grant agreement no. PIRSES-GA-2013-610692 (INNOSTORAGE). This work was partially funded by the Spanish Government (ENE2011-28269-C03-01 and ULLE10-4E-1305). The authors would like to thank the Catalan Government for the quality accreditation given to their research group (2009 SGR 534) as well as to the city hall of Puigverd de Lleida.

NOMENCLATURE

Δt^+	temperature difference [°C]
τ	time [h]
η_{PCM}	PCM storage efficiency [%]
C_{eq}	equivalent specific heat capacity [kJ/(kg K)]
E_{st}	stored energy [kWh]
FA	frequency of activation [%]
FTC	frequency of thermal comfort [%]
ITD	intensity of thermal discomfort [°C h]
L	latent heat [kJ/kg]
M	PCM mass [kg]
P	period [h]
q_L^+	heat flux entering the wall from the left side [kW]
q_R^+	heat flux entering the wall from the right side [kW]
T	temperature [°C]

SUBINDEX

D	discomfort conditions
lim	comfort limit
lower	lower comfort limit
m	Melting
op	operative

PC   phase-change process
PCM   phase-change material
upper   upper comfort limit
s   solidification

REFERENCES

1. Directive 2010/31/EU of the European parliament and of the council of 19 May 2010 on the energy performance of buildings. 20 June 2014. Available from http://www.epbd-ca.eu.
2. L.F. Cabeza, A. Castell, M. Medrano, I. Martorell, G. Pérez, I. Fernández, Experimental study on the performance of insulation materials in Mediterranean construction, *Energy and Buildings* 42 (2010) 630–636.
3. I. Axaopoulos, P. Axaopoulos, J. Gelegenis, Optimum insulation thickness for external walls on different orientations considering the speed and direction of the wind, *Applied Energy* 117 (2014) 167–175.
4. A. Gagliano, F. Patania, F. Nocera, C. Signorello, Assessment of the dynamic thermal performance of massive buildings, *Energy and Buildings* 72 (2014) 361–370.
5. C. Di Perna, F. Stazi, A.U. Casalena, M. D'Orazio, Influence of the internal inertia of the building envelope on summertime comfort in buildings with high internal heat loads, *Energy and Buildings* 43 (2011) 200–206.
6. N. Aste, A. Angelotti, M. Buzzetti, The influence of the external walls ther- mal inertia on the energy performance of well insulated buildings, *Energy and Buildings* 41 (2009) 1181–1187.
7. D.E.M. Bond, W.W. Clark, M. Kimber, Configuring wall layers for improved insulation performance, *Applied Energy* 112 (2013) 235–245.
8. L.F. Cabeza, C. Castellón, M. Nogués, M. Medrano, R. Leppers, O. Zubillaga, Use of microencapsulated PCM in concrete walls for energy savings, *Energy and Buildings* 39 (2007) 113–119.
9. A. Castell, I. Martorell, M. Medrano, G. Pérez, L.F. Cabeza, Experimental study of using PCM in brick constructive solutions for passive cooling, *Energy and Buildings* 42 (2010) 534–540.
10. F. Kuznik, J. Virgone, J.J. Roux, Energetic efficiency of room wall containing PCM wallboard: a full-scale experimental investigation, *Energy and Buildings* 40 (2008) 148–156.
11. F. Kuznik, J. Virgone, Experimental assessment of a phase change material for wall building use, *Applied Energy* 86 (2009) 2038–2046.
12. S. Behzadi, M.M. Farid, Experimental and numerical investigations of the effect of using phase change materials for energy conservation in residential buildings, *HVAC and R Research* 17 (2011) 366–376.
13. I. Mandilaras, M. Stamatiadou, D. Katsourinis, G. Zannis, M. Founti, Experimental thermal characterization of a Mediterranean residential building with PCM gypsum board walls, *Building and Environment* 61 (2013) 93–103.
14. P. Lamberg, J. Jokisalo, K. Sirén, The effects on indoor comfort when using phase change materials with building concrete products. *Proceedings of Healthy Buildings 2000*, 2 (2000) 751–756, SIY Indoor Air Information OY.
15. J. Jokisalo, P. Lamberg, K. Sirén, Thermal simulation of PCM structures with TRNSYS, in: *Terrastock 2000*, Stuttgart, Germany, 2000.
16. H. Schranzhofer, P. Puschnig, A. Heinz, W. Streicher, Validation of a TRNSYS sim- ulation model for PCM energy storages and PCM wall construction elements, in: *ECOSTOCK 2006—10th International Conference on Thermal Energy Storage*, Pomona, NJ, 2006.

17. M. Ibán̄ez, A. Lázaro, B. Zalba, L.F. Cabeza, An approach to the simulation of PCMs in building applications using TRNSYS, *Applied Thermal Engineering* 25 (2005) 1796–1807.
18. http://www.trnsys.com/tess-libraries/individual-components.php.
19. F. Kuznik, J. Virgone, K. Johannes, Development and validation of a new Trnsys Type for the simulation of external building walls containing PCM, *Energy and Buildings* 42 (7) (2010) 1004–1009.
20. F. Kuznik, J. Virgone, Experimental investigation of wallboard containing phase change material: data for validation of numerical modeling, *Energy and Buildings* 41 (5) (2009) 561–570.
21. C.O. Pedersen, Advanced zone simulation in EnergyPlus: incorporation of variable properties and phase change material (PCM) capability, in: *Proceedings: Building Simulation*, Beijing, China, 2007.
22. P.C. Tabares-Velasco, Verification and validation of EnergyPlus phase change material model for opaque wall assemblies, *Building and Environment* 54 (2012) 186–196.
23. A. Tardieu, S. Behzadi, J.J.J. Chen, M.M. Farid, Computer simulation and experimental measurements for an experimental PCM-impregnated office building, in: *Proceedings of Building Simulation 2011: 12th Conference of International Building Performance Simulation Association*, Sydney, Australia 2011.
24. G. Evola, L. Marletta, F. Sicurella, A methodology for investigating the effective- ness of PCM wallboards for summer thermal comfort in buildings, *Building and Environment* 59 (2013) 517–527.
25. ASHRAE, ASHRAE terminology, in: ASHRAE Handbook CD, ASHRAE, 1999–2002.
26. EN Standard, Indoor environmental input parameters for design and assessment of energy performance of buildings addressing indoor air quality, thermal environment, lighting and acoustics, in: EN Standard 15251, 2007.
27. F. Sicurella, G. Evola, E. Wurtz, A statistical approach for the evaluation of thermal and visual comfort in free-running buildings, *Energy and Buildings* 47 (2012) 402–410.

3.4 Peak Load Shifting with Energy Storage and Price-Based Control System

Reza Barzin, John J. J. Chen, Brent R. Young, and Mohammed Farid
University of Auckland

3.4.1 INTRODUCTION

High electricity usage at certain times of the day, known as peak load, introduces stress to the grid as supplied electricity is inadequate during the high peak demand period. To satisfy such demand, expensive peak power generation must be brought online during the peak period [1]. Also, variability of power generation based on renewable energy such as solar and wind has a huge impact on the electricity supply [2]. Peak load shifting is a possible solution, with electricity being stored during low load periods for use in peak load periods [3]. Because heating, cooling and air conditioning in many developed countries are responsible for almost 30% of the total electricity consumption [4], storing heat (or cold) could contribute significantly to peak load shifting.

Variable electricity rate or so-called ToU (time of use) electricity, is one of the tools used to encourage people to use electricity during off-peak periods. Many researchers have suggested using thermal energy storage (TES) to store heat or cold during off-peak periods to be used during the peak period [5]. Usually in TES, energy is stored in form of sensible heat, latent heat [6] and sorption [7]. Sensible heat storage materials have low thermal storage density which leads to large storage volume. Phase change materials (PCMs) however, offer a good thermal storage capacity because of their high storage density and they are used in applications where it is necessary to store heat or cold [8,9]. A large number of these studies can be found on the application of PCM in hot water cylinders [10,11], fridge and freezers [12], solar power plants [13], and buildings [14].

A number of studies have been carried out on the application of PCM in refrigerators, freezers [15] and cold storages [16]. The main objective of these studies was to improve the performance of refrigerators and freezers by introducing thermal inertia to reduce temperature fluctuation [17]. Azzouz et al. used PCM to improve COP of a household refrigerator by 12% [18]. Gin and Farid also used PCM to prevent significant temperature rise in a freezer through power failure and frequent door openings. These studies showed that the application of PCM in a freezer can considerably improve the performance of the freezer [19].

There are many reported studies on the use of PCM in buildings in both active [20] and passive systems [21]. An active system refers to storage systems in which an additional fluid loop is used to charge and discharge the stored energy to supply heating or cooling. On the other hand, a passive system does not involve any additional heat exchanger. Chilled water tanks and ice storage tanks are some of the most common active TES equipment [22]. Underfloor heating using PCM and PCM wallboards are examples of passive systems [23]. The application of PCM in building materials such as wallboards provides a large surface area for heat transfer, it is very easy to install and does not require any additional space in the building. These characteristics make PCM interesting alternatives to storing energy for passive applications [24].

There *have* been a number of studies carried out on the application of PCM for peak load application using active systems such as ice storage units [25,26] which resulted in successful peak load shifting. They have also applied a number of optimal control studies on active systems based on the published official variable electricity rates to minimize the electricity cost. The use of published official variable electricity rates can, however, cause a number of problems. The main problem is that the expected electricity load might not match the actual demand of the day [27]. Peak load shifting through consumers also can cause serious problems; for instance, DR (demand response) approach frequently failed in many countries as it requires people to adjust their consumption according to a dynamic electricity tariff. Given this approach, a sudden need to make frequent active consumption decisions may lead consumers to grow tired of keeping track of rates and usage. For instance, DR was tested in Salt Lake City, Utah, and Puget Sound, Washington and up to 98% of consumers gave up TOU tariffs and returned to fixed electricity rates [28]. In contrast, some researchers have suggested DSM (demand-side management), using DLC (direct load control) as a better solution for peak load shifting problem. However, many DLC projects using RACs (Remote Appliance Controllers) have failed because of their large impact on the users' comfort level [29].

To be successful with peak load shifting, a suitable energy storage needs to be incorporated during peak load periods (when the appliance is turned off because of high load) to have a minimum impact on consumers' comfort. In this article, the application of PCM was investigated to achieve a successful peak load shifting (based on RAC) while minimizing its effect on consumers' comfort level. A price-based control strategy is also proposed which enables the use of online electricity prices and prevents any mismatch between the predicted load and the actual load. The proposed strategy has been tested on a PCM-incorporated domestic freezer as well as on experimental test huts equipped with DuPont PCM wallboards.

3.4.2 METHODOLOGY

To prove the concept, a domestic freezer and experimental huts provided with heat pumps were used as two different case studies. The proposed price-based method was applied and the results with and without using PCM were compared.

3.4.2.1 PRICE-BASED METHOD

The proposed method shares many common features with the DLC method using RAC, but with some differences. In the proposed method, electricity load is monitored by the electricity provider and, as soon as the electricity consumption peaks, the electricity provider increases the electricity price and then users who are using the price-based control system will automatically stop using electricity. To use this method, the electricity provider also needs to set a price constraint for each day.

To prove the concept, a program was developed using the LabVIEW software to enable the controller to read the electricity price from any given website when the website address is provided. The developed program reads the electricity price from the Electricity Information website which publishes the wholesale electricity rates in New Zealand [30]. Price constraints need to be suggested by the electricity provider but, since this information is not available yet, a price constraint was selected based on the previous trends.

The importance of a correct price constraint for the day was discussed in the results section. If the electricity company provides a price constraint for each day on the webpage, the same developed program could be used to read the price constraint of the day. Thus, in the later stage of the experiment, to mimic a real situation in the future, both price and price constraint were fed directly to the control system remotely from an office located at the University of Auckland's city campus, 12 km away from the experimental huts. The experimental huts are situated at the Tamaki campus, Auckland, New Zealand.

3.4.2.2 THE EXPERIMENTAL SETUP

3.4.2.2.1 Domestic Freezer

In the first part of the experimental study, a 153 L vertical freezer (model Elba E150, Fisher & Paykel) with a nominal power of 150 W was used. An aqueous ammonium chloride solution (19.5wt% NH_4Cl) with a phase-change temperature of −15.4°C was used as the PCM because of its high latent heat (250 kJ/kg) and its peak melting point [12].

The freezer cabinet temperature was measured using T-type thermocouples. All thermocouples were calibrated against a reference thermometer (Ebro TFX430) from −20°C to 20°C in a stirred flask of methyl alcohol to an accuracy of 0.1°C. The PCM was contained in anodized aluminium containers because of its high thermal conductivity. The anodized layer protects against corrosion. Seven of these panels were placed against the three walls in the freezer, covering 26% of the surface area of the walls, and occupying only 3% by volume of the freezer storage space.

3.4.2.2.2 Data Acquisition and Control in the Freezer

As shown in Figure 3.4.1, the controller reads electricity price from the electricity provider and the price constraint of the day from the office situated at the University of Auckland, city campus. As shown in the flowchart in Figure 3.4.2, the controller compares the OP (online price) and the PC (price constraint). If the OP is less than the PC, the controller keeps the freezer running to charge the PCM.

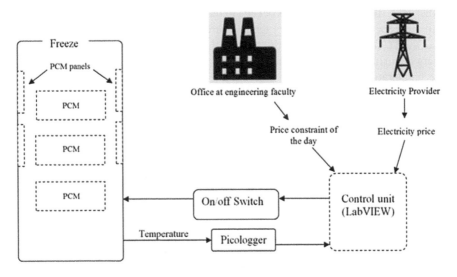

FIGURE 3.4.1 Schematic diagram for the freezer experimental setup.

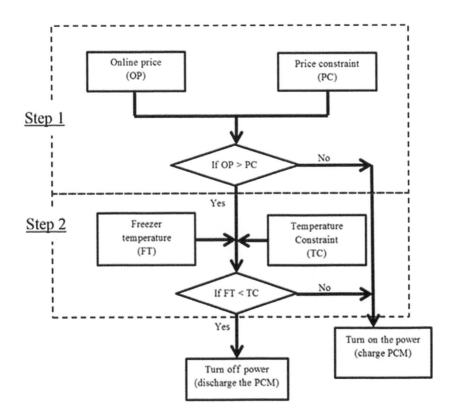

FIGURE 3.4.2 Flowchart of the price-based method applied for the freezer experiment.

Otherwise, the control system proceeds to the next step. In the next step, the control unit receives data on the cabinet temperature of the freezer (FT) and if it is lower than the TC (temperature constraint), the compressor is turned off to use the stored energy. During the high peak period, if the temperature of the freezer rises above the temperature constraint of $-12°C$ (FT $\geq$ TC), the controller ignores the high price and turns on the compressor to prevent any possible food damage, as shown in the flow chart.

3.4.2.2.3   Experimental Hut

The proposed method was tested using a full-scale test facility consisting of two identical test huts and a computer office situated at the University of Auckland Tamaki campus. These huts were used in this study to test the proposed method for space heating applications. Both huts were constructed using standard light-weight materials, were elevated above the ground, and had a north-facing window (Figure 3.4.3a). The first hut, referred to as Hut 1 or the "reference hut", has interior walls and ceiling finished with ordinary 13 mm thick gypsum wallboards and is used as the base case for all the experiments. In Hut 2, the 13 mm gypsum boards were replaced with 10 mm gypsum board plus a layer of DuPont Energain® wallboard. The DuPont Energain® sheets had a thickness of 5.2 mm and the PCM contained in it had a peak melting point of 21.7°C, and a latent heat storage capacity of 70 kJ/kg [31].

3.4.2.2.4   Data Acquisition and Control in the Hut Experiment

A CompactRIO was used as the control unit for hut experiments because of its ability to acquire all the required data and processing them as well as sending the required commands to the heating unit. As shown in Figure 3.3.4, the CompactRIO was programmed to receive the OP, PC and the temperature setpoint using the internet connection. Temperatures within each hut were measured using T-type thermocouples connected to a 16-channel thermocouple input module (NI 9214) enabling measurement sensitivity up to 0.02°C and measurement accuracy up to ±0.45°C. Room

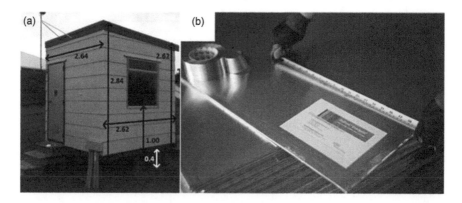

FIGURE 3.4.3 Experimental hut at Tamaki campus (a) exterior view and (b) DuPont Energain® wallboards used in Hut 2.

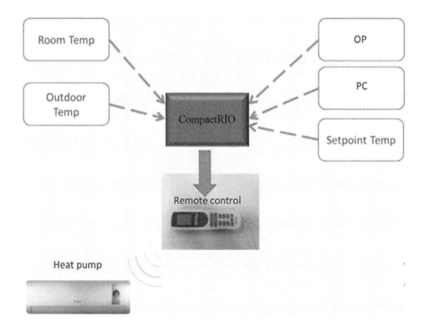

FIGURE 3.4.4 Schematic diagram showing the data acquisition and control used in the space heating experiment.

temperature was measured using four different temperature sensors, each installed on one of the walls. For this study, the thermocouple installed on the west wall was used as the room temperature. Additionally in Hut 2, one thermocouple in each wall was used to measures the wall temperature. Electricity consumption in both huts was measured using Cerrel Electrade LP-1W1 with accuracy up to ±0.5%. Solar irradiance also was measured using a VAEQ0SE pyranometer. The Embedded controller in CompactRIO received all the information and sent signals to the remote control of the heat pump; the remote control sent the signal to the heat pump to maintain the room within the desired temperature range.

As shown in Figure 3.4.5, during the low peak period when the electricity price is low, the controller keeps the room temperature between 21°C and 23°C. This temperature range is referred to as the "low price range" in which 21°C and 23°C are referred to as the "low price lower constraint" and "low price upper constraint", respectively. As soon as the price rises above PC, the controller switches to discharge mode, in which the room temperature kept within the temperature range of 17°C–19°C (referred to as high peak temperature range shown by the grey area) to discharge the stored energy in PCM and also use the least possible energy during the high peak period. Temperatures 17°C and 19°C are referred to as "high price lower constraint" and "high price upper constraint", respectively. As mentioned earlier, DuPont wallboards have a peak melting point of 21.7°C, which is between the low price and the high price temperature range. This enables successful charge and discharge of the PCM and consequently uses the full potential of the PCM.

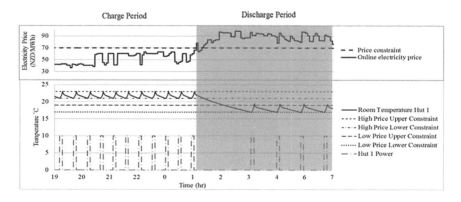

FIGURE 3.4.5   Room temperature setpoint during high peak and setpoint period.

3.4.3   RESULTS AND DISCUSSION

3.4.3.1   FREEZER EXPERIMENT

To test the benefit of using PCM in a freezer an experiment was conducted first with no PCM, and the results are shown in Figure 3.4.6. Part A of the figure presents the online electricity price (OP) and the PC (price constraint) of the day. During the period that the OP is lower than the PC of the day (low peak period), the controller sends a 5-V analogue signal to the switch to keep the freezer running and thus operates within its normal cycling range (between −21°C and −17°C). As soon as the electricity price rises above the price constraint of the day (peak period), the controller sends a zero volt signal to the switch to turn off the freezer during this period until the electricity price drops below the price constraint.

However, at any time during this process, if the freezer's temperature rises beyond the temperature constraint, the controller ignores the electricity price and turns the freezer on to prevent further increase in the temperature. This is to keep the freezer temperature low enough to prevent any possible food damage.

Figure 3.4.7 shows the experimental results in which the price-based control strategy was applied to the freezer with and without PCM. The experiment has been carried out for identical price data and using the same price constraint of 75 NZD/MWh and the same set temperature of −12°C.

The results show that the freezer with no PCM had a more rapid temperature increase from −22°C to the freezer temperature constraint of −12°C during the price peak time. Following that, the controller turned on the power to prevent further temperature rise, and kept the temperature close to the temperature constraint during the peak period (shown by C). Once the electricity price dropped below PC, the controller allowed the freezer to use electricity normally and bring the freezer temperature back to −22°C. The results showed that the freezer could be turned off for a maximum of 40 min during the peak period.

On the other hand, using the same strategy for the freezer with PCM, it can be clearly seen in Figure 3.4.7 that, despite power being turned off, the temperature stayed below the temperature constraint limit during the high peak period.

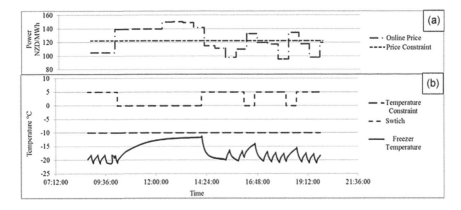

FIGURE 3.4.6   (a and b) Price-based control for freezer without PCM.

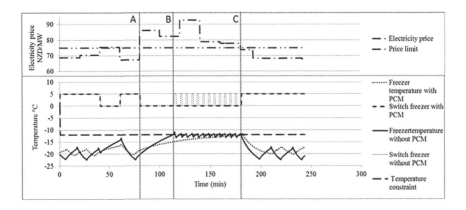

FIGURE 3.4.7   Price-based control of the freezer filled with PCM panels.

The freezer power was switched off for more than 150 min until the price dropped below the PC. Cost saving using this method varies depending on the electricity price profile (the higher the price during the cut-off period, the more money can be saved). For this experiment, the application of PCM resulted in a 16.5% reduction in the total cost of electricity use. This experiment showed that using PCM can enable successful application of the price-based strategy that enables the user to stop using electricity for a period three times longer than when no PCM is used.

It should be noted that the temperature constraint of −12°C for the freezer was chosen to illustrate the effect of PCM in the freezer and, if a lower desired temperature is to be applied, a PCM with a lower melting point needs to be used.

3.4.3.2   HUT EXPERIMENT

As described earlier, the test huts at the University of Auckland Tamaki campus were used to study the application of PCM to assist the proposed price-based method.

The proposed method was applied under various conditions and results for both huts were compared in terms of energy and electricity cost saving.

3.4.3.2.1 Space Heating Using Price-Based Control Method

Figure 3.4.8 shows the interior temperature of Hut 1 and Hut 2 when both huts were kept at a temperature range of 21°C–23°C until 7:30 in the morning (charge period). The controller stopped supplying heat to the huts during the early morning high peak period (shown by the grey shaded area). However, because of the solar irradiance during the day, the room temperature in both huts remained above 21°C. As a result, the PCM in Hut 2 remained fully charged until evening. During the high peak period of early evening, the room temperature dropped sharply and reached the minimum temperature constraint at 17:27 while the room temperature for Hut 2 remained within the setpoint range until 22:17, using the stored energy for almost 5 h longer compared to Hut 1.

Based on the above discussion, this method of control can be very successful for most winter days with relatively low solar irradiance. However, during days with high solar radiation, this method can cause overheating during the day, as shown in Figure 3.4.9. As can be observed, during the charge period both huts are heated and consequently the PCM in the wallboards is fully charged in the morning. During the day, high solar radiation causes a temperature rise in both huts, which brings the room temperature in both to an uncomfortable range. This is because the PCM in Dupont wall boards is fully charged, and cannot absorb any further energy to prevent temperature rise in Hut 2.

To address this problem, the controller was programmed to keep the room temperature within the range of 17°C–19°C throughout the night when the next day's weather was forecast to be sunny, even if the price is low. The results in Figure 3.4.10 show that the controller kept the room temperature for both huts between 17°C and

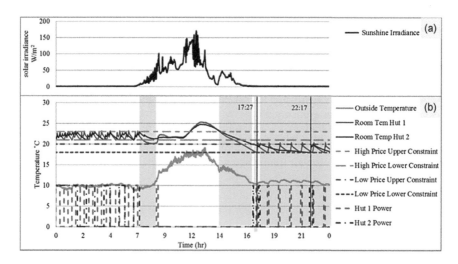

FIGURE 3.4.8 Hut experiment for a day with low solar radiation. (a) Solar irradiance and (b) room temperature for Hut 1 and Hut 2.

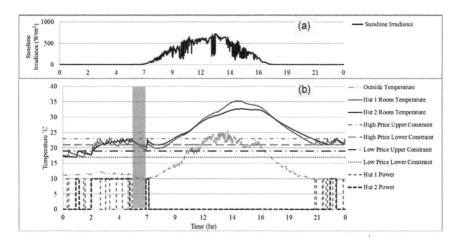

FIGURE 3.4.9 Hut experiment for a day with high solar radiation. (a) Solar irradiance and (b) room temperature for Hut 1 and Hut 2.

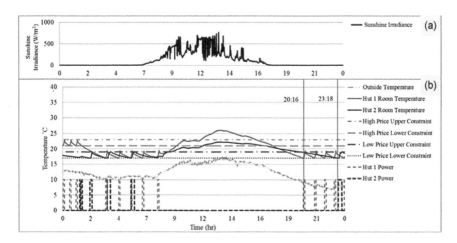

FIGURE 3.4.10 Hut experiment for a day with high solar radiation, using weather forecast. (a) Solar irradiance and (b) room temperature for Hut 1 and Hut 2.

19°C until the room temperature increased because of solar radiation. The room temperature for Hut 1 rose above 25°C, while the temperature in Hut 2 did not reach 23°C, which clearly shows that the PCM in Hut 2 was getting charged during that period by capturing a good amount of solar energy. The charged energy was then used during the evening peak period to keep the room temperature within the limit of 17°C–19° C until 23:18, while Hut 1 required the use of electricity 4 h earlier at 20:16. This method of control requires accurate weather forecasting, and unfortunately because of unstable weather during the winter period in Auckland, New Zealand, an accurate weather forecast is not possible. This makes it very difficult to use weather forecasting as a reliable method.

3.4.3.2.2 Power Consumption

To better present the possible savings using the proposed price-based method, the room temperatures in both huts were compared over a 6-day period, as shown in Figure 3.4.11. During this period, the total electrical energy consumption and corresponding costs were compared. The control system used the price- based method throughout this period and the price constraint was kept constant at 70 NZD/MWh. During Day 6, the price-based method was used in combination with the weather forecast.

Figure 3.4.11 shows room temperature for Hut 1 and Hut 2 over the 6-day period. Over this period, a total electricity saving of 21.5% was achieved with a corresponding electricity cost saving of 26.7%. A wrong assumption of the price constraint can significantly deteriorate performance of the proposed price-based method, as can be observed in Day 3. Table 3.4.1 shows the electrical and cost savings achieved for each day.

As shown in Table 3.4.1, the highest saving achieved in Day 5 was 60.08%, followed by Day 6 and Day 2 with 36.50% and 32.03%, respectively. However, Hut 2 consumed 9.39% more energy compared to Hut 1 in Day 3 because of the unavailability of a suggested price constraint by the electricity provider.

Figure 3.4.12 shows Day 2 where the electricity price stayed below the price constraint all night, allowing the control system to successfully charge the PCM during the night. The stored energy then used during the high peak period kept Hut 2 within the setpoint temperature range until midnight. This resulted in a 32% saving in electrical energy consumption and 45% in electricity cost.

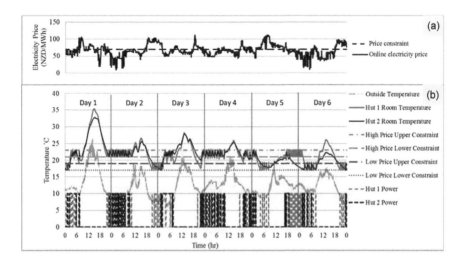

FIGURE 3.4.11   Hut experiment over 6 days. (a) Electricity price and (b) room temperature for Hut 1 and Hut 2.

TABLE 3.4.1

Daily Power Saving and Cost Saving over the 6-Day Period

	Day 1	Day 2	Day 3	Day 4	Day 5	Day 6	Total
Power saving	0.24%	32.03%	−9.39%	15.94%	60.08%	36.50%	21.5%
Cost saving	1.43%	45.08%	−6.81%	16.02%	62.64%	44.00%	26.7%

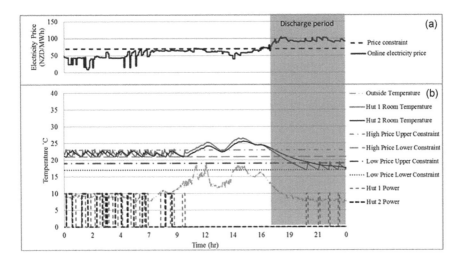

FIGURE 3.4.12 Hut experiment Day 2. (a) Electricity price and (b) room temperature for Hut 1 and Hut 2.

As shown in Figure 3.4.13, during Day 5, the controller stopped using electricity during the early hours of the morning (2:00 am onwards) because of the high electricity price. As a result, Hut 2 used the stored energy until sunrise, while Hut 1 had to use electrical energy to keep the room temperature within the setpoint range. During the evening, despite the low electricity price, Hut 2 did not use electricity to charge PCM as the next day was forecast to be sunny, while Hut 1 used heating to ensure that it would have enough stored heat in case of any electricity peak. As a result, the heater only turned on 4 cycles compared to 22 cycles in Hut 1, which resulted in a 60% power saving. As forecast, solar radiation charged the PCM during Day 6, storing enough energy to supply heat during high peak period in the evening. Because of this, very little heating was required in Hut 2 which resulted in a power saving of 36% during Day 6.

As shown in Table 3.4.1, Hut 2 consumed more energy compared to Hut 1 in Day 3. As can be seen in Figure 3.4.14, both huts were charged in the early morning but as a result of low price in the evening, the stored energy could not be used. However, in this case, it only affected the daily energy consumption rates and did not affect the overall energy consumption, as the consumed energy was later used during

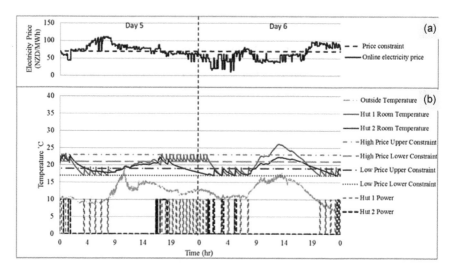

FIGURE 3.4.13 Hut experiment Days 5 and 6. (a) Electricity price and (b) room temperature for Hut 1 and Hut 2.

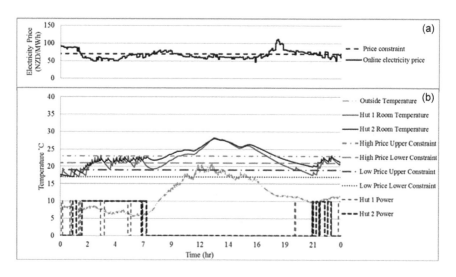

FIGURE 3.4.14 Hut experiment Day 3. (a) Electricity price and (b) room temperature for Hut 1 and Hut 2.

the next day. As it is clear, Hut 2 required a higher amount of energy to get charged compared to Hut 1, which resulted in a more energy consumption for Hut 2.

As mentioned earlier, the price constraint of the day is one of the most important parameters in the proposed methodology. Looking closer on Day 4, the price trend shows that the price constraint is too high for that day (Figure 3.4.15a). As a result, the controller maintained the room temperature within the low peak temperature

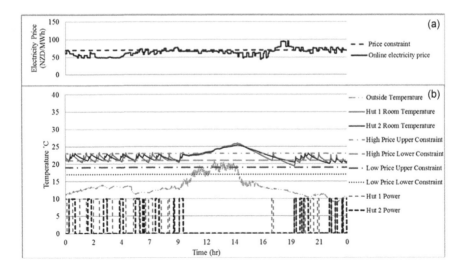

FIGURE 3.4.15 Hut experiment Day 4. (a) Electricity price and (b) room temperature for Hut 1 and Hut 2 using price constraint equal to 75 NZD/MWh.

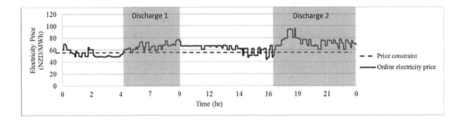

FIGURE 3.4.16 Day 4 electricity price using price constraint equal to 57 NZD/MWh.

range almost all day. This means that the controller did not benefit from price variation to charge when the price is cheaper and use when it is more expensive, and thus could not contribute to the peak load shifting. Also more energy was consumed in both huts to keep the room temperature in a higher temperature range for a longer period. This naturally also increased heat losses from walls (because of higher temperature difference of the room temperature and outdoor temperature).

Using a lower price constraint such as 57 NZD/MWh on Day 4 could significantly improve the performance of the system by storing energy during the period of the day with a lower electricity price and using the stored energy during the period with a higher price (Figure 3.4.16). As a result, the stored energy at night could be discharged during the morning peak (shown as discharge 1) and use solar energy to store energy again for the evening peak (shown as discharge 2). However, because of using the price constraint equal to 70 NZD/MWh, the controller assumed the electricity price was cheap almost all day.

As suggested earlier, to successfully apply the price-based control method, the electricity providers needed to provide both online electricity prices and the price

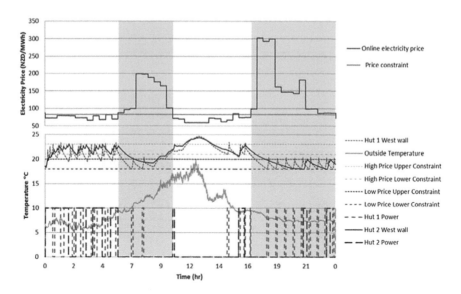

FIGURE 3.4.17   Application of price-based method using suitable price constraint.

constraint of the day. The results in Figure 3.4.17 show that selecting a suitable price constraint enables the controller to better use the variable price. It should be noted that having a price constraint lower than 60 NZD/MWh in Figure 3.4.17 would not allow the controller to charge the PCM throughout the day. The experimental results highlight the importance of using a suitable price constraint for the proposed method.

3.4.4   CONCLUSIONS

The price-based control method was experimentally applied on a domestic freezer and two experimental huts to elucidate the concept. PCM was used in both applications to improve the performance of the proposed method. Experimental results for the freezer experiment showed a cost saving of 16.5% and also showed that, using PCM, the freezer can stop using electricity three times longer than would a freezer with no PCM. The application of the price-based method on experimental huts also showed a significant improvement in both energy and cost saving, reaching up to 62.62% cost saving per day with a corresponding 60% energy saving. The application of the price-based method shown resulted in a successful peak load shifting for both freezer and space heating applications. The results also showed the importance of selecting a suitable price constraint to maximize the benefits, suggesting that the electricity distributer needed to provide the price constraint data for the consumers.

ACKNOWLEDGEMENTS

The work presented in this publication was supported by MBIE of New Zealand, which has funded the project UOAX0704-CR-3 and also both the European Union and the Royal Society of New Zealand for funding the project INNOSTORAGE IRSES-610692.

ABBREVIATIONS

AC:	air conditioner
Cop:	coefficient of performance
DLC:	direct load control
DR:	demand response
DSC:	differential scanning calorimetry
FT:	freezer temperature
LHTES:	latent heat thermal energy storage
OP:	online price
PC:	price constraint
PCM:	phase change materials
RAC:	remote applicant control
RT:	room temperature
TC:	temperature constraint
TES:	thermal energy storage
ToU:	time of use

REFERENCES

1. Peterson RB. A concept for storing utility-scale electrical energy in the form of latent heat. *Energy* 2011;36(10):6098–109.
2. Lund H. Large-scale integration of wind power into different energy systems. *Energy* 2005;30(13):2402–12.
3. Packer MB, Glicksman LR. An assessment of thermal energy storage in conjunction with heat pumps for residential heating and cooling. *Energy* 1979;4(3):393–9.
4. Qureshi WA, Nair NKC, Farid MM. Impact of energy storage in buildings on electricity demand side management. *Energy Convers Manag* 2011;52(5):2110–20.
5. Li G, Qian S, Lee H, Hwang Y, Radermacher R. Experimental investigation of energy and exergy performance of short term adsorption heat storage for residential application. *Energy* 2014;65(0):675–91.
6. Li G, Hwang Y, Radermacher R. Experimental investigation on energy and exergy performance of adsorption cold storage for space cooling application. *Int J Refrig* 2014;44(-0): 23–35.
7. Ii G. *Review of Thermal Energy Storage Technologies and Experimental Investigation of Adsorption Thermal Energy Storage for Residential Application*. College Park, MD: University of Maryland; 2013.
8. Martin V, He B, Setterwall F. Direct contact PCM-water cold storage. *Appl Energy* 2010;87(8): 2652–9.
9. Medrano M, Yilmaz MO, Nogues M, Martorell I, Roca J, Cabeza LF. Experi mental evaluation of commercial heat exchangers for use as PCM thermal storage systems. *Appl Energy* 2009;86(10):2047–55.
10. Nabavitabatabayi M, Haghighat F, Moreau A, Sra P. Numerical analysis of a thermally enhanced domestic hot water tank. *Appl Energy* 2014;129(0):253–60.
11. Nkwetta DN, Vouillamoz P-E, Haghighat F, EI-Mankibi M, Moreau A, Daoud A. Impact of phase change materials types and positioning on hot water tank thermal performance: using measured water demand profile. *Appl Therm Eng* 2014;67(1–2):460–8.
12. Liu M, Saman W, Bruno F. Development of a novel refrigeration system for refrigerated trucks incorporating phase change material. *Appl Energy* 2012;92(0):336–42.

13. Mathur A, Kasetty R, Oxley J, Mendez J, Nithyanandam K. Using encapsulated phase change salts for concentrated solar power plant. *Energy Procedia* 2014;49(0):908–15.

14. Castell A, Farid MM. Experimental validation of a methodology to assess PCM effectiveness in cooling building envelopes passively. *Energy Build* 2014;81(0):59–71.

15. Li G, Hwang Y, Radermacher R, Chun H-H. Review of cold storage materials for sub-zero applic ations. *Energy* 2013;51 (0):1–17.

16. Li G, Liu D, Xie Y. Study on thermal properties of TBAB-THF hydrate mixture for cold storage by DSC. *J Therm Anal Calorim* 2010;102(2):819–26.

17. Gin B, Farid MM. The use of PCM panels to improve storage condition of frozen food. *J Food Eng* 2010;100(2):372–6.

18. Azzouz K, Leducq D, Gobin D. Performance enhancement of a household refrigerator by addition of latent heat storage. *Int J Refrig* 2008;31(5):892–901.

19. Gin B, Farid MM, Bansal PK. Effect of door opening and defrost cycle on a freezer with phase change panels. *Energy Convers Manag* 2010;51(12):2698–706.

20. Li G, Hwang Y, Radermacher R. Review of cold storage materials for air conditioning application. *Int J Refrig* 2012;35(8): 2053–77.

21. Qureshi WA, Nair NKC. Farid MM. Demand side management through efficient thermal energy storage using phase change material. *Energy Convers Manag* 2011;52(5):2110–20.

22. Morgan S, Moncef K. Field testing of optimal controls of passive and active thermal storage. *ASHRAE Trans* 2010;116:134–46.

23. Farid M, Kong WJ. Underfloor heating with latent heat storage. *Proc Inst Mech Eng Part A J Power Energy* 2001;215(5):601–9.

24. Zhou D, Zhao CY, Tian Y. Review on thermal energy storage with phase change materials (PCMs) in building applications. *Appl Energy* 2012;92:593–605.

25. Henze GP. An overview of optimal control for central cooling plants with ice thermal energy storage. *J Sol Energy Eng Trans ASME* 2003;125(3):302–9.

26. Henze GP. Impact of real-time pricing rate uncertainty on the annual performance of cool storage systems. *Energy Build* 2003;35(3):313–25.

27. Henze GP, Le TH, Fiorita AR, Felsmann C. Sensitivity analysis of optimal building thermal mass control. *J Sol Energy Eng Trans ASME* 2007;129(4): 473–85.

28. Kim J-H, Shcherbakova A. Common failures of demand response. *Energy* 2011;36(2):873–80.

29. Sepulveda A, Paull L, Morsi WG, Li H, Diduch CP, Chang L. A novel demand side management program using water heaters and particle swarm optimization, *EPEC 2010-IEEE Electrical Power and Energy Conference: "Sustainable Energy for an Intelligent Grid"*, Halifax.

30. WITS website, Online electricity price for Auckland, New Zealand. Available from: URL: http://www.elect ricityinfo.co.nz.

31. Liu H, Awbi HB. Performance of phase change material boards under natural convection. *Build Environ* 2009;44(9):1788–93.

3.5 Application of Weather Forecast in Conjunction with Price-Based Method for PCM Solar Passive Buildings – An Experimental Study

Reza Barzin, John J. J. Chen, Brent R. Young, and Mohammed Farid
University of Auckland

3.5.1 INTRODUCTION

As defined by the International Energy Agency (IEA), current energy consumption trends and their environmental consequences are raising great concern [1]. Increased CO_2 emissions from high energy consumption have resulted in serious environmental problems worldwide. In the EU and U.S., buildings are responsible for about 40% of total energy consumption in the country, and about 40% of greenhouse gas emissions are attributed to energy use in buildings. It is interesting to note that roughly 50% of this energy is used for space heating and cooling applications. This clearly shows that a reduction in building heating and cooling demand can significantly affect total energy consumption in buildings and consequently the emission of greenhouse gases [2,3].

Renewable energy resources such as solar or wind can contribute to a significant reduction in the emission of greenhouse gases. However, the intermittent nature of these energy sources requires energy storage because the energy availability usually does not match the time of use [4–6]. Energy storage in buildings is of benefit when there is a mismatch between the time of the energy's availability [7] and its use and can assist the flexibility of energy demand [8]. Batteries are the simplest example of energy storage devices, as they store electrical energy for use when required [9]. The same concept can be applied to store available heat and cold for use when needed.

Thermal energy storage (TES) material is considered as one of the potential solutions to the energy management problem. TES incorporates materials with high energy storage capacity in the form of sensible heat (sensible heat storage, SHS),

241

latent heat (latent heat storage, LHS) or thermochemical energy (thermo- chemical heat storage, TCHS). Phase change materials have the ability to store large amounts of energy as the latent heat of melting [7]. These materials can provide constant energy delivery because of their isothermal nature [10].

3.5.1.1 APPLICATION OF PCM IN SOLAR PASSIVE BUILDINGS

Application of the phase change material (PCM) technology to store solar energy for space heating purposes can be traced back to just after World War II [5]. However, the application of PCMs to store energy faces some difficulties such as the low thermal conductivity of these materials. As a result, there are some difficulties associated with the design of PCM heat exchangers. To address this problem, various storage designs such as the application of plate type and tubular heat exchangers, heat exchangers with the addition of high thermal conductivity materials have been suggested [11–14]. However, the high cost of some of these methods prevented their commercialization.

The application of PCM in building components, however, is advantageous by having a large surface area, which is needed to charge and discharge the PCM. Additionally, this can be done at very low cost and does not need any additional space. These characteristics make PCM interesting alternatives to storing energy for passive applications. In passive heating, solar energy can be stored in latent heat storage materials incorporated in the construction components during daytime and later released passively into the room's air as the PCM solidify [15,16].

For instance, Chen et al. [17] studied the application of a new phase change material in solar passive buildings to absorb the excessive solar radiation during winter days and use it during the night when heating is required. Simulation results showed that the application of PCM can reduce peak temperature during the day by 1.5°C. However, the amount of cost/energy saving achieved using this method was not discussed in the study.

In another study, Shilei et al. [18] experimentally studied the application of PCM-impregnated gypsum wallboards in a solar passive building to reduce electricity costs as well as the electrical energy needed for heating. They concluded that, because of low variation of outdoor temperature, little saving could be achieved; however, this still encourages an application of this method to save energy in a building during winter. A similar simulation study performed by Zhou et al. [19] resulted in a 46% and 56% reduction in indoor temperature swings with the application of gypsum board and shape-stabilized PCM, respectively. Athienitis et al. [20] also performed numerical and experimental studies on the application of PCM-impregnated gypsum board in a passive solar test room during winter. The results showed up to a 4°C reduction in the room peak temperature and close to 15% reduction in heating at night. Khudhair et al. also used PCM-impregnated gypsum board to improve thermal behavior of lightweight buildings. The results showed significant reduction in temperature variation [21]. A similar study by Kuznik et al. [22] also showed significant enhancement in the thermal comfort of the building.

Several studies have used active thermal energy storage for peak load shifting; for instance, Bourne and Novoselac have used a compact PCM based thermal energy storage for peak load shifting [23]. Similarly, Mankibi has used PCM-air heat exchanger to perform peak load shifting achieving up to 10% peak load shifting [24]. Peak load shifting also can be done through passive application of PCM as demonstrated by Jin et al. [25]. They showed that the PCM impregnation into building walls can results in up to 2 h of peak load shifting. Qureshi et al. also used PCM-impregnated gypsum boards in a lightweight experimental hut and resulted into 31% energy saving over 12 days period [26]. To improve this, Barzin et al. proposed a price-based method to store energy in PCM when the price of electricity drops (during the off-peak period) and use the store electricity when the electricity price is high [27]. They experimentally tested this method using two identical test huts, one of them was equipped with DuPont sheets and resulted in a successful peak load shifting with an energy saving of up to 60%. They have also experimentally studied the price-based method in combination with a number of PCM designs such as the application of PCM underfloor heating in combination with DuPont wallboards for space passive space heating [28] or PCM-impregnated gypsum boards in combination with night ventilation for space cooling [29] which resulted in successful peak load shifting and significant energy saving of up to 35% and 92% per day, respectively.

3.5.1.2 Application of Weather Forecasts in Energy Management in Buildings with PCM

The application of weather forecasts to improve the efficiency of a process is a well-established method. For instance, many studies can be found in the field of electrical power demand forecasting. Similar methods can be used to forecast weather in the building sector [30].

Some studies have used weather forecasts for buildings with no thermal energy storage. For instance, Argiriou et al. [31] experimentally tested the use of weather forecasts to predict ambient temperature and solar radiation for the following day and consequently to predict the heating requirements of the test cell the next day. The results showed an annual saving equal to 7.5% in the case of the test cell studies. Similar results were obtained using weather and solar irradiance forecasts in the case of a full-scale office building; this resulted in a saving of 15% [32].

Application of weather forecasting for buildings using active TES also has been investigated to optimize the process. For instance, Candanedo et al. [33] used ice storage in combination with a chiller to perform peak load shifting in a building. They used weather forecasts and internal gains forecasts to calculate the required cooling demand for the next day. Through the application of a model-based control algorithm, they achieved a 20%–30% energy saving.

Weather forecasting, however, plays a crucial role if it is included and an inaccurate forecast can adversely affect the performance of the system significantly. Henze et al. [34] showed in simulation studies that, assuming the weather forecast and building models are perfect, there can be a significant saving; however, in real applications, prediction uncertainty can dramatically affect the performance of the system [35].

According to the literature, application of phase change materials in passive solar buildings is a well-practiced method. Weather predictions also have been successfully tested to reduce the electricity consumption of large buildings as well as buildings with active thermal storage materials. However, a review of the literature reveals an absence of applications of weather prediction for passive solar buildings.

In this study, the application of PCM-impregnated gypsum boards in passive solar buildings was investigated for heating in winter, using two identical test huts. Weather forecasting was also used to minimize the energy consumption in the huts. Problems regarding inaccurate weather forecasts were also discussed in detail. One of the test huts was used as the reference while the other hut was retrofitted with PCM-impregnated gypsum board. The control system considers real-time electricity pricing whereby low-price electricity is used optimally to charge PCM and the stored heat is used during the high peak electricity price period of the day.

3.5.2 METHODOLOGY

3.5.2.1 EXPERIMENTAL SETUP

Experimental studies were performed at the Tamaki Campus of the University of Auckland, New Zealand. Two fully instrumented test huts were used for the experimental studies, as shown in Figure 3.5.1. Both huts were constructed using standard lightweight materials with internal dimensions of $2.4 \times 2.4 \times 2.4$ m. The huts have north-facing windows with dimensions of 1×1 m, and both were heavily insulated using 100 mm thick pink glass wool in the walls and ceiling and 60 mm of polystyrene foam under the floor [28].

The first hut was used as reference for all experiments through-out this study and referred to as Hut 1. The interior walls and ceiling of Hut 1 were finished using ordinary 13 mm thick gypsum board. However in the second hut, the ordinary gypsum

FIGURE 3.5.1 Experimental huts at Tamaki Campus, University of Auckland, New Zealand.

boards were replaced with PCM-impregnated gypsum boards. The second hut is referred to as Hut 2. Each hut was equipped with a heat pump to provide heating demands. The heat pump also was used to increase the room temperature during the low price period to charge the PCM.

To focus on the application of weather forecast, internal gains were not included in this study. This could introduce error in total energy consumption in each hut which is why the total energy consumption was not reported in this study. However, this has minimal effect when the results for both huts are compared due to the fact that internal gains were not included in both huts, which means a constant load were deducted from both cases.

3.5.2.2 THERMAL ENERGY STORAGE

In this study, PT 20 from PureTemp was impregnated in gypsum board using the direct immersion method. The PCM used has a latent heat storage capacity of 180 J/g and peak melting point of 20°C [36]. The results obtained from previous studies suggested that PT20 can be a suitable PCM to be used in this study due to the climate of the region which is why this PCM was used in this study [21,26–29].

To impregnate the gypsum boards with PCM, two metal trays with dimensions of $0.7 \times 0.7 \times 0.09$ m and $0.9 \times 0.9 \times 0.15$ m were used as shown in Figure 3.5.2. The large tray was filled with water while the smaller tray was filled with PT20 and left floating inside the larger tray. Water inside the large tray was heated to 55°C to heat the PT20 inside the smaller tray. After reaching a steady temperature, the gypsum boards, which were cut into 0.6×0.6 m pieces, were immersed in the molten PCM one by one, each for a period of 10 min. Weighing of the gypsum boards before and after impregnation showed an approximate PCM impregnation of 22 wt%.

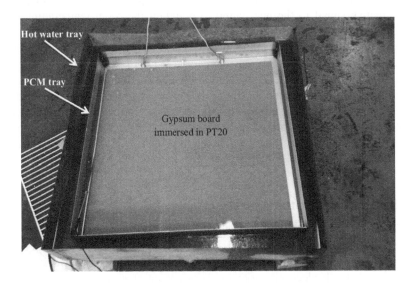

FIGURE 3.5.2   Impregnation of PT20 using the direct immersion method.

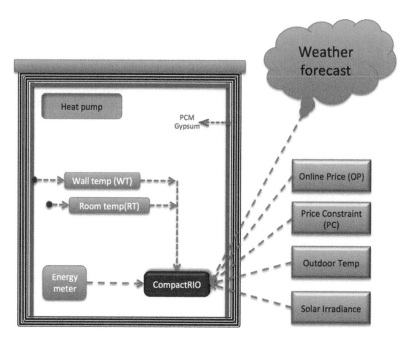

FIGURE 3.5.3 Schematic diagram for data acquisition and control system in the experimental huts.

3.5.2.3 DATA ACQUISITION

The experimental huts were designed to operate independently. A CompactRIO was used in each hut to collect all the required data, process them and finally send the corrective signal to the heat pump to maintain the room temperature within the desired range. As shown in Figure 3.5.3, the CompactRIO was programmed to receive weather forecast data, price constraints (PC), online prices (OP), outdoor temperatures, solar irradiance, wall temperatures (WT), room temperatures (RT) and electricity consumption. To measure temperature, T-type thermocouples were connected to a 16-Channel thermocouple input module (NI 9214) enabling measurement sensitivity up to 0.02 and measurement accuracy up to ±0.45°C. All thermocouples were calibrated against a referenced thermometer (Ebro TFX430) from 0°C to 35°C. Electrical consumption was measured using Carrel Electrade LP-1w1 with an accuracy of up to ±0.5%.

3.5.2.4 CONTROL SYSTEM

LabVIEW software was used to read the online electricity price from any given website when its address was provided. As a result, the developed code could be used in any other country to read online electricity prices from any given website. In this study, the wholesale electricity price for Auckland city region (obtained from http://www.electricityinfo.co.nz) was used to obtain the online price (OP). Similarly, the

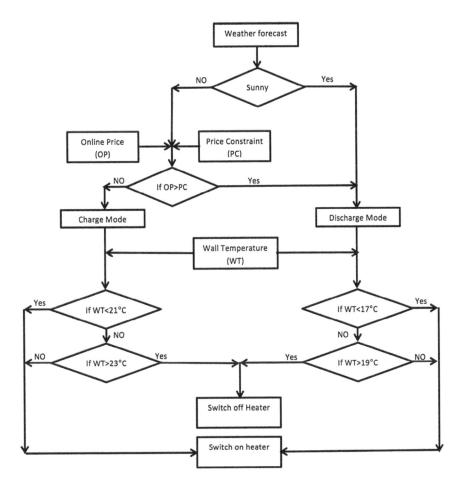

FIGURE 3.5.4 Flow chart of the control system to use price-based method in conjunction with weather forecast.

www.metservice.com [37] website was used to obtain weather forecast data for the location of the experimental huts.

The control system throughout this study operates the heating unit based on the flowchart shown in Figure 3.5.4. The control system receives the weather forecast for the next day and, if it is not forecast to be sunny, uses the price-based method; otherwise, it proceeds to the discharge mode. In the price-based method, the control mode is used based on the relative magnitudes of OP and PC. The controller proceeds to the charge mode if the OP is smaller than PC and the wall temperature is kept within the "low peak temperature range". The limits of this range are 23°C and 21°C, which are referred to as the "low price upper constraint" or LPU and "low price lower constraint" or LPL, respectively. This temperature range was selected to be above the peak melting point of PCM to ensure PCM melting during the charge period. Choosing the charge mode, the controller receives wall temperature data and keeps the wall temperature within the low peak temperature range using the heat pump. In this study, wall temperature was used instead of room temperature to improve the charge and discharge processes, as will be discussed further in Section 3.5.3. As soon as the electricity price rises above the PC value, the controller switches to discharge mode, in which a lower temperature range is applied; this is referred to as the "high price temperature range". In this mode, the control system uses a lower temperature range of 19°C–17°C. These limits are referred to as the "high price upper constraint" or HPU and "high price lower constraint" or HPL, respectively.

If the weather is forecast to be sunny, the control system automatically switches to discharge mode regardless of the electricity price and maintains the room temperature within the "high price temperature range" to prevent any energy consumption for charging of the PCM. This enables the control system to use solar energy instead of the heat pump to charge the PCM. The temperature constraints during the low and high price periods are presented in Table 3.5.1.

In the case of a misleading weather forecast, the controller switches to the price-based method during the day to use the heat pump to charge the PCM during the low peak price period of the day and use the stored energy during the evening peak price period.

Figure 3.5.5 shows an example of the application of price-based control where room temperature was maintained within a low peak price temperature range of

TABLE 3.5.1

Temperature Constraints during Low and High Price Periods

	Lower Temperature Constraint (°C)	Upper Temperature Constraint (°C)
Low price period	21	23
High price period (grey shaded area)	17	19

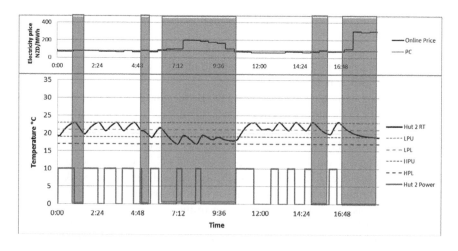

FIGURE 3.5.5   Application of price-based method using low peak and high peak temperature range.

21°C–23°C during the off-peak period when the electricity price is below the price constraint. Every time the electricity price rises above the PC value, the control system stops using electricity for heating and maintains the room temperature within the high peak temperature range of 17°C–19°C as long as the electricity price is higher than the price constraint. The high peak period is shown by the gray-shaded areas.

3.5.3   RESULTS AND DISCUSSION

In the first stage of this study, room temperature was used as the control variable (CV), as shown in Figures 3.5.5 and 3.5.6. However, because of the small interior area of the experimental huts, and the high power of the heating units, a quick indoor temperature rise was observed. In fact, room temperature reached the upper temperature constraint quickly while the PCM temperature remained below its melting point. As shown in the figure, during a low peak price period, the room temperature rose to 23°C, while the PCM temperature did not rise above 20°C during this period.

This clearly shows that the PCM was not charged during the low peak period. Figure 3.5.7 shows experimental results over a 4-day period using the room temperature as CV. As can be observed, during the early morning peak (shown by gray-shaded area), the drop in Hut 2 room temperature is very similar to that in Hut 1, which is evidence of the fact that very little PCM charging was achieved. As shown in Table 3.5.2, Hut 2 achieved an energy saving of only 8% with a corresponding cost saving of 13% during this period.

To overcome this problem, wall temperature was used as a control variable in both huts, as shown in Figure 3.5.8. During the charging period, the room temperature rises above the LPU range and consequently raises the PCM temperature above its melting temperature within a shorter period as shown in the figure. After the first heating cycle, very little temperature difference can be observed between the room temperature and the wall temperature during the charge period. As can be seen in

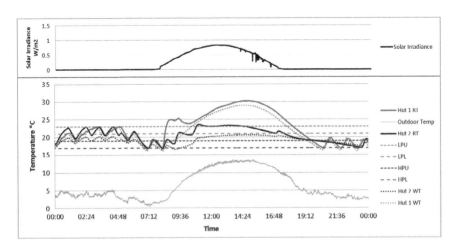

FIGURE 3.5.6   Application of room temperature as control variable to charge and discharge the PCM.

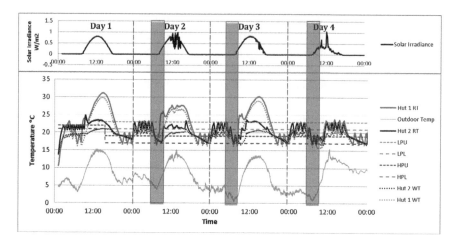

FIGURE 3.5.7 Experimental results for Huts 1 and 2 over a period of 4 days using the room temperature as the CV.

TABLE 3.5.2

Electrical Energy and Cost Savings over a Period of 4 Days Using the Room Temperature as the CV

	Day 1	Day 2	Day 3	Day3	Day 4
Electrical saving (%)	−12	26	13	7	8
Price saving (%)	−25	42	18	13	13

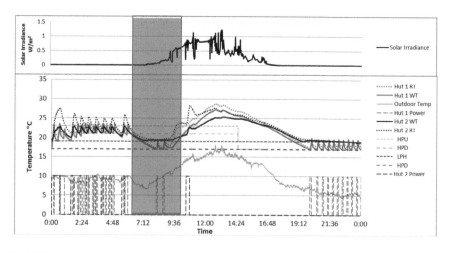

FIGURE 3.5.8 Room and wall temperatures of Huts 1 and 2 using price-based control and the wall temperature as CV.

the figure, room temperature and wall temperature are almost identical during the discharge process because the temperature change during the discharge process is very slow. Successful charging of the PCM during the charging period is one of the most important criteria, which was achieved using wall temperature instead of room temperature as the CV. The application of wall temperature as the CV provides better indication of the status of the PCM whether it is solid or liquid, and consequently, the control system can provide a more effective control action.

As shown in Figure 3.5.8, the PCM temperature was maintained within the low price peak temperature range most of the night. During the early morning price peak, both huts stopped using electricity for heating and as a result, the room temperature in Hut 1 dropped fast and heating was required to be applied, while Hut 2 did not require any heating because of the successful PCM charging during the low price peak period. Later during the day, the room temperature remained above the melting point of the PCM due to solar irradiation. After sunset, during the high price period, the room temperature in Hut 1 dropped sharply and reached 17°C at 19:23 while the room temperature in Hut 2 remained above 17°C until midnight, achieving an electrical saving equal to 14%, with a corresponding cost saving of 32%. The method presented shows acceptable performance during days with medium or low solar radiation. However application of the same method on days with high solar radiation can be of a problem, as Figure 3.5.9 shows.

The PCM temperature in Figure 3.5.9 remained above its melting point during the low price period at night, making the PCM fully charged. High solar radiation during the day caused a sharp temperature rise in Hut 2 similar to that in Hut 1 to above 29°C because the PCM in Hut 2 was almost fully charged and could not absorb further amounts of heat. Additionally, because there was very little heating demand during the evening price period, Hut 2 could not discharge all the stored energy and consequently consumed 30% more electricity compared to Hut 1. This could be even worse if both Huts did not require any heating until the next day (due to high

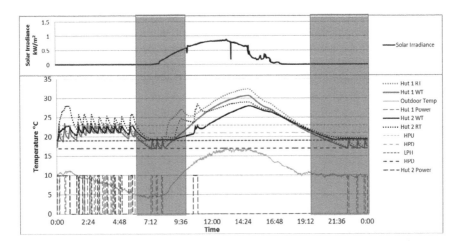

FIGURE 3.5.9 Room and wall temperatures of Huts 1 and 2 using the price-based method on a sunny day.

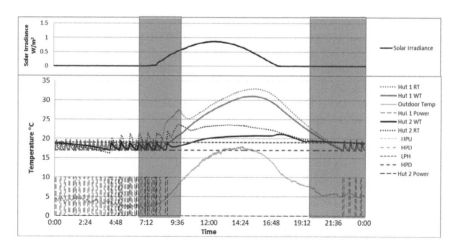

FIGURE 3.5.10 Room and wall temperatures of Huts 1 and 2: application of weather forecasting in conjunction with the price-based method.

temperature during the day). Hence, using the same strategy during warmer periods of the year may result in an inferior performance for Hut 2.

To prevent this excessive temperature rise and to increase savings in energy, weather forecast data was used to predict solar radiation and outdoor temperature for the next day. Figure 3.5.10 shows the experimental result for a day, which was predicted to be sunny. As shown in the figure, despite the low electricity price, the controller used the high price temperature range (and maintained the temperature between 17°C and 19°C) through the night to prevent charging of the PCM. During the day, the high solar radiation raised the room temperature sharply in Hut 1 to above 30°C, while the room temperature in Hut 2 remained below 24°C all day. This clearly indicates that the PCM in Hut 2 was getting charged during the day by capturing sufficient solar energy. As can be seen in the figure, the PCM temperature remained constant for almost 5 h from 12:00 to 17:00 while absorbing heat from the interior space of the hut. After sunset, the room temperature in Hut 1 dropped to 17°C and used electricity for heating from 10 pm onward while the room temperature in Hut 2 remained above 19°C until midnight using the stored energy during the day. As a result of the proposed method, solar energy can be used to charge the PCM passively and use the stored energy during the evening high price period. A significant energy saving, equal to 41%, was achieved using the proposed method.

Figure 3.5.11 shows the experimental results over an 11-day period to further investigate the proposed method. During this period, the proposed method was tested against various conditions. A total energy saving of 31%, with a corresponding 40% cost saving, was achieved, as can be seen in Table 3.5.2.

Table 3.5.3 shows the energy and price savings achieved over the period of 11 days shown in Figure 3.5.11. As shown in the table, the energy saving for most days was significant. The highest energy saving in Hut 2 was achieved during Day 7, with 90% less energy consumption than Hut 1 followed by Day 2 and Day 1 with 82% and 63% energy saving, respectively. However, during this period, Hut 2 used more energy

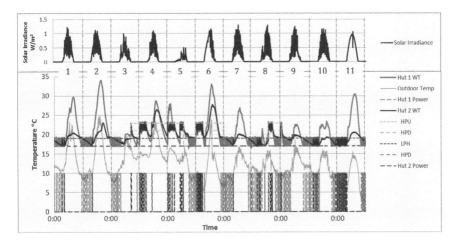

FIGURE 3.5.11   Room and wall temperatures of Huts 1 and 2: application of weather forecasting in conjunction with the price-based method over a period of 11 days.

TABLE 3.5.3

Electrical Energy and Cost Saving over a Period of 11 Days Using Weather Forecasting in Conjunction with the Price-Based Method

	Day 1	Day2	Day3	Day 4	Day 5	Day 6	Day 7	Day 8	Day 9	Day 10	Day 11	Total
Energy saving (%)	63	82	64	−73	22	−8	90	−11	53	41	37	31
Price saving (%)	47	60	71	−77	46	7	79	16	67	55	41	40

during Days 4, 6 and 8 because of inaccurate weather forecast. Despite that, the electricity cost in Hut 2 was less that Hut 1 throughout this period, except for Day 4, due to application of the price-based method. It is also interesting to see that, despite inaccurate weather forecasting, the total price and energy saving were equal to 40% and 31%, respectively, over the test period of 11 days.

Inaccurate weather predictions can result in inappropriate strategy selection, as can be seen during Day 4 and Day 6 (Figure 3.5.12). These days were predicted to be cloudy with little solar radiation, and consequently, the controller decided to charge the PCM through the night and use the stored energy during the peak period throughout the day. However, the forecast was incorrect and the high solar radiation raised the room temperature and caused overheating in both huts. During the high price period of Day 4, no heating was required in Hut 1 or Hut 2, and as a consequence, no heat was discharged from the PCM. This means that all the energy stored through the night before was not used during the same day, which resulted in 73% more energy consumption in Hut 2 than in Hut 1.

Day 6 was slightly better. Because of the cold outdoor temperature, Hut 1 needed to be heated during the high price period and at the same time some stored energy

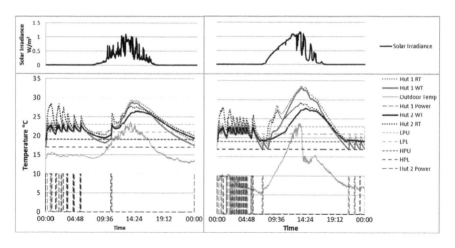

FIGURE 3.5.12   Room temperature and wall temperature of Huts 1 and 2 during Days 4 and 6: effect of inaccurate weather forecasting.

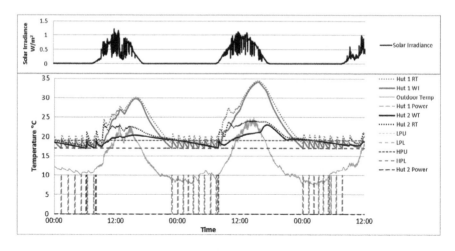

FIGURE 3.5.13   Room and wall temperatures of Huts 1 and 2 during Days 1 and 2: accurate weather forecasts.

in Hut 2 could be used to prevent a temperature drop in the hut. However, this still resulted in 8% more energy consumption in Hut 2 than in Hut 1.

Figure 3.5.13 shows the experimental results during the first 2 days where high energy saving was achieved. Mainly because of the weather forecast accuracy, the controller benefitted from the solar energy to charge the PCM and use the stored energy during the high price period. The stored energy in the PCM was sufficient to keep the room temperature above the HPL all night until the next day. As can be observed, Hut 2 used only three cycles of heating during this period, while hut 1 used more than 20 cycles. As a result of the accurate weather forecast and high solar

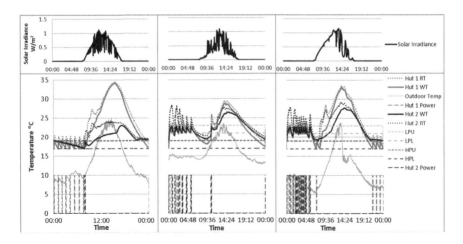

FIGURE 3.5.14   Room temperature and wall temperature of Huts 1 and 2: effect of accurate weather forecasting on performance of the control system.

irradiance, an energy saving of 63% and 82%, respectively (compared to Hut 1) was achieved during these 2 days.

Comparison of Days 2, 4 and 6 shows the significance of accurate weather forecasts, as shown in Figure 3.5.14. Looking closely at these days, it can be observed that the outdoor temperature profile and the solar irradiance are very similar. However, choosing an appropriate heating control strategy resulted in 82% energy saving in Day 2, while using an inappropriate strategy based on an inaccurate weather forecast resulted in 73% and 8% more energy consumption in Hut 2 for Days 4 and 6, respectively.

For days with low solar radiation, and if the outdoor temperature is predicted to be relatively high during the evening, the controller did not allow charging the PCM at night; instead, it uses the low price period in the afternoon to charge the PCM for a short period, as can be seen during Days 3 and 8 (Figure 3.5.15). This can ensure the availability of a sufficient amount of stored energy for use during the high price period, while minimizing the heat losses from the PCM.

The experimental results suggest that application of the price-based method in conjunction with weather forecast data can significantly improve energy storage in buildings incorporating PCM. However, due to the geographic location of Auckland City, there are some uncertainties associated with the weather forecasts, which could significantly worsen the performance of storage materials [35]. Thus, this mode of control will result in higher savings in regions with more stable weather conditions that can be forecasted more accurately.

3.5.4   CONCLUSION

Application of PCM-impregnated gypsum boards in a passive solar building was investigated experimentally. Initially, the price-based control method showed a noticeable energy saving of 14% with a corresponding cost saving of 32%. To

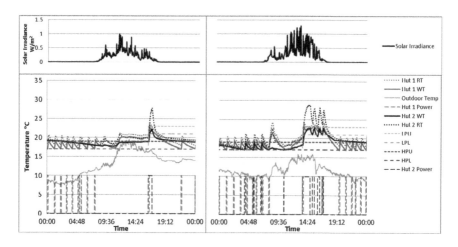

FIGURE 3.5.15   Room and wall temperatures of Huts 1 and 2: controller approach for Days 3 and 8 with low solar radiation and high outdoor temperature.

achieve further savings and to prevent overheating during days with high solar radiation, weather forecast data were incorporated into the control strategy. The results obtained showed a significant energy saving, reaching up to 90% in 1 day and 31% energy savings over-all, with corresponding 40% cost savings over an 11-day period despite inaccurate weather forecasts on some of the days. This work therefore provides strong justifications for the application of the proposed method in areas where reliable weather forecasts are possible.

ACKNOWLEDGEMENTS

The work presented in this publication was supported by MBIE of New Zealand, which has funded the project UOAX0704-CR-3 and also both the European Union and the Royal Society of New Zealand for funding the project INNOSTORAGE IRSES-610692.

ABBREVIATIONS

AC:	air conditioner
CV:	control variable
DSC:	differential scanning calorimetry
HPL:	high price lower constraint
HPU:	high price upper constraint
LHS:	latent heat storage
LHTES:	latent heat thermal energy storage
LPL:	low price lower constraint
LPU:	low price upper constraint
OP:	online price
PC:	price constraint

PCM: phase change materials
RT: room temperature
SHS: sensible heat storage
TCHS: thermochemical heat storage
TES: thermal energy storage
WT: wall temperature.

REFERENCES

1. Diakaki C, Grigoroudis E, Kolokotsa D. Towards a multi-objective optimization approach for improving energy efficiency in buildings. *Energy Build* 2008;40 (9):1747–54.
2. Lee KO, Medina MA, Raith E, Sun X. Assessing the integration of a thin phase change material (PCM) layer in a residential building wall for heat transfer reduction and management. *Appl Energy* 2015;137:699–706.
3. Soares N, Costa JJ, Gaspar AR, Santos P. Review of passive PCM latent heat thermal energy storage systems towards buildings' energy efficiency. *Energy Build* 2013;59:82–103.
4. Farid MM, Khudhair AM, Razack SAK, Al-Hallaj S. A review on phase change energy storage: materials and applications. *Energy Convers Manage* 2004;45 (9–10):1597–615.
5. Khudhair AM, Farid MM. A review on energy conservation in building applications with thermal storage by latent heat using phase change materials. *Energy Convers Manage* 2004;45(2):263–75.
6. El-Sawi A, Haghighat F, Akbari H. Centralized latent heat thermal energy storage system: model development and validation. *Energy Build* 2013;65:260–71.
7. Dutil Y, Rousse D, Lassue S, Zalewski L, Joulin A, Virgone J, et al. Modeling phase change materials behavior in building applications: comments on material characterization and model validation. *Renew Energy* 2014;61:132–5.
8. Kuznik F, Arzamendia Lopez JP, Baillis D, Johannes K. Design of a PCM to air heat exchanger using dimensionless analysis: application to electricity peak shaving in buildings. *Energy Build* 106 (0):65-73.
9. Candanedo JA, Athienitis AK. Investigation of anticipatory control strategies in a net-zero energy solar house. *ASHRAE Trans* 116 (1):246–59.
10. Kuznik F, Virgone J. Experimental assessment of a phase change material for wall building use. *Appl Energy* 2009;86(10):2038–46.
11. Morrison DJ, Abdel-Khalik SI. Effects of phase-change energy storage on the performance of air-based and liquid-based solar heating systems. *Sol Energy* 1978;20(1):57–67.
12. Marshall R, Dietsche C. Comparisons of paraffin wax storage subsystem models using liquid heat transfer media. *Sol Energy* 1982;29(6):503–11.
13. Lacroix M. Numerical simulation of a shell-and-tube latent heat thermal energy storage unit. *Sol Energy* 1993;50(4):357–67.
14. Velraj R, Seeniraj RV, Hafner B, Faber C, Schwarzer K. Heat transfer enhancement in a latent heat storage system. *Sol Energy* 1999;65(3):171–80.
15. Zhang Y, Zhou G, Lin K, Zhang Q, Di H. Application of latent heat thermal energy storage in buildings: state-of-the-art and outlook. *Build Environ* 2007;42(6):2197–209.
16. Xiao W, Wang X, Zhang Y. Analytical optimization of interior PCM for energy storage in a lightweight passive solar room. *Appl Energy* 2009;86(10):2013–8.
17. Chen C, Guo H, Liu Y, Yue H, Wang C. A new kind of phase change material (PCM) for energy-storing wallboard. *Energy Build* 2008;40(5):882–90.
18. Shilei L, Neng Z, Guohui F. Impact of phase change wall room on indoor thermal environment in winter. *Energy Build* 2006;38(1):18–24.

19. Zhou G, Zhang Y, Wang X, Lin K, Xiao W. An assessment of mixed type PCM- gypsum and shape-stabilized PCM plates in a building for passive solar heating. *Sol Energy* 2007;81(11):1351–60.

20. Athienitis AK, Liu C, Hawes D, Banu D, Feldman D. Investigation of the thermal performance of a passive solar test-room with wall latent heat storage. *Build Environ* 1997;32(5):405–10.

21. Khudhair AM, Farid MM, Chen JJJ, Bansal PK. Thermal energy storage in buildings using PCM: computer simulation. pp. 257–62.

22. Kuznik F, Virgone J, Johannes K. In-situ study of thermal comfort enhancement in a renovated building equipped with phase change material wallboard. *Renew Energy* 2011;36(5):1458–62.

23. Bourne S, Novoselac A. Compact PCM-based thermal stores for shifting peak cooling loads. *Build Simul* 2015;8(6):673–88.

24. Mankibi ME, Stathopoulos N, Rezaï N, Zoubir A. Optimization of an Air-PCM heat exchanger and elaboration of peak power reduction strategies. *Energy Build* 2015;106:74–86.

25. Jin X, Medina MA, Zhang X. On the importance of the location of PCMs in building walls for enhanced thermal performance. *Appl Energy* 2013;106:72–8.

26. Qureshi WA, Nair NKC, Farid MM. Impact of energy storage in buildings on electricity demand side management. *Energy Convers Manage* 2011;52 (5):2110–20.

27. Barzin R, Chen JJJ, Young BR, Farid MM. Peak load shifting with energy storage and price-based control system. *Energy* 2014.

28. Barzin R, Chen JJJ, Young BR, Farid MM. Application of PCM underfloor heating in combination with PCM wallboards for space heating using price based control system. *Appl Energy* 2015;148:39–48.

29. Barzin R, Chen JJJ, Young BR, Farid MM. Application of PCM energy storage in combination with night ventilation for space cooling. *Appl Energy* 2015;158:412–21.

30. Dovrtel K, Medved S. Weather-predicted control of building free cooling system. *Appl Energy* 2011;88(9):3088–96.

31. Argiriou AA, Bellas-Velidis I, Balaras CA. Development of a neural network heating controller for solar buildings. *Neural Networks* 2000;13(7):811–20.

32. Argiriou AA, Bellas-Velidis I, Kummert M, André P. A neural network controller for hydronic heating systems of solar buildings. *Neural Networks* 2004;173:427–40.

33. Candanedo JA, Dehkordi VR, Stylianou M. Model-based predictive control of an ice storage device in a building cooling system. *Appl Energy* 2013;111:1032–45.

34. Henze GP, Felsmann C, Knabe G. Evaluation of optimal control for active and passive building thermal storage. *Int J Therm Sci* 2004;43(2):173–83.

35. Henze GP, Kalz DE, Felsmann C, Knabe G. Impact of forecasting accuracy on predictive optimal control of active and passive building thermal storage inventory. *HVAC R Res* 2004;10(2):153–78.

36. PureTemp. http://www.puretemp.com/technology_docs.

37. weather forecast data, MetService New Zealand. www.metservice.com.

3.6 Application of PCM Energy Storage in Combination with Night Ventilation for Space Cooling

Reza Barzin, John J. J. Chen, Brent R. Young, and Mohammed Farid
University of Auckland

3.6.1 INTRODUCTION

In recent years, with dramatic increases in energy consumption in developed and developing countries, the shortage of energy sources and environmental issues has become increasingly important. The building sector is responsible for almost 40% of energy consumption in Europe and 36% worldwide [1,2]. Space heating and cooling are responsible for roughly 50% of this energy consumption in buildings, which is why energy saving in space heating and cooling can significantly reduce total energy consumption and consequently would reduce greenhouse gases emissions [3].

To reduce energy consumption and to assist with the smoothing of diurnal variation in energy demand, energy-efficient buildings have gained much attention since the 1970s [4]. Incorporation of thermal energy storage (TES) into traditional buildings is not only considered an effective method of minimizing energy consumption but also it is a useful method to reduce the mismatch between energy supply and demand [5–7]. Energy can be stored in the form of sensible heat and latent heat [8]. The former depends on the mass and specific heat of the storage material. This method has been used for centuries to store/release thermal energy passively, but today it is difficult to use this method as charge and discharge induces a temperature variation and also requires a large amount of space [9]. As an alternative, latent heat storage is based on the absorption/release of energy associated with a change of state. Such latent heat storage materials can store a much larger amount of energy, as the latent heat of fusion is large compared to sensible heat. Additionally, there is only a very small temperature change during heat charge and discharge [10,11].

These days, lightweight buildings are becoming more popular due to their ease of construction and architectural flexibility for retrofitting purposes [12]. However, having a low thermal mass is the main disadvantage of these buildings. External cooling, heating and solar heating loads largely affect their interior temperature. The application of phase change material (PCM) can significantly reduce temperature fluctuation and consequently reduce energy consumption in these buildings [9]. PCM stores 5–14 times more heat per unit volume than sensible heat-based storage materials such as water, masonry or rock [13]. Several researchers have investigated the possibility of increasing the storage capacity of lightweight buildings using energy storage materials including PCMs. However, this is not limited to lightweight construction as Ascione et al. [10] have applied the concept in a well-insulated heavyweight building in Europe and showed improvement in the thermal behavior of the building.

Traditionally, nocturnal ventilative cooling or night ventilation has been used to dissipate heat from the building structure using convective heat loss, by allowing the outdoor cooler air to pass through the building at night. In this method, the building mass is cooled down during the night and in fact building structures act as a heat storage medium, with the stored energy being used to prevent overheating the buildings the following day [14]. This can significantly improve thermal comfort without increasing electricity demand [15]. Previous monitoring studies on real buildings concluded that the use of night ventilation in buildings can reduce 20%–25% of their air-conditioning demand or, when air-conditioning is not used, it can reduce peak indoor temperature by up to 3°C [16–18]. However, such benefits cannot be seen in lightweight buildings because of their low thermal mass.

For lightweight buildings however, "free cooling" in which air available at a lower temperature at night may be used. It could also be used to store the coolness through external heat storage. The main difference between night ventilation and "free cooling" is their storage components [19]. As mentioned earlier, nocturnal ventilation uses the building structure as the storage while in "free cooling", a storage medium is used to store the cold at night. The advantage of "free cooling" over the night ventilation is that the stored cold in the free cooling method can be extracted from the storage medium whenever it is needed [19].

The storage medium for "free cooling" can be in the form of sensible or latent heat or a combination of the two [20]. Latent heat energy storage is preferred over sensible heat storage due to its high energy storage density and isothermal characteristics during heat charge and discharge [21,22]. A large number of simulations and experimental studies have been performed on "free cooling" using PCM storage units. For instance, Takeda et al. [23] used a PCM packed bed thermal storage in a ventilation system operating using the "free cooling" concept. Results showed a significant reduction in the ventilation load (up to 62%) of the building. Similar works done on different PCM storage designs using PCM in various shaped products such as aluminum panels [24], spherically encapsulated PCM [25], PCM granules [26] and many more. All showed reductions in the cooling demand of the building. However, the relatively high cost of this method compared to its storage capacity, the fact that they do not contribute to the thermal inertia of the building, and the additional space requirements are the main disadvantages of these systems. To address this, Álvarez et al. [27] encapsulated PCM in cylindrical containers and located in an air chamber of the building for "free cooling" application. De Garcia et al. [28] also used PCM in ventilated double skin façade and showed a significant improvement in the thermal behavior of that building. As a result, they were able to use an active storage system with forced convection, while using this storage in the air cavity helped to increase the thermal inertia of the building as well.

Incorporating PCM into building materials (such as wallboards, gypsum boards, and bricks) is commonly used as a passive method for its ease of application, low cost, lack of additional space requirement and provision of a large heat transfer area [29–31]. Kuznik et al. [32] experimentally investigated the application of PCM wallboards, showing a significant enhancement in the thermal comfort of the building. Many researchers investigated this method under free-floating conditions, showing a significant reduction in the peak interior temperature in summer [33–35] and a significant reduction also in the electricity needed for heating and cooling [36]. However, to fully use the storage capacity, PCM needs to be fully charged and discharged at each cycle, otherwise the full potential of PCM will not be achieved. For example, considering PCM building envelope construction, which intends to take advantage of PCM to prevent over-heating of the building during hot summer days; the amount of energy that can be absorbed during the day depends on the amount of PCM which has been solidified during the night before. If the PCM remains liquid during the night, it can only absorb heat in the form of sensible heat, which will be negligible. In other words, PCM needs to be solidified at night to be able to absorb a good amount of heat in the form of latent heat and consequently prevents over-heating of the building the following day.

Under free-floating conditions, PCM incorporated into the interior of the building may not fully solidify at night. This is because building insulation prevents energy losses and, as a result, interior temperatures stay higher than outdoor temperature throughout the night. Hence, the interior temperature might remain above PCM's freezing point most of the night, preventing PCM from solidifying. To remove heat from the building, Zhou et al. [37] used night ventilation to reduce interior temperature and consequently solidify the PCM. They performed numerical analysis on the application of shape-stabilized PCM in combination with night ventilation, resulting in an application of PCM with peak melting temperature of 26°C, which could reduce peak room temperature (RT) by 2°C. Zhou et al. [38] also numerically investigated the application of the same shape-stabilized PCM in an office environment with night ventilation and active cooling during office hours to maintain RT below 28°C. They showed that, using this technique, 76% of daytime cooling energy consumption could be saved.

To the authors' knowledge, the above-mentioned two research publications are the only published articles on the application night ventilation in combination with PCM wallboards. Also, literature shows an absence of experimental works on the use of PCM-impregnated gypsum boards in buildings combined with night ventilation.

In this study, the application of PCM-impregnated gypsum boards was experimentally investigated for cooling application using two identical test huts. One test hut was used as the control while the other was used as a retrofit hut containing PCM-impregnated gypsum boards as a storage medium. Initially, air-conditioning units were used to solidify PCM using lower-priced electricity during the night and use the stored coolness during the peak period when the electricity price is more expensive. The price-based control method used in this study was further discussed in Barzin et al. [39] and Barzin et al. [40]. In the next stage, a night ventilation system was used to solidify the PCM instead of the AC unit to increase energy saving.

3.6.2 METHODOLOGY

3.6.2.1 EXPERIMENTAL SETUP

The field tests were performed using two identical fully instrumented test huts with interior dimensions of $2.4 \times 2.4 \times 2.4$ m and designed to operate independently [40]. The test huts are located at Tamaki Campus, University of Auckland, New Zealand (Figure 3.6.1). Both huts were constructed using standard lightweight materials, were elevated 0.4 m above the ground and have north facing single glazed windows with dimensions of 1×1 m. Both huts were insulated using 0.1 m thick pink glass wool in the walls and ceiling and 0.06 m of polystyrene foam under the floor to reduce heat loss from the huts [39]. The first, hut referred to as "Hut 1," is used as the base case for all experiments throughout this study and has interior walls and ceiling finished with ordinary 13 mm thick gypsum boards. In second hut, referred to as "Hut 2," the 13 mm gypsum boards were replaced with PCM-impregnated gypsum boards. Each hut was equipped with an AC unit (installed on the south wall) to provide cooling demand as required and a power meter to monitor the electricity consumption. A 20 W ventilation fan was installed only in Hut 2 (no ventilation fan was installed in Hut 1). This is because Hut 1 didn't require any cooling at night and its RT easily drops below 17°C at night. T-type thermocouples were used for all temperature measurements inside and outside the experimental huts. Walls and PCM temperatures are measured; however, only the west wall was reported in this study. All thermocouples were calibrated against a reference thermometer (Ebro TFX430) from 0°C to 35°C.

3.6.2.2 PCM SELECTION AND IMPREGNATION

PT20 from PureTemp was used as the energy storage material in this study due to its unique characteristics. The PCM is 100% renewable, with a narrow melting range,

FIGURE 3.6.1 Experimental test huts at Tamaki Campus, University of Auckland, New Zealand.

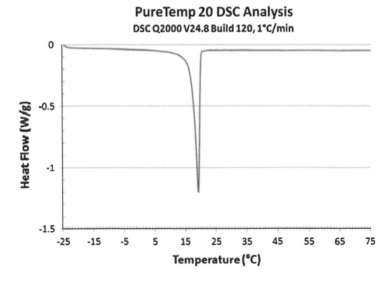

FIGURE 3.6.2 Differential scanning calorimetry (DSC) of PT20 as reported by PureTemp [41].

produced from agricultural sources. PT20 has heat storage capacity and peak melting point equal to 180 J/g and 20°C, respectively [41] (Figure 3.6.2).

Imbibing or direct immersion technique was used to impregnate gypsum boards. To impregnate the gypsum boards, two metal trays with dimensions of $0.9 \times 0.9 \times 0.15$ m and $0.7 \times 0.7 \times 0.09$ m were used, as shown in Figure 3.6.3. The large tray was filled with water while the smaller tray was left floating inside the large tray. Water inside the large tray was heated to 55°C to melt the PT20 contained in the small tray. After reaching steady temperature, the boards were immersed in the molten PCM one by one, each for a period of 10 min. Weight measurement of the gypsum boards before and after impregnation showed roughly a PCM impregnation of 22 wt%.

3.6.2.3 CONTROL SYSTEM

A code was developed using the LabVIEW software to read electricity prices from any given website when its address is provided. As a result, this code can be used at any country to read online electricity prices from any website. In this study, wholesale electricity prices for Auckland city region (obtained from http://www.electricityinfo.co.nz) were used. The control system uses this information to monitor the electricity price and uses electricity as long as its price is low. As soon as the electricity price peaks, the control unit stops using electricity for cooling in the building as long as the RT is below the temperature constraint.

A CompactRIO was used as the control system in each hut to acquire all the necessary data, process them and send corrective commands to the AC unit or ventilation fan. As shown in Figure 3.6.4, the CompactRIO was programmed to receive price constraint (PC), online price (OP), solar radiation, outdoor temperature, RT, power consumption of the AC unit and ventilation system. The embedded controller

FIGURE 3.6.3 Impregnation of PT 20 into the gypsum board using the imbibing method.

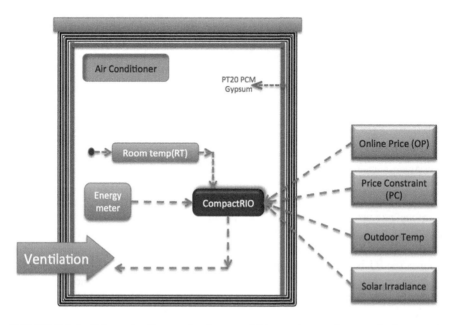

FIGURE 3.6.4 Schematic diagram for data acquisition and control system in the experimental huts.

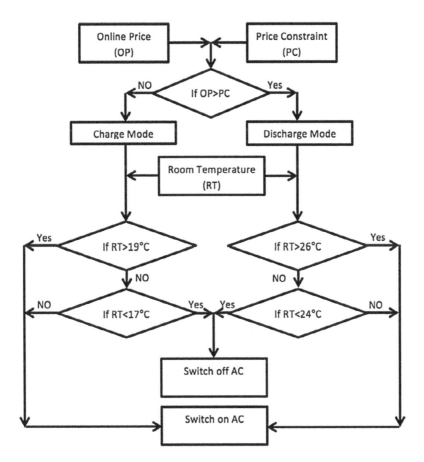

FIGURE 3.6.5   Flow chart of the control system to operate AC unit.

in CompactRIO receives all this information, sends signals to the remote control of the AC unit, and the remote control sends the signal to the heat pump to maintain the room within the desired temperature range. The control unit also controls operation period of the ventilation fan using an electric switch.

Throughout this study, the control system operates the AC unit based on the flowchart shown in Figure 3.6.5, which shows that the operation mode depends on the relative magnitudes of OP and PC. If the OP is smaller than the PC, the electricity price is considered low and the controller proceeds to the "charge mode". In the charge mode, the RT needs to be maintained within the "low peak temperature range", between 17°C and 19°C; which are referred to as the "low price lower constraint" or LPL and "low price upper constraint" or LPU, respectively. This temperature range was selected to solidify PCM during the charge period. Choosing the charge mode, the controller receives the RT data and uses the AC unit to keep the RT within the low peak temperature range. If the OP rises above the PC, the controller switches to the "Discharge mode" in which the controller uses a higher temperature range of 24°C and 26°C, referred to as "high price temperature range" to reduce cooling load

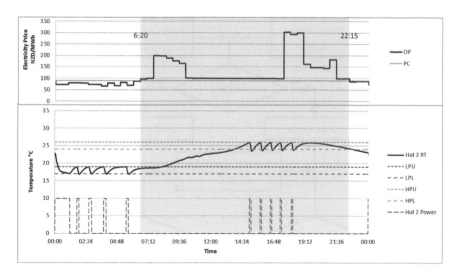

FIGURE 3.6.6 Application of price-based control method using low peak and high peak temperature range.

and also allows the encapsulated PCM to absorb heat during this period. The 24°C and 26°C are referred to as "high price lower constraint" or HPL and "high price upper constraint", or HPU, respectively.

Figure 3.6.6 presents an example to further clarify the price-based method. A price constraint equal to 90 NZD/MWh was set for this experiment. During the night and early morning period until 6:20, electricity price was below 90 NZD/MWh (Low Price Period). As a result, the RT was maintained within the range of 17°C–19°C, as can be observed in the figure. As soon as the electricity price rises above the price constraint, the control system changes the temperature set point to the high price temperature range of 24°C–26°C to use less energy until the electricity price drops below the temperature constraint once again at 22:15. The high price period is shown by the gray-shaded area in the figure.

3.6.2.4 NIGHT VENTILATION

To provide night ventilation, a 20 W fan was installed on a 0.25 × 0.25 m opening at the bottom of the entrance door of Hut 2 to blow outdoor air into the hut (with nominal flow rate of 300 m³/h) during the night period as required. The control system automatically operates the ventilation fan and outside the ventilation hours it is fully closed. The ventilation system operates solely during night period between 9 pm and 7 am.

3.6.3 RESULTS

3.6.3.1 APPLICATION OF AC UNIT

In the first part of the study, an AC unit was used in each hut to supply the cooling demand. As shown in Figure 3.6.7, Hut 1 does not require any cooling at night due

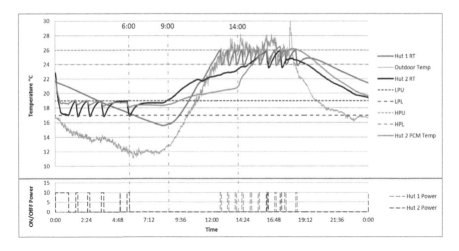

FIGURE 3.6.7 Thermal behavior of the impregnated PCM during a charge and discharge period.

to its low thermal mass and low outdoor temperature. As can be observed, the RT for Hut 1 easily falls below 17°C at night without using any additional cooling. This means further cooling is not required. However, due to the presence of the PCM, the RT in Hut 2 remains above 19°C and needs to be cooled down using the AC unit to charge the impregnated PCM.

To have a better understanding of the charge and discharge process of the PCM-impregnated gypsum boards, both PCM and RT are presented in Figure 3.6.7 over a 24 h period. As can be observed, the RT was maintained within the low price temperature range, and despite cycling of the RT between 19°C and 17°C, the PCM temperature remained at 19°C for more than 5 h. At 6:00, the PCM temperature dropped below 19°C, which indicates that the PCM has partially solidified (charged). At 9:00, the RT rose above 19°C due to solar radiations and quickly reached 22°C at around 11:30 while the PCM temperature remained relatively constant below 21°C until 14:00 when the PCM was fully discharged. Shortly after the discharge, the PCM temperature reached the RT and consequently the RT rose faster and reached the temperature constraint of 26°C. During charge and discharge periods, PCM temperature remained fairly constant at 19°C and 21°C.

Figure 3.6.8 shows that the RT in Hut 2 was maintained within the low price temperature range of 17°C–19°C (using the AC unit) to solidify the PCM for the next day while Hut 1 did not require any cooling. During the day, the RT in Hut 1 increased sharply due to solar radiation and reached the upper temperature constraint of 26°C at 13:03, whereas in Hut 2, the RT did not reach the upper constraint until 16:04; almost 3 h after Hut 1, consuming 60% less energy during the peak period due to the stored coolness in the PCM-impregnated gypsum boards. However, as a result of not using any electricity for cooling Hut 1 during night, Hut 1 consumed an overall 15% less electrical energy compared to Hut 2. Despite this negative impact of the PCM, a cost saving of 5% was achieved in Hut 2 through the use of electricity during the low

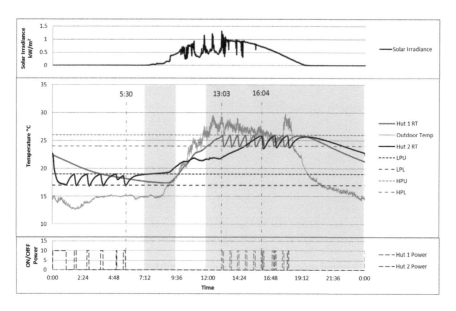

FIGURE 3.6.8   Thermal performance of Hut 2 in comparison with Hut 1 using AC unit for cooling, during a summer day with high solar radiation.

peak period. This can be even less energy saving in the case of rainy days or days with low solar radiation as shown in Figure 3.6.9.

Figure 3.6.9 shows the RT in both huts over a 6-day period. During sunny days, as shown earlier in Figure 3.6.8, Hut 2 uses slightly more electricity compared to Hut 1. Hut 2 uses even more electricity in case of rainy days as, because of the limited solar radiation and low outdoor temperature, no cooling is required through the day, and as a result, all stored energy is wasted through heat loss to the environment. This can be seen during days 3 and 4 in Figure 3.6.9 where Hut 2 used AC through the night, and the following day, no cooling was required and all stored energy was wasted during the day. During these 2 days, Hut 1 used no electrical energy while Hut 2 used AC for almost 5 h per night. A similar performance can be seen for days with low solar radiation as can be observed during Day 5, where Hut 2 used almost nine times more electricity compared to Hut 1 for similar reasons. This is because that the stored energy at night was more than what is required energy for the next day. As presented in Table 3.6.1, application of AC to charge the PCM at night can be very risky: without accurate weather forecast information, this can result in higher electricity consumption rather than energy saving. Electricity cost saving was only achieved for Days 1, 2 and 6, when high solar radiation resulted in full discharge of the PCM. During Days 3, 4 and 5, Hut 1 used AC for a few minutes while Hut 2 used more than 14 h. During 6 days, Hut 2 consumed twice the electrical energy and increased electricity cost by 30%, as shown in Table 3.6.1.

The negative numbers presented in the table shows that Hut 2 used more energy than Hut 1. For instance, 890% during Day 5 shows that Hut 2 has used almost nine times more energy compared to Hut 1. The abovementioned negative impact of using PCM in combination with night air-conditioning encouraged us to test the idea of

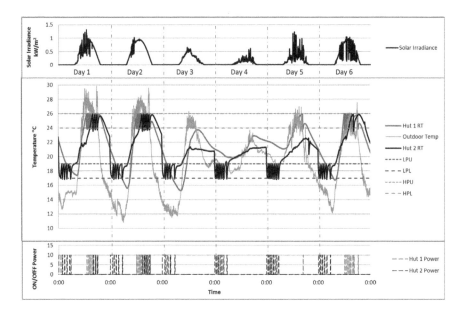

FIGURE 3.6.9 Room-temperature comparison between Huts 1 and 2 in a period of 6 days using an AC unit for cooling.

TABLE 3.6.1

Energy and Cost Saving/Losses Achieved through the Use of AC to Charge PCM during Low Peak Period

	Day 1(%)	Day 2(%)	Day 3(%)	Day 4(%)	Day 5(%)	Day 6(%)	Total (%)
Energy saving	−21	−6	–	–	−890	−6	−100
Cost saving	5	22	–	–	−631	35	−30

using night cooling as shown in Figure 3.6.10. Figure 3.6.10 shows that during the first day, Hut 2 was charged during the night period and as a consequence Hut 2 did not need to use any cooling during the day until 16:17, almost four hours later than Hut 1. The next day however, the PCM in Hut 2 was not charged using AC at night and, as a result, both huts performed almost identically during the day. This shows that, when not charging the PCM, both huts perform very similar. One concludes that the observed difference during the first day was solely due to the stored energy in the PCM and no thermal mass difference was involved. The experiment also shows that, despite a low outdoor temperature of below 15°C, the RT did not drop below 20 and as a result no significant charging of PCM was achieved.

As demonstrated by numerous research studies, successful application of PCM in buildings can produce savings in energy and electricity costs. However, the above analysis showed that the improper use of PCM could lead to higher power

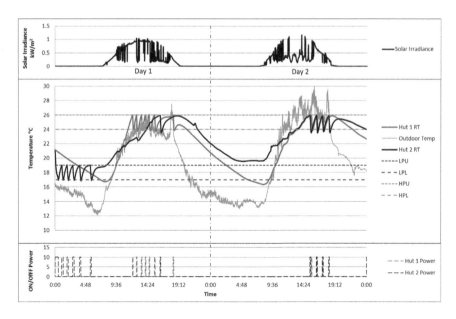

FIGURE 3.6.10 Performance of Hut 2 in comparison with Hut 1: effect of charging of the PCM.

consumption instead. It can be observed in Figures 3.6.11 that Hut 2 was not charged, and as a result, the RT in Hut 2 stayed above 20°C at all times. This means that the PCM in Hut 2 was never solidified during this period; as a consequence, it could only store energy in the form of sensible heat which is very limited. Experimental results showed a 3% higher power consumption in Hut 2 compared to Hut 1 during this period due to higher RT in Hut 2 before sunrise.

3.6.3.2 APPLICATION OF NIGHT VENTILATION IN COMBINATION WITH AC IN HUT 2

As shown in Figure 3.6.9, using an AC unit through the night to charge the PCM can cause an increase in electricity consumption in Hut 2. On the other hand, not using any cooling at night to charge the PCM can lead to similar problem, as demonstrated in Figure 3.6.11. However, night ventilation should provide a significant benefit if night air temperature is lower than the PCM melting point. For instance, in Figure 3.6.11, during the whole 4-day period, out-door temperature at night was lower than the melting point of the PCM. To use night ventilation, a 20 W fan was used in Hut 2 to blow fresh air into the room during night period. Figure 3.6.12 shows the experimental results when night ventilation was used in Hut 2 during the night period. As can be observed, due to low outdoor temperature at night, the application of night ventilation helped to decrease the RT and the PCM temperature. As shown in the figure, the PCM temperature dropped as a result of night ventilation until it reached 19°C and plateaued for almost 4 h until it got fully charged. During the next day, the RT in Hut 1 rose sharply due to high solar radiation and reached

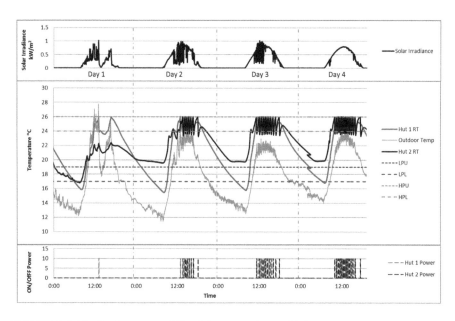

FIGURE 3.6.11 Performance of Hut 2 in comparison with Hut 1 over a period of 4 days without using any cooling at night (not using AC or night ventilation).

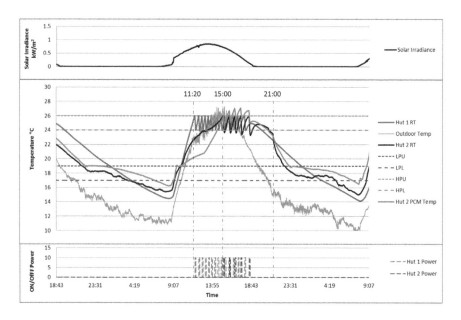

FIGURE 3.6.12 Performance of Hut 2 in comparison with Hut 1 using night ventilation in Hut 2.

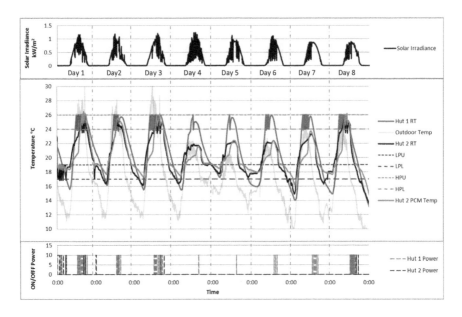

FIGURE 3.6.13 Hut experiment using night ventilation over a period of 8 days.

26°C at 11:20. However as a result of successful implementation of night ventilation, Hut 2 remained at an acceptable level without using electricity for air-conditioning until 15:00, 3 h and 40 min longer than Hut 1, which resulted in 66% energy saving in Hut 2 compared to Hut 1. The night ventilation system was activated again at 21:00 as shown in the figure, causing a sudden temperature drop in the RT and PCM temperature to charge the PCM for the next day. The main reason for this improvement was utilization of the free cooling by using outside fresh air having sufficiently low temperature to solidify the PCM and store enough energy for the next day.

Figure 3.6.13 presents a period of 8 days in which AC was replaced with night ventilation in Hut 2. As shown in the figure, AC was used to charge the PCM in Hut 2 only during day 1 to ease the visual comparison between the use of AC and night ventilation. During Day 1, Hut 2 consumed 0.5% more electricity compared to Hut 1; however a cost saving of 44% was achieved in Hut 2 as a result of using the price-based control strategy. However, the application of night ventilation instead of the AC unit resulted in an impressive total electricity saving of 73%, with corresponding electricity cost saving of 67% over the period of a week (Days 2–8).

The control system is programmed to use the AC unit if the next day is forecast to be sunny and the outdoor temperature is not low enough to charge the PCM. As can be observed in the early hours of Day 2, the AC unit in Hut 2 was turned on for a short period to assist the ventilation system, reducing the RT and assisting charging of the PCM.

Application of weather forecasts in the control system plays a critical role in electricity saving as can be observed on Days 4, 5 and 6 shown in Figure 3.6.13. As shown in the figure, during early hours of these days, both the room and outdoor temperatures were high during the charging period. Not using weather forecasts could lead to application of AC to charge the PCM for any possible high temperature during the next day. However, weather forecast data suggested very little cooling demand in the following day, and consequently, the controller decided not to use the AC unit. Not using this information in the control strategy could easily lead to higher electricity consumption in Hut 2 compared to Hut 1 during these 3 days.

The best results can be observed when the outdoor temperature is low enough to bring the PCM temperature below its melting point and ensure the PCM is fully charged for use for the following day. During Days 7 and 8, when PCM has been fully charged due to low outdoor temperature in the previous night, high energy savings of 92% and 76%, respectively, were achieved. On Day 7, PCM temperature reached 17°C at night and, as a consequence, no heating was required during the next day in Hut 2, while AC was used in Hut 1 for 3 h and 30 min. Full charge was also achieved on Day 8 as PCM temperature reached 17.5°C at night and the RT in Hut 2 remained below 26°C, the following day, for almost 4 h longer than Hut 1, achieving 73% energy saving as detailed in Table 3.6.2.

As shown in Table 3.6.2, Hut 2 did not need to use AC for four consecutive days and solely used electricity to operate the 20 W fan. The lowest energy saving achieved during this period was 48% on Day 5 due to the very low cooling demand in Hut 1.

3.6.4 CONCLUSION

Application of PCM-impregnated gypsum boards in lightweight buildings was experimentally tested using two identical test huts at Tamaki Campus, University of Auckland. The experimental results showed that if PCM is not used with proper control strategy, it may lead to an increase in the air-conditioning energy required. For the PCM to work efficiently, a combination of "night ventilation" and "free cooling" method should be applied in which low temperature outdoor air at night is used to charge the PCM inside the building. Using this method to charge the PCM with coolness instead of using AC has significantly improved energy saving. A combination of the proposed method with weather forecast data resulted in an impressive electricity saving of 73% over a 1-week period. This method of energy storage is very suitable for office application where comfort temperature need to be maintained during office hours. Night ventilation could create inconvenience for occupants if applied in residential buildings.

TABLE 3.6.2

Energy and Electricity Cost Saving Achieved Using Night Ventilation System

	Day 2 (%)	Day 3 (%)	Day 4 (%)	Day 5 (%)	Day 6 (%)	Day 7 (%)	Day 8 (%)	Total (%)
Energy saving	57	64	55	48	83	92	76	73
Electricity cost saving	67	34	58	52	85	93	73	67

ACKNOWLEDGEMENTS

The work presented in this publication was supported by MBIE of New Zealand, which has funded the project UOAX0704-CR-3 and also both the European Union and the Royal Society of New Zealand for funding the project INNOSTORAGE IRSES-610692.

REFERENCES

1. Diarce G, Campos-Celador Á, Martin K, Urresti A, García-Romero A, Sala JM. A comparative study of the CFD modeling of a ventilated active façade including phase change materials. *Appl Energy* 2014;126(0):307–17.
2. Barreneche C, de Gracia A, Serrano S, Elena Navarro M, Borreguero AM, Inés Fernández A, et al. Comparison of three different devices available in Spain to test thermal properties of building materials including phase change materials. *Appl Energy* 2013;109(0):421–7.
3. Lee KO, Medina MA, Raith E, Sun X. Assessing the integration of a thin phase change material (PCM) layer in a residential building wall for heat transfer reduction and management. *Appl. Energy.* 137, 699–706.
4. Paksoy HÖ, Beyhan B. 22-Thermal energy storage (TES) systems for greenhouse technology. In: Cabeza LF, editor. *Advances in Thermal Energy Storage Systems*. Woodhead Publishing; Cambridge, UK, 2015. pp. 533–48.
5. Zhou D, Shire GSF, Tian Y. Parametric analysis of influencing factors in Phase Change Material Wallboard (PCMW). *Appl Energy* 2014;119(0):33–42.
6. Cabeza LF, Martorell I, Miró L, Fernández AI, Barreneche C. 1-Introduction to thermal energy storage (TES) systems. In: Cabeza LF, editor. *Advances in Thermal Energy Storage Systems*. Woodhead Publishing; Cambridge, UK, 2015. pp. 1–28.
7. Stritih U, Osterman E, Evliya H, Butala V, Paksoy H. Exploiting solar energy potential through thermal energy storage in Slovenia and Turkey. *Renew Sustain Energy Rev* 2013;25(0):442–61.
8. Bas çetinçelik A, Öztürk HH, Paksoy HÖ, Demirel Y. Energetic and exergetic efficiency of latent heat storage system for greenhouse heating. *Renew Energy* 1999;16(1–4):691–4.
9. Kuznik F, Virgone J. Experimental assessment of a phase change material for wall building use. *Appl Energy* 2009;86(10):2038–46.
10. Ascione F, Bianco N, De Masi RF, de'Rossi F, Vanoli GP. Energy refurbishment of existing buildings through the use of phase change materials: energy savings and indoor comfort in the cooling season. *Appl Energy* 2014;113(0):990–1007.
11. Cabeza LF, Castell A, Barreneche C, De Gracia A, Fernández AI. Materials used as PCM in thermal energy storage in buildings: a review. *Renew Sustain Energy Rev* 2011;15(3):1675–95.
12. Hoes P, Trcka M, Hensen JLM, Hoekstra Bonnema B. Investigating the potential of a novel low-energy house concept with hybrid adaptable thermal storage. *Energy Convers Manage* 2011;52(6):2442–7.
13. Sharma A, Chen CR, Vu Lan N. Solar-energy drying systems: a review. *Renew Sustain Energy Rev* 2009;13(6–7):1185–210.
14. Ramponi R, Gaetani I, Angelotti A. Influence of the urban environment on the effectiveness of natural night-ventilation of an office building. *Energy Build* 2014;78(0):25–34.
15. Pfafferott J, Herkel S, Wambsganß M. Design, monitoring and evaluation of a low energy office building with passive cooling by night ventilation. *Energy Build* 2004;36(5):455–65.

16. Kolokotroni M, Aronis A. Cooling-energy reduction in air-conditioned offices by using night ventilation. *Appl Energy* 1999;63(4):241–53.

17. Kubota T, Chyee DTH, Ahmad S. The effects of night ventilation technique on indoor thermal environment for residential buildings in hot-humid climate of Malaysia. *Energy Build* 2009;41(8):829–39.

18. Santamouris M, Kolokotsa D. Passive cooling dissipation techniques for buildings and other structures: the state of the art. *Energy Build* 2013;57(0):74–94.

19. Waqas A, Ud Din Z. Phase change material (PCM) storage for free cooling of buildings—a review. *Renew Sustain Energy Rev* 2013;18(0):607–25.

20. Sharma A, Tyagi VV, Chen CR, Buddhi D. Review on thermal energy storage with phase change materials and applications. *Renew Sustain Energy Rev* 2009;13(2):318–45.

21. Zhou D, Zhao CY, Tian Y. Review on thermal energy storage with phase change materials (PCMs) in building applications. *Appl Energy* 2012;92:593–605.

22. Konuklu Y, Unal M, Paksoy HO. Microencapsulation of caprylic acid with different wall materials as phase change material for thermal energy storage. *Sol Energy Mater Sol Cells* 2014;120(Part B):536–42.

23. Takeda S, Nagano K, Mochida T, Shimakura K. Development of a ventilation system utilizing thermal energy storage for granules containing phase change material. *Sol Energy* 2004;77(3):329–38.

24. Lazaro A, Dolado P, Marin JM, Zalba B. PCM-air heat exchangers for free- cooling applications in buildings: empirical model and application to design. *Energy Convers Manage* 2009;50(3):444–9.

25. Arkar C, Medved S. Free cooling of a building using PCM heat storage integrated into the ventilation system. *Sol Energy* 2007;81(9):1078–87.

26. Nagano K, Takeda S, Mochida T, Shimakura K, Nakamura T. Study of a floor supply air conditioning system using granular phase change material to augment building mass thermal storage—heat response in small scale experiments. *Energy Build* 2006;38(5):436–46.

27. Álvarez S, Cabeza LF, Ruiz-Pardo A, Castell A, Tenorio JA. Building integration of PCM for natural cooling of buildings. *Appl Energy* 2013;109:514–22.

28. de Gracia A, Navarro L, Castell A, Ruiz-Pardo Á, Alvárez S, Cabeza LF. Experimental study of a ventilated facade with PCM during winter period. *Energy Build* 2013;58:324–32.

29. Biswas K, Lu J, Soroushian P, Shrestha S. Combined experimental and numerical evaluation of a prototype nano-PCM enhanced wallboard. *Appl Energy* 2014;131:517–29.

30. Kuznik F, Johannes K, David D. 13-Integrating phase change materials (PCMs) in thermal energy storage systems for buildings. In: Cabeza LF, editor. *Advances in Thermal Energy Storage Systems*. Woodhead Publishing; Cambridge, UK, 2015. pp. 325–53.

31. Kuznik F, David D, Johannes K, Roux J-J. A review on phase change materials integrated in building walls. *Renew Sustain Energy Rev* 2011;15(1):379–91.

32. Kuznik F, Virgone J, Johannes K. In-situ study of thermal comfort enhancement in a renovated building equipped with phase change material wallboard. *Renew Energy* 2011;36(5):1458–62.

33. Khudhair AM, Farid MM. A review on energy conservation in building applications with thermal storage by latent heat using phase change materials. *Energy Convers Manage* 2004;45(2):263–75.

34. Castell A, Martorell I, Medrano M, Pérez G, Cabeza LF. Experimental study of using PCM in brick constructive solutions for passive cooling. *Energy Build* 2010;42(4):534–40.

35. Osterman E, Tyagi VV, Butala V, Rahim NA, Stritih U. Review of PCM based cooling technologies for buildings. *Energy Build* 2012;49(0):37–49.

36. Qureshi WA, Nair N-KC, Farid MM. Impact of energy storage in buildings on electricity demand side management. *Energy Convers Manage* 2011;52 (5):2110–20.

37. Zhou G, Yang Y, Wang X, Zhou S. Numerical analysis of effect of shape- stabilized phase change material plates in a building combined with night ventilation. *Appl Energy* 2009;86(1):52–9.

38. Zhou G, Yang Y, Xu H. Energy performance of a hybrid space-cooling system in an office building using SSPCM thermal storage and night ventilation. *Sol Energy* 2011;85(3):477–85.

39. Barzin R, Chen JJJ, Young BR, Farid MM. Application of PCM underfloor heating in combination with PCM wallboards for space heating using price based control system. *Appl Energy* 2015;148:39–48.

40. Barzin R, Chen JJJ, Young BR, Farid MM. Peak load shifting with energy storage and price-based control system. *Energy* 92:505–14.

41. PureTemp. <http://www.puretemp.com/technology_docs>.

3.7 Application of PCM Underfloor Heating in Combination with PCM Wallboards for Space Heating Using Price-Based Control System

Reza Barzin, John J. J. Chen, Brent R. Young, and Mohammed Farid
University of Auckland

3.7.1 INTRODUCTION

High energy demand, and excessive consumption of natural resources, has become one of the most critical global challenges. Energy consumption in buildings is responsible for 40% of global energy in the European Union, contributing to production of up to 35% of greenhouse gases [1]. It is expected that between 2008 and 2030, the world energy demand will increase by about 50% [2]. More than 30% of this energy consumption is caused by heating, cooling and ventilation systems demonstrating the huge potential for improving energy efficiency of buildings [3]. This can be done through modifying the construction systems and using new technologies such as thermal energy storage (TES) for more efficient energy usage in buildings. Latent heat thermal energy storage (LHTES) is preferred for its high energy storage density [4]. Many researchers have suggested the use of phase change materials (PCMs) in buildings to improve their performance [5].

Active and passive storage systems are two types of TES systems that can be used in buildings. Active systems, such as ice storage [6], require an additional fluid loop to charge and discharge the storage tank and to deliver cooling to the existing chilled water loop. PCM also can be used in a storage tank and uses an additional liquid to charge and discharge the PCM. In one example, Martin et al. [7] used direct contact PCM water storage and managed to obtain between 30 and 80 kW/m³ of storage. Solar-assisted PCM storage tanks are also another example of active storage systems which can be used for heating application, as demonstrated through modeling and experimental study [8–10] (similar works can be found in [11–13]). Passive systems, however, do not require any additional heat exchanger to extract heat or cold from the storage [14,15]. Passive TES systems may use the thermal mass of the building to store energy in the form of sensible heat [16,17] or may use PCM in building envelopes to store energy in the form of latent heat, with the objective of shifting and reducing peak cooling loads[18]. PCM can be incorporated into various building components such as walls [19], floors [20], and window glazing [21].

Impregnation of PCM into construction materials has become a popular area of research in recent decades. This is mainly because walls and floors offer large heat transfer areas within the building, enabling a high heat transfer rate [22]. PCM wallboard [23–25], bricks [26], concrete [27–29], and PCM-impregnated gypsum boards [30,31] are among the common construction elements used for passive storage of energy.

In 2005, a German research group used microencapsulated PCM (peak melting point of 25°C) in two similar rooms of a full-size building. Results showed that PCM had a very little effect on the room temperature (RT) because of the high melting point of the PCM chosen. The study also suggested that an adequate ventilations system was required for the solidification of the PCM during the night [32]. Voelker et al. (2008) used microencapsulated PCM (melting range between 25°C and 28°C) in gypsum boards. The boards were used in two identical lightweight chambers. PCM-modified gypsum board reduced the temperature fluctuation in the chamber by 3°C. However, because of the high melting point of the PCM, the RT increased to 35°C, well above the comfort level [33]. Numerous experimental studies using microencapsulation or gypsum board impregnated with PCM have shown similar results [34].

In [35], the ENERGAIN PCM board (DuPont) was used, which contained 60 wt% PCM compared to 25%–30% in the case of gypsum boards. Results showed that using ENERGAIN PCM boards reduced the RTs by up to 4.2°C compared to a room with normal gypsum board. Further studies focusing mainly on characterization of ENERGAIN boards can be found in [36,37].

The main focus in all these studies has been to increase the thermal mass of lightweight buildings. There has been very little work published on the application of PCM wallboards for peak load shifting application. Qureshi et al. [38] used the same facilities built by [34] consisting of office-size construction containing gypsum board impregnated by RT21 as PCM to facilitate peak load shifting. PCM gypsum wallboards (RT21 used as the PCM) were used to store heat and use it for peak load shifting. The results obtained showed a significant reduction in electricity consumption during peak periods and allowed a total of 31% less energy consumption during a period of 12 days [38].

Unlike the application of PCM in walls, there is limited literature regarding the application of PCM in underfloor heating systems. More than a decade ago, Farid and Kong used a salt hydrate PCM (peak melting point about 28°C) in concrete slabs to reduce temperature fluctuations of the slab surface. The application of PCM was shown to significantly reduce temperature fluctuations of the slab surface, indicating the potential future use of this method for peak load shifting applications. However, the study was done on a small concrete slab and did not explore what the effect of the proposed method would be in a real building. Later in 2005, a Chinese research group used a high melting point PCM in an underfloor heating system (with phase transition temperature of 52°C). The results showed that enough energy can be stored in the PCM for the following day. However, the reported floor temperature in the study was above 40°C, which is too high for human comfort. According to the European standard, floor temperature should not exceed 35°C [39].

Despite a large number of reported studies on the application of PCM in building materials, very little work has been reported on the use of PCM (passively) for price-based peak load shifting. This article presents the application of PCM wallboards in combination with a PCM underfloor heating system to perform a successful peak load shifting. The research uses a price-based peak load shifting scheme to demonstrate how application of PCM can significantly improve both energy and cost savings.

3.7.2 METHODOLOGY

3.7.2.1 PRICE-BASED CONTROL

In the proposed price-based method, the electricity provider monitors electricity load and, as soon as the electricity consumption peaks, the electricity provider increases the unit price. Then users who are using the price-based control system make a decision whether to continue using electricity and pay the higher price, or turn it off. A program was developed using LabVIEW software [40] to read the electricity price from a website when the website address is provided. In this study, wholesale electricity prices for the Auckland city region (obtained from http://www.electricityinfo.co.nz/) were used as the electricity price variable.

To mimic the real situation, both price and price constraints (PCs) were fed directly to the control system remotely from the University of Auckland city campus, 12 km away from the experimental huts located at Tamaki Campus of the University.

3.7.2.2 EXPERIMENTAL SETUP

The proposed method for space heating applications was tested using two identical test huts, Hut 1 and Hut 2, situated at the University of Auckland Tamaki Campus. Both huts were constructed using standard lightweight materials, were elevated above the ground, and had a north-facing window (Figure 3.7.1). Hut 1 was used as the reference in this study. Its interior walls and ceiling were finished with ordinary 13 mm thick gypsum board. It is referred to as Hut 1 or the "reference hut" and is used as the base-case in this study. These experimental Huts are similar to those used by [34,38], with the exception of the way PCM and the heating is done and controlled, as described in the following section.

3.7.2.2.1 Underfloor Heating System

In the first part of this study, Hut 2's interior walls and ceiling were finished with gypsum wallboards, similar to Hut 1. However, the floor was covered with 10 mm PCM-impregnated gypsum board used in combination with an underfloor heating

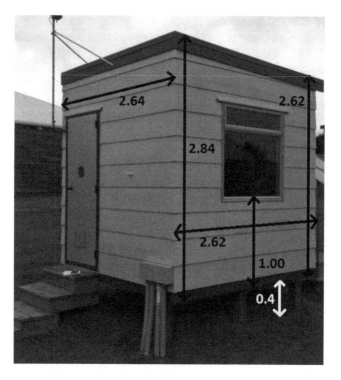

FIGURE 3.7.1   Experimental hut at Tamaki Campus, University of Auckland.

system. A paraffin-based PCM was provided specially to this project by Shell in Malaysia (peak melting point of 28°C and latent heat storage capacity of 120 J/g) was used. Five Devi™ electric foil heaters were used in the hut as the underfloor heating element, as shown in Figure 3.7.2a. Each foil had a heated area of approximately 32 cm × 240 cm with a power rating of 145 W. The whole unit has a total power rating of 725 W. The layout of the underfloor heating system is presented in Figure 3.7.2b.

The recommended floor surface temperature depends upon the surface material and on whether the occupants are wearing shoes or walking barefoot. According to ASHARE standards, a floor temperature between 19°C and 29°C is recommended in the occupied zone (giving less than 10% discomfort) [41]. In the European standard, it is acceptable to use 35°C outside the occupied zone. This study uses the recommended floor temperature range of 19°C–29°C for day time and 34°C for the midnight period when occupants are sleeping, to be able to both charge the PCM and maintain the RT within the set point range, as described in Table 3.7.1.

3.7.2.2.2   PCM Underfloor Heating System in Combination with PCM Wallboard

In the second part of the study, the 13 mm gypsum boards in Hut 2 walls were replaced with 10 mm gypsum board plus a layer of DuPont Energain® wall board.

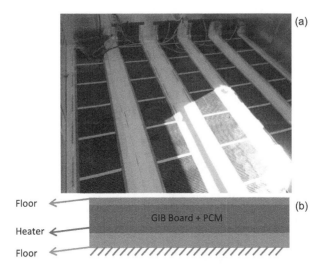

FIGURE 3.7.2 The underfloor heating system used in Hut 2. (a) Devi foil heaters and (b) underfloor heating system layout.

TABLE 3.7.1
Temperature Set Point Ranges during High and Low Price Periods

	Lower Temperature Constraint (°C)	Upper Temperature Constraint (°C)
High price period	17	19
Low price period	21	23

The DuPont Energain® sheets had a thickness of 5.2 mm; the PCM contained in it had a peak melting point of 21.7°C and a latent heat storage capacity of 70 kJ/kg [36].

3.7.2.3 Data Acquisition and Control

As illustrated in Figure 3.7.3, a CompactRIO unit was programmed to receive RT, outdoor temperature, solar radiation, power consumption of the heating unit, online electricity price (OP) and PC. PC needs to be obtained from the electricity provider. But since this information is not yet available, a PC was selected based on previous trends for this study. As shown in the figure, the heat pump (AC) was used to supply heat to Hut 1 while underfloor heating elements were used to heat Hut 2.

As shown in Figure 3.7.4, the control mode used was dependent on the relative magnitudes of OP and PC. If OP is smaller than PC, the controller proceeds to "Charge mode" and RT is kept within the "low peak temperature" range of 21°C and 23°C. These

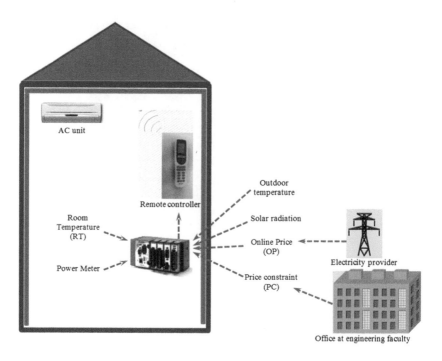

FIGURE 3.7.3 Data acquisition and control system in the experimental huts.

are referred to as the "low price lower constraint" and "low price upper constraint", respectively. As mentioned earlier, DuPont wallboards have a peak melting point of 21.7°C. Keeping the room within the low peak temperature range enables charging the wallboards during the low peak period. In the next step, the control unit receives RT data and uses the heating unit to keep the RT within the range of 21°C–23°C. As soon as OP rises above PC, the controller switches to "Discharge mode". This implies a lower temperature range, the "high price temperature range", of 17°C–19°C to use less energy and capture the stored energy in the wallboards during this period. The 17°C and 19°C temperatures are referred to as "high price lower constraint" and "high price upper constraint", respectively. Temperature constraint values are presented in Table 3.7.1.

RT was measured using T-type thermocouples connected to a 16-channel thermocouple input module (NI 9214) enabling measurement sensitivity up to 0.02°C and measurement accuracy up to ±0.45°C. All thermocouples were calibrated against a reference thermometer (Ebro TFX430) from 0°C to 35°C. Electrical consumption of the heating units in both huts was measured using Carrel Electrade LP-1W1 with accuracy up to ±0.5%.

3.7.3 RESULTS AND DISCUSSION

In the following sections, two types of experimental arrangements were made. The first was based on introducing an underfloor heating system incorporating PCM, while the second includes the use of DuPont wallboards in combination with the underfloor heating system.

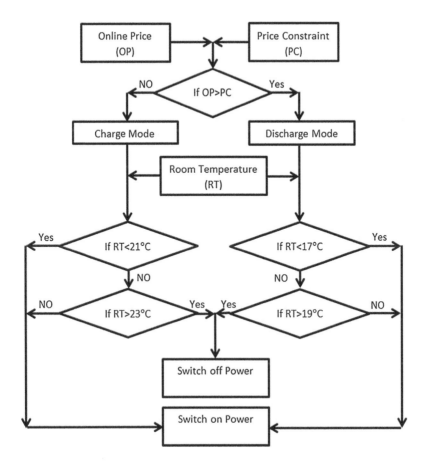

FIGURE 3.7.4 Flowchart of the price-based method applied to the experimental huts.

3.7.3.1 UNDERFLOOR HEATING SYSTEM

Figure 3.7.5 shows the charge/discharge process of the PCM in the floor gypsum board over a 24-h period in Hut 2. As shown, during the night, RT was kept within the high price temperature range until 6:00 am. During this period, PCM impregnated in the underfloor heating system remained fully charged since its temperature is sustained above its melting point of 28°C. After the morning peak at 6:00 am, the controller stopped supplying electricity to the heater, and, as a consequence, the temperature of the PCM dropped sharply from 40°C to 27°C and remained relatively constant for almost 3 h until all PCM solidified (shown as "Discharge" in the figure). At 20:13, the RT reached 17°C and the controller turned on the underfloor heating elements to prevent further temperature drop in the hut. During the heating process, phase change of the PCM from solid to liquid can be observed as a gradient change (shown as "Discharge" in the figure).

Figure 3.7.6 shows the interior temperature of Hut 1 and Hut 2 when both were heated during the night using an AC unit and the under-floor heating system, respectively. The

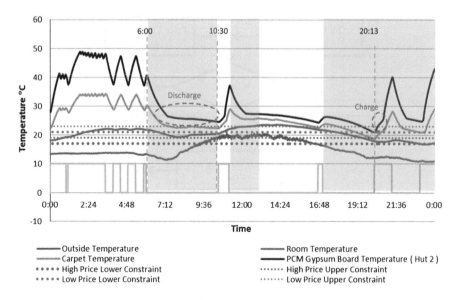

FIGURE 3.7.5   Hut 2 room temperature, charge and discharge of the PCM underfloor heating system.

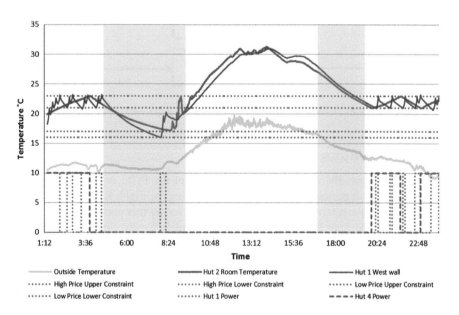

FIGURE 3.7.6   Room temperature of Huts 1 and 2: application of PCM underfloor heating system for morning peak.

underfloor heating PCM was fully charged. The controller stops supplying heat to the huts during the early morning high peak period (shown by the gray shading) and, as a consequence, the interior temperature of Hut 1 drops sharply and reaches the minimum temperature constraint at 7:30. The RT for Hut 2 remains above the temperature constraint until the room is heated as a result of solar radiation. This shows clearly that the PCM underfloor heating system can successfully be used for early morning peaks.

However, despite the good performance of the underfloor heating system in the early morning peak, it does not have any effect during the evening peak, as shown in Figure 3.7.6 (the temperature drops similarly in both huts during the evening period). This is because there is usually no need for heating during the day, and, as a consequence, the stored energy in the underfloor system is fully discharged. This can be clearly observed in Figure 3.7.7, which shows that the underfloor heating system was not turned on during the day and the PCM temperature at 17:00 dropped to 24°C, suggesting that the PCM is fully discharged and cannot contribute any more to the heating of the hut.

To prevent excessive floor temperature, the underfloor heating system was programmed not to allow the carpet temperature to exceed 29°C during daytime and not to exceed 34°C from 1:00 am until 6:00 am (when occupants are sleeping) as shown in Figure 3.7.8. The controller stops supplying heat as soon as the carpet temperature exceeds these carpet temperature constraints. PCM temperature in either of these cases can easily rise above 30°C, which ensures that the PCM is being fully charged. During the discharge period, carpet temperature remains around 23°C until the PCM is fully discharged.

3.7.3.2 Underfloor Heating in Combination with PCM Wallboards

As shown in previous section, underfloor heating can contribute mainly to early morning peaks and not to evening electricity peaks. To improve this, DuPont wallboards

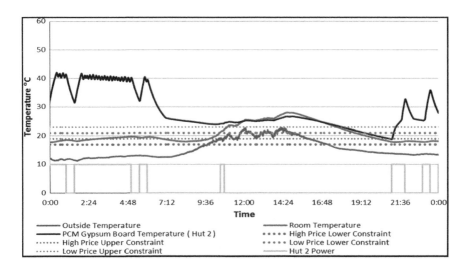

FIGURE 3.7.7 Application of PCM underfloor heating system for evening peak.

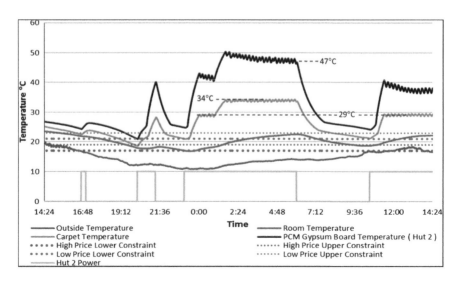

FIGURE 3.7.8 Underfloor heating system: Carpet temperature and PCM temperature during charge and discharge period in Hut 2.

were installed on the wall of Hut 2 in addition to the PCM underfloor heating system to deal with possible evening peaks. Hut 1 was heated using an AC unit and Hut 2 by the underfloor heating system.

Figure 3.7.9 shows the indoor temperatures for Huts 1 and 2 when they were both heated and maintained at 22°C during the low peak period at night. As can be observed in the figure, both PCM used in underfloor system and wallboards were fully charged. Around 7:00 am, the electricity price rose and consequently the controller stopped supplying heat to the hut during the peak period. As a result of solar radiation, the RT of both huts remained above 22°C during the day, which ensures that the PCM in Hut 2 remained fully charged. During the high peak period in the evening, indoor temperature in Hut 1 dropped and reached the minimum temperature constraint at around 20:00 h while RT in Hut 2 remained within the set point range until the next day, using the stored energy for almost 4 h longer.

As demonstrated above, this method can be successful in most winter days with low solar gains. However, on a sunny day with high solar radiation, this method can cause overheating during the day. As shown in Figure 3.7.10, both huts are heated during the night and consequently PCM in both the underfloor system and wall-boards are fully charged in the morning. High solar radiation causes a temperature rise in Hut 2 that is very similar to Hut 1, and this brings the interior temperature to an unacceptable level. This is because the PCM is fully charged, cannot absorb any further energy and therefore cannot prevent a temperature rise inside Hut 2.

To prevent this excessive heating during the day, the RT was set to 20°C instead of the 21°C–23°C range in Figure 3.7.10. As a result, the PCM for the underfloor heating system was fully charged but not the walls. In other words, changes were made to use electricity solely to charge underfloor heating PCM and use solar energy to charge

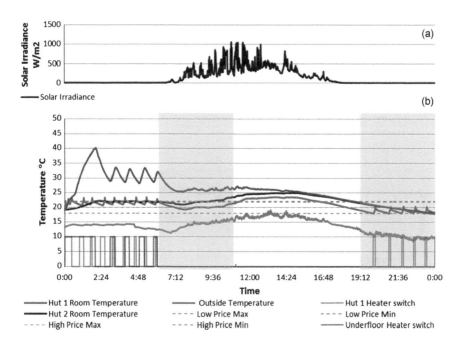

FIGURE 3.7.9   Hut experiment for a day with Low solar radiation (a) solar irradiance, (b) room temperature for Huts 1 and 2.

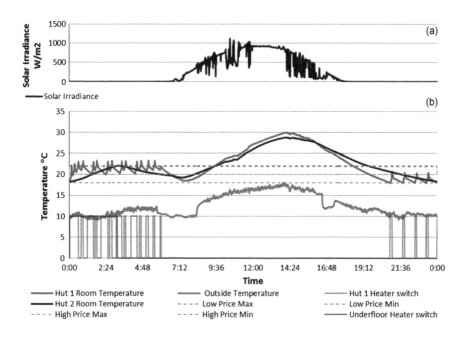

FIGURE 3.7.10   Hut experiment for a day with high solar radiation. (a) Solar irradiance and (b) room temperature for Hut 1 and Hut 2.

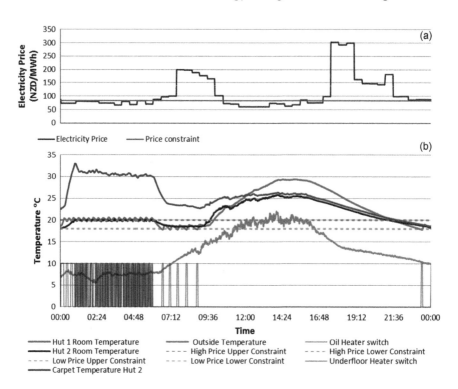

FIGURE 3.7.11 Experiment results for a day with high solar radiation using lower set point. (a) Electricity price and (b) room temperature (RT) for Huts 1 and 2.

the wall PCM. To keep the RT closer to the set point, an oil heater was used instead of the AC unit in Hut 1.

In Figure 3.7.11, RT was kept at 20°C during the night to prevent melting the PCM in the DuPont wallboard, while the floor temperature remained at 31°C to ensure the PCM in the underfloor heating system was fully charged. As a consequence, during the morning peak, RT for Hut 2 remained above the temperature constraint for a 3-h period longer than Hut 1 as a result of stored energy in the underfloor heating system. Later, during the midday, RT in Hut 1 rose to 30°C because of high solar radiation, while in Hut 2 it remained below 25.5°C. This clearly shows that the PCM in Hut 2 was getting charged during that period by capturing a good amount of solar energy. The stored solar energy during the day was also discharged during the evening peak as the RT remained above 18°C, using the stored energy.

To apply this strategy, the set temperatures of 20°C and 18°C were used as the low and high peak temperature set points, respectively. Additionally, the control system monitors solar radiation and, when it is a rainy or a cloudy day, the controller raises the low peak temperature set point from 20°C to 23°C to charge the PCM in the wallboard for the evening peak. As shown in Figure 3.7.12, due to very little solar radiation in the morning period, the controller at 10:38 am raises the temperature set point from 20°C to 23°C to charge the PCM during low peak periods of the day, resulting in storing enough energy for use during the high peak period in the evening. Because

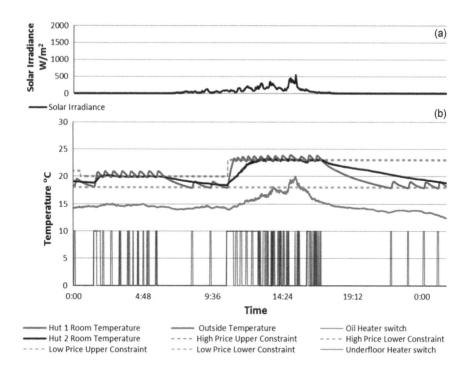

FIGURE 3.7.12   Set point change for day with low solar radiation. (a) Solar irradiance and (b) room temperature for Huts 1 and 2.

of the stored energy, RT in Hut 2 was successfully held above 18°C for almost 4 h longer than Hut 1.

Figure 3.7.13 shows solar radiation and RT in Huts 1 and 2 over a 5-day period. As presented in Table 3.7.2, very little electricity saving was achieved in Day 1 because of the low solar radiation during the day; however, using the price-based control system, a 17.2% cost saving was possible. This is mainly as a result of using electricity during low peak periods to store energy and use it during the high peak period. High solar gains caused overheating in Hut 1 during Days 2 and 5, while excessive heat was captured by the PCM walls in Hut 2, which discharged it during the evening peak. The stored solar energy in the wallboards then was used to maintain the RT above 18°C during the evening peak, which resulted in energy savings of 25% and 15% for Days 2 and 5, respectively.

As shown in Table 3.7.2, the highest energy saving of 35%, with a corresponding power saving of 44% was achieved in Day 3. This was mainly because the stored energy was not fully discharged during Day 2, and consequently, Hut 2 used less energy during the early hours of Day 3. This shows that the electricity saving achieved in each day comes from solar irradiance and is affected by outdoor temperature. Moreover, the previous day's conditions also significantly affect the energy consumption of the next day.

Based on the obtained results presented in Table 3.7.2, and using the proposed method, a total electricity saving of 18.8% was possible, with a corresponding

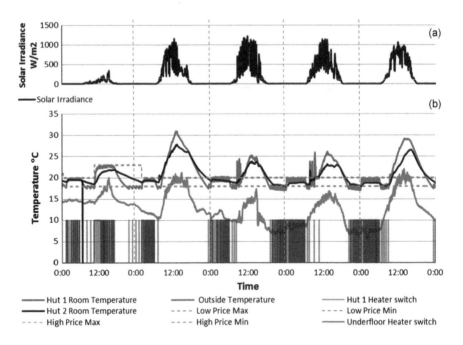

FIGURE 3.7.13 Hut experiment over 5 days. (a) Solar irradiance and (b) room temperature for Huts 1 and 2.

TABLE 3.7.2

Power and Cost Saving Achieved Using the Proposed Energy Storage System

	Day 1 (%)	Day 2 (%)	Day 3 (%)	Day 4 (%)	Day 5 (%)	Total (%)
Energy saving	5.8	25.0	35.0	16.2	14.6	18.8
Cost saving	17.2	27.7	44.4	22.8	28.8	28.7

electricity cost saving of 28.7%. As mentioned earlier, all energy consumption measurements were performed with a maximum error of 0.5%.

3.7.3.3 FURTHER COMMENTS AND DISCUSSIONS

PCM selection is one of the most important steps toward successfully using full potential of PCM. The selected PCM should undergo full charge and discharge in each cycle, otherwise latent heat storage capacity of the PCM will not be fully used. The PCM used in this study for the underfloor heating system has a peak melting temperature of 28°C, which can easily get fully charged and discharged, as can be seen in Figure 3.7.6. Failing to fully charge and discharge at each cycle can significantly deteriorate the performance of the PCM, as can be seen in [32,33,39].

As mentioned earlier, the application of a PCM underfloor heating system and the application of DuPont wallboard to absorb heat for heating purposes have been studied earlier in [35–36,39]. However, the combination of a PCM underfloor heating system with DuPont wallboards in this study resulted in getting benefits from both of the systems and consequently reducing electricity consumption while maintaining the RT within the desired range. The application of this method resulted in sizable saving of 18.8% over a 5-day period, with a maximum saving of 35%. The amount of saving mainly depends on the solar radiation of the day and the outdoor temperature, as is concluded in [38]. This method can benefit from the solar radiation and variable electricity prices during the winter period to reduce electricity consumption while performing peak load shifting. The proposed method in this research can be applied in any locality that offers a variable electricity price. It should be noted that the melting point of the PCM needs to be selected according to the thermal comfort requirements.

3.7.4 CONCLUSIONS

A series of experiments were conducted in two identical test huts at the Tamaki Campus, University of Auckland. A price-based control method was tested initially using a PCM underfloor heating system, which showed a successful morning peak load shifting. However, very little improvement was observed during the evening peak load period. DuPont Energain® wallboards were installed on the interior walls to provide energy storage for evening peak. Experimental results showed that the application of the underfloor heating system in combination with PCM wallboard, enables very efficient energy usage. A total energy saving and electrical cost saving equal to 18.8% and 28.7%, respectively, were achieved over a period of 5 days. The highest energy saving achieved during this period was 35%, with a corresponding cost saving of 44.4%.

ACKNOWLEDGEMENT

The authors would like to acknowledge MBIE for funding this project.

ABBREVIATIONS

PCM: phase change materials
TES: thermal energy storage
LHTES: latent heat thermal energy storage
DSC: differential scanning calorimetry
RT: room temperature
OP: online price
PC: price constraint
AC: air conditioner.

REFERENCES

1. El-Sawi A, Haghighat F, Akbari H. Assessing long-term performance of centralized thermal energy storage system. *Appl Therm Eng* 2014;62(2): 313–21.
2. Menoufi K et al. Evaluation of the environmental impact of experimental cubicles using life cycle assessment: a highlight on the manufacturing phase. *Appl Energy* 2012;92:534–44.

3. Castell A et al. Life cycle assessment of alveolar brick construction system incorporating phase change materials (PCMs). *Appl Energy* 2013;101: 600–8.
4. Chiu JNW, Martin V. Multistage latent heat cold thermal energy storage design analysis. *Appl Energy* 2013;112:1438–45.
5. Navarro L et al. Thermal loads inside buildings with phase change materials: experimental results. *Energy Proc* 2012;30:342–9.
6. Morgan S, Moncef K. Field testing of optimal controls of passive and active thermal storage. *ASHRAE Transac* 2010;116(1):134–46.
7. Martin V, He B, Setterwall F. Direct contact PCM–water cold storage. *Appl Energy* 2010;87(8):2652–9.
8. Esen M, Ayhan T. Development of a model compatible with solar assisted cylindrical energy storage tank and variation of stored energy with time for different phase change materials. *Energy Convers Manage* 1996;37(12): 1775–85.
9. Esen M, Durmuş A, Durmuş A. Geometric design of solar-aided latent heat store depending on various parameters and phase change materials. *Sol Energy* 1998;62(1):19–28.
10. Esen M. Thermal performance of a solar-aided latent heat store used for space heating by heat pump. *Sol Energy* 2000;69(1):15–25.
11. Heier J, Bales C, Martin V. Combining thermal energy storage with buildings – a review. *Renew Sustain Energy Rev* 2015;42:1305–25.
12. Chiu JNW, Gravoille P, Martin V. Active free cooling optimization with thermal energy storage in Stockholm. *Appl Energy* 2013;109:523–9.
13. Bruno F, Tay NHS, Belusko M. Minimising energy usage for domestic cooling with off-peak PCM storage. *Energy Build* 2014;76:347–53.
14. Henze GP. Energy and cost minimal control of active and passive building thermal storage inventory. *J Sol Energy Eng Trans ASME* 2005;127(3):343–51.
15. Álvarez S et al. Building integration of PCM for natural cooling of buildings. *Appl Energy* 2013;109: 514–22.
16. Bahadori MN, Haghighat F. Weekly storage of coolness in heavy brick and adobe walls. *Energy Build* 1985;8(4):259–70.
17. Arteconi A, Hewitt NJ, Polonara F. State of the art of thermal storage for demand-side management. *Appl Energy* 2012;93:371–89.
18. Farid M, Kong WJ. Underfloor heating with latent heat storage. *Proc Inst Mech Eng Part A: J Power Energy* 2001;215(5):601–9.
19. Castell A et al. Experimental study of using PCM in brick constructive solutions for passive cooling. *Energy Build* 2010;42(4):534–40.
20. Eicker U. Cooling strategies, summer comfort and energy performance of a rehabilitated passive standard office building. *Appl Energy* 2010;87(6): 2031–9.
21. Long L et al. Performance demonstration and evaluation of the synergetic application of vanadium dioxide glazing and phase change material in passive buildings. *Appl Energy* 2014;136: 89–97.
22. Khudhair AM, Farid MM. A review on energy conservation in building applications with thermal storage by latent heat using phase change materials. *Energy Convers Manage* 2004;45(2):263–75.
23. Bastani A, Haghighat F, Kozinski J. Designing building envelope with PCM wallboards: design tool development. *Renew Sustain Energy Rev* 2014;31:554–62.
24. Biswas K et al. Combined experimental and numerical evaluation of a prototype nano-PCM enhanced wallboard. *Appl Energy* 2014;131:517–29.
25. Zhou G, Yang Y, Xu H. Performance of shape-stabilized phase change material wallboard with periodical outside heat flux waves. *Appl Energy* 2011;88(6): 2113–21.
26. Cheng R et al. A new method to determine thermophysical properties of PCM-concrete brick. *Appl Energy* 2013;112:988–98.

27. Arce P et al. Use of microencapsulated PCM in buildings and the effect of adding awnings. *Energy Build* 2012;44:88–93.
28. Darkwa J. Mathematical evaluation of a buried phase change concrete cooling system for buildings. *Appl Energy* 2009;86(5):706–11.
29. Li H et al. Development of thermal energy storage composites and prevention of PCM leakage. *Appl Energy* 2014;135:225–33.
30. Borreguero AM et al. Thermal testing and numerical simulation of gypsum wallboards incorporated with different PCMs content. *Appl Energy* 2011;88(3):930–7.
31. Zhou D, Zhao CY, Tian Y. Review on thermal energy storage with phase change materials (PCMs) in building applications. *Appl Energy* 2012;92:593–605.
32. Schossig P et al. Micro-encapsulated phase-change materials integrated into construction materials. *Sol Energy Mater Sol Cells* 2005;89(2–3):297–306.
33. Voelker C, Kornadt O, Ostry M. Temperature reduction due to the application of phase change materials. *Energy Build* 2008;40(5):937–44.
34. Khudhair AM, Farid MM. Use of phase change materials for thermal comfort and electrical energy peak load shifting: experimental investigations. In: *ISES Solar World Congress 2007*, ISES 2007; Beijing, China, 2007.
35. Kuznik F, Virgone J. Experimental assessment of a phase change material for wall building use. *Appl Energy* 2009;86(10):2038–46.
36. Liu H, Awbi HB. Performance of phase change material boards under natural convection. *Build Environ* 2009;44(9):1788–93.
37. Kuznik F, Virgone J. Experimental investigation of wallboard containing phase change material: data for validation of numerical modeling. *Energy Build* 2009;41(5):561–70.
38. Qureshi WA, Nair NKC, Farid MM. Impact of energy storage in buildings on electricity demand side management. *Energy Convers Manage* 2011;52(5): 2110–20.
39. Olesen BW. Radiant floor heating in theory and practice. *ASHRAE J* 2002;44(7):19–26.
40. National Instruments. *LabVIEW Version 11*, National Instruments, Austin, TX; 2011.
41. Thermal environmental conditions for human occupancy. ASHRAE Standard; 2010(STANDARD 55). pp. 1–44.

3.8 Analysis of Energy Requirements versus Comfort Levels for the Integration of Phase Change Materials in Buildings

Martin Vautherot and François Maréchal
Ecole Polytechnique Fédérale de Lausanne

Mohammed Farid
The University of Auckland

3.8.1 INTRODUCTION

The use of phase change materials (PCMs) as a mean for thermal energy storage in buildings is complex but can provide multiple advantages. The interest of PCMs is that by going through a phase change, large amounts of heat can be stored and released. Building materials store thermal energy through sensible heat, which requires a temperature difference while a PCM involves the latent heat at a selected temperature, therefore requiring lower temperature difference for the heat storage. These materials have been investigated and still are for their properties which can be compared using existing reviews [1–3].

A convenient integration of PCMs in buildings is through the use of impregnated gypsum boards, which are commonly added as an internal layer inside a building. This low-cost component provides a suitable structure for PCM containment, which is great for both new constructions and retrofitting. One of the main interests of the gypsum board is that it is the innermost layer for most constructions. The replacement of an existing conventional gypsum board is therefore an easy task [4,5]. There are three methods of how PCM can be incorporated into the construction material:direct incorporation, immersion and encapsulation. Microencapsulation prevents problems associated with PCM volume change and provides greater heat exchange area, which increases heat transfer rate. Microencapsulation is most appropriate when using gypsum boards; it prevents PCM leakage and shows good cycling stability [6].

The main interest when using PCMs in buildings is that they should have the ability to reduce indoor temperature swings without any external help. To do so, the indoor temperature must vary across the phase-change temperature range of the PCM. When the material goes from solid phase to liquid phase through a melting process, it absorbs large amount of heat and therefore slows down the temperature rise that would otherwise occur inside the building. When the ambient temperature drops, the PCM goes through a solidification process and releases heat which has the effect of slowing down the decrease in temperature inside the building [7].

Not only can the integration of PCM in buildings reduce the energy requirements, but it can also enhance comfort. By reducing the indoor temperature variations, buildings rely less on heating, ventilation and air conditioning (HVAC) system. PCMs can also reduce the period of heating and cooling as temperature peaks can be avoided. At the moment, the emphasis in most investigations on PCMs is mainly about the energy consumption and not about comfort levels. However, the goal of an HVAC system is to provide comfortable conditions for people; therefore, a better understanding on how these two parameters vary in the presence of PCMs is necessary since it has not been studied in the literature before. Also, the question of the optimum choice PCM phase-change temperature range is not assessed in the literature and turns out to be of great importance as investigated in this article.

The objectives of this work are: (a) to show how the HVAC set points influence both energy requirements and comfort; (b) to investigate how the phase-change temperature range of PCM influences the energy consumption and the level of comfort; and (c) to show that appropriate PCM designs depend on the trade-off between energy requirements and comfort levels.

3.8.2 METHODOLOGY OF THE INVESTIGATION FOR A TYPICAL HOUSE USING COMPUTER SIMULATION

3.8.2.1 DEVELOPMENT OF A BUILDING SIMULATION MODEL

Computer-based simulations show several advantages when it can be validated against past experimental measurements. An important fact is that they save time and can be used to perform a study over a year-round within a few hours of simulation. It also allows great flexibility to show the influence of few parameters, while keeping all the other factors constant. To perform the calculations, the interface Design Builder which is based on the other software Energy Plus is used [8]. The latter allows the integration of PCMs and has been validated in previous studies [9–11].

The University of Auckland built two identical offices (with and without PCM) provided with a data collection system. The offices are built in Tamaki near Auckland where the climate is temperate. The two offices have been modelled, and the results of the simulations are compared with the experimental data collected over a week as shown in Figure 3.8.1. The thin dashed line shows the indoor air temperature as obtained from the simulation, the black solid bold line shows the measured indoor temperature, and the light green solid line shows the outside temperature.

The office with the PCM-impregnated gypsum boards is the one which was modelled and shown in Figure 3.8.1. It can be observed that the gap between the

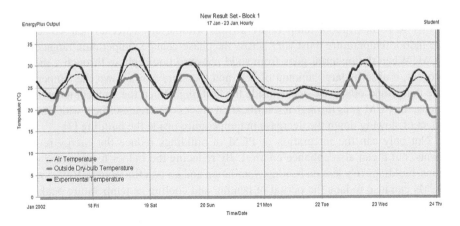

FIGURE 3.8.1 Validation of the simulation software with experimental data over a week. (For interpretation of the references to colour in this figure the reader is referred to the web version of this article.)

experimental and simulated curves is reasonably small. This confirms the fact that simulation software can be trusted as it takes well into account the integration of PCM. The deviation from the experimental measurements may have come from the difficulty of measuring some of the parameters such as infiltration rate.

3.8.2.2 Modelling of a Typical House

3.8.2.2.1 Geometry and Materials

Following model validation, simulations were conducted for a typical two-story house in Auckland. The construction of the house is based on real construction plans and a typical materials structure. It is a two-story family house for five people. The geometry was therefore added in the simulation software and is shown in Figure 3.8.2. A few simplifications have been done to speed up the simulations while keeping the results relevant. The two "Master bedrooms" are merged with their respective "En suite". The "Living room" and the "Hall" of the first floor are merged together. Finally, the three "Bedrooms" of the first floor are merged together. The black arrow in the figure points towards North. This North-facing orientation is explained by the fact that the house is situated in the Southern Hemisphere. Every bedroom and living room has two windows to let as much light as possible in.

The characteristics of the materials provided to the software and every type of structure (roof, walls, partitions etc...) are given in Tables 3.8.1 and 3.8.2. It must be kept in mind that the PCM is added to the gypsum board; therefore, it has a significant influence on the house's thermal mass. The total surface area of gypsum board is $810.5\,m^2$ for the whole house with a floor surface area of $256\,m^2$.

3.8.2.2.2 HVAC

To run a realistic study for the house, several assumptions had to be made to define the overall system. The simulations only give the energy loads needed, and no HVAC

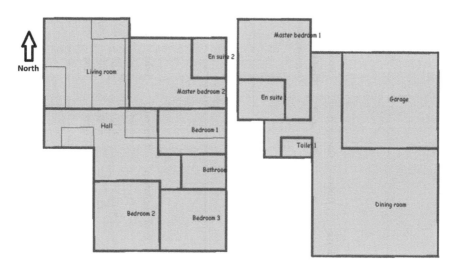

FIGURE 3.8.2 View of the geometry of the simulated house (first floor on the left and ground floor on the right).

system was defined. The study aimed to observe the heating and cooling loads needed and not how to provide them. A zoning in the house is assumed so that every heated or cooled room receives the appropriate amount of heating or cooling needed to meet the set points. The loads are assumed to be variable so that the temperatures remain constant once the set points are reached, and the power adapts. To have a realistic design, all main rooms are assumed to have an HVAC system. Therefore, only the bathroom, toilet and garage are left with no HVAC.

3.8.2.2.3 Occupancy

Regarding the occupancy, the house is assumed to be occupied by a family of five people. The occupancy schedule is of a great importance when doing this type of simulation and was kept as realistic as possible. On weekdays, people leave the house at 8 am and come back at 5 pm. Then they spend their time between the bedrooms and the common rooms (kitchen, living room and hall) until 10 pm when everybody goes to their respective bedrooms. On the weekend, people stay in the house the whole day and spend their time between the bedrooms and common rooms from 9 am to 10 pm, after that they go to their respective bedrooms. The HVAC schedule matches with the room occupancy and is therefore assumed to operate when there are people in the rooms. Meaning that when the people are not present the HVAC system is off, and the people start the HVAC system when they arrive at home. Another aspect of PCMs is the ability to perform peak load shifting, but this is not investigated in this article [11].

3.8.2.3 Inputs in the Model

The two variable parameters in this study are the heating set point and the phase-change temperature range of the PCM used in the gypsum board. In this study,

TABLE 3.8.1

Construction of the House with the Different Materials

	Exterior wall	Interior wall	Ground floor	Interior floor	Roof	Windows	Doors	Garage door	Roof floor
Outer layer	Brick Air gap 45mm Wall insulation	Gypsum board Air gap 90mm	Concrete Carpet	Gypsum board Air gap 240mm	Roof tiles	Double glazing	Wood	Steel sheet	Roof insulation
Inner layer	Gypsum board	Gypsum board		Gypsum board					Gypsum board

TABLE 3.8.2

Properties of the Construction Materials

	Unit	Steel Sheet	Carpet	Wood	Gypsum Board	Wall Insulation	Roof Insulation	Concrete	Brick	Tiles
Thickness	[m]	0.002	0.001	0.035	0.013	0.09	0.17	0.085	0.07	0.014
Conductivity	[W/m K]	13.8	0.056	0.14	0.25	0.04	0.04	1.13	0.84	0.73
Density	[kg/m³]	7817	213	650	600	12	12	2000	1700	2500
Specific heat	[J/kg K]	460	1400	1200	1089	840	840	1000	800	773

five heating set points and six types of PCMs were tested. This yields a total of 30 combinations.

3.8.2.3.1 Heating Set Point

The heating set point goes from 20°C to 24°C while the cooling set point is kept constant at 25°C. The choice to vary the heating set point is done because in New Zealand, most of the energy requirements come from heating. The set point defines the target that the air temperature must reach. A minimum set point temperature of 18°C is chosen for nighttime in the bedrooms. Therefore, the HVAC system changes its target after 10 pm to reduce the energy requirements and maintain a comfortable minimum sleeping temperature [12].

3.8.2.3.2 Types of Gypsum Boards

Six types of gypsum boards are tested: five using PCMs with different melting ranges and one conventional gypsum board having no PCM. The gypsum board has a thickness of 13 mm and a density of 600 kg/m³ while the PCM has a density of 770 kg/m³ which leads to a density of 643 kg/m³ for the PCM-impregnated gypsum board [13]. The latent heat storage density of the PCM-impregnated gypsum board is 33.5 kJ/kg (25% weight of pure PCM with a latent heat of 134 kJ/kg) and takes effect only during the phase change. The latent heat is assumed to be distributed linearly throughout the phase-change process. It is chosen to keep a constant phase-change temperature range of 5°C, which is a realistic assumption for most commercial PCMs. A narrower range would increase significantly the cost of PCM. Hence, the six different gypsum boards are:

No PCM/PCM 18–23°C/PCM 19–24°C/PCM 20–25°C/PCM 21–26°C/PCM 22–27°C. Even though the economic aspects are not at the heart of the study, it must be said that all of the suggested PCMs are assumed to have the same estimated cost of $3 NZD/kg, which is an expected future price [14]. One scenario is therefore defined as a combination of both inputs. For example, "PCM 21–26 22°C" refers to the scenario with a heating set point of 22°C and with a PCM having a phase-change temperature range between 21°C and 26°C.

The total storage capacity of the PCM in the whole house is given in Equation (3.8.1):

$$\dot{Q}_{storage} = A_{GB} e_{CB}\, \rho_{PCMGB}\ \text{wt}\%_{PCM}\ h_{lat,PCM} = 227\ \text{MJ} \qquad (3.8.1)$$

where $\dot{Q}_{storage}$ in [MJ] is the storage capacity of PCM in the house; $A_{GB} = 810.5\,\text{m}^2$ is the total surface area of gypsum board in the house; $e_{GB} = 0.013\,\text{m}$ is the thickness of the gypsum board; $\rho_{PCMGB} = 643\,\text{kg/m}^3$ is the density of the PCM-impregnated gypsum board; $\text{wt}\%_{PCM} = 0.25$ is the weight ratio of PCM in the gypsum board; $h_{lat,PCM} = 134$ kJ/kg is the latent heat of the PCM.

3.8.2.4 OUTPUTS OF THE MODEL

The goal of this study is to observe two outputs: comfort level and energy requirement. Both are factors of high importance, and they go together creating a trade-off between energy savings and comfort. However, some scenarios show better combination and the target is to see what are these scenarios.

3.8.2.4.1 Comfort Level

The comfort level is measured by the number of discomfort hours, i.e. the number of hours during which the comfort conditions as defined by the standard ASHRAE 55-2004 are not met [15]. The standard states that if there is someone in the room and the conditions of temperature and humidity depicted in Figure 3.8.3 are not satisfied, there is discomfort. It considers both the operative temperature and humidity ratio when deciding whether or not there is comfort. The humidity levels are taken from the weather file of the city and are assumed not to be controlled during the simulation.

When measuring the comfort level based on the standard, it is the operative temperature which is considered and not the air temperature which is the target of the HVAC. The operative temperature is an average between the air and radiant temperature, the latter being dependent on the wall temperature. The radiant temperature has a strong influence on the feeling of warmth or coldness for a person inside a room. The addition of thermal mass with the PCM-impregnated gypsum board reduces the temperature swings of the building's envelope, and therefore influences the feeling of comfort. The simulation software provides the comfort level as the total number of discomfort hours for the rooms usage as depicted in Table 3.8.3. The garage is assumed to be unoccupied and rooms of the same type are merged together as explained before.

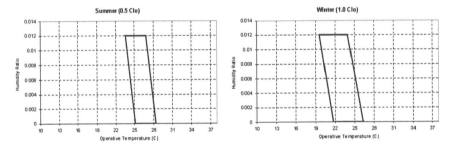

FIGURE 3.8.3 Comfort zones according to ASHRAE Standard 55-2004 used in the simulation software [15].

TABLE 3.8.3

Total Number of Hours of Occupancy per Zone in the House over a Year

	Occupancy [hour/year]
Ground floor: Master Bedroom 1	6150
Ground floor: Dining room	2918
Ground floor: Toilet	365
First floor: Master Bedroom 2	6150
First floor: Hall and Living room	2918
First floor: Bathroom	730
First floor: Other Bedrooms	6150
Total	25,381

It has to be kept in mind, however, that the concept of comfort is a subjective one and therefore may vary with external parameters such as culture, activity, age, and gender.

3.8.2.4.2 Energy Requirements

Energy requirement is the main focus of most studies involving the integration of PCMs and is the most critical parameter. The energy requirements define the consumption (varying with the HVAC system). From that, the annual costs of the consumed energy can be calculated. This decides if an investment in PCMs is economically viable and determines people's decisions to use PCM in their house. In this investigation, however, only the heating and cooling ideal loads are given. A distinction is made between heating and cooling loads as it will be seen that they vary differently with the addition of PCM in the construction.

3.8.3 RESULTS AND DISCUSSION

Table 3.8.4 summarises the outputs of the annual 30 year-round simulations based on the same year's meteorological data. The results are sorted by constant heating set points to make relevant comparisons depending on the targeted temperature.

By keeping the heating set point constant, energy savings can be compared for the designs with PCM-impregnated gypsum boards. Therefore, a negative value in the savings means that instead of bringing better conditions, PCM makes it worse.

This way of presenting the data is to emphasise the fact that there are several acceptable solutions.

The discomfort savings index is defined as in Equation (3.8.2) and is based on the number of hours when there are people in the rooms while the comfort conditions are not met.

$$\text{Discomfort savings} = \frac{\left(T_{\text{Discomfort, No PCM}} - T_{\text{Discomfort, PCM}}\right)}{T_{\text{Discomfort, No PCM}}} 100 \; [\%] \qquad (3.8.2)$$

where Discomfort savings in [%] is the comfort improvement from the integration of PCMs; $T_{\text{discomfort, PCM}}$ and $T_{\text{discomfort, No PCM}}$ are, respectively, the yearly number of hours of discomfort for a case with and without PCM, respectively, at a specific heating set point.

The results shown in Table 3.8.4 can be used to assist a designer to select a suitable combination of heating set point and PCM melting range based on energy consumption and comfort. It also allows better understanding on whether the savings are from the heating or cooling load. This is depicted in Figure 3.8.4 and clearly shows that most of the savings occur during cooling in summer. In summer, PCM stores large quantity of coolness at night, while in winter PCM stores only the limited excess energy coming from solar radiation. In summer, the PCM slows down any rise in temperature which otherwise would involve a need for cooling. The heat stored can be released freely later when the temperature drops at night. Regardless of the heating set point, the cooling load decreases sharply when the melting temperature range of the PCM increases.

TABLE 3.8.4

Results for the 30 Year-Round Simulations Investigating the Comfort and Energy Requirements

Design	Heating Load [kWh]	Cooling Load [kWh]	Total Load [kWh]	Discomfort Hours [hours]	Energy Savings [%]	Discomfort Savings [%]
No PCM 20°C	3171	1410	4581	15,822		
PCM 18–23 20°C	2926	597	3523	16,790	23	−6
PCM 19–24 20°C	2734	450	3184	16,236	31	−3
PCM 20–25 20°C	2745	321	3066	14,953	33	5
PCM 21–26 20°C	2815	251	3067	14,546	33	8
PCM 22–27 20°C	2868	254	3122	14,921	32	6
No PCM 21°C	4020	1432	5452	13,799		
PCM 18–23 21°C	4407	604	5011	14,335	8	−4
PCM 19–24 21°C	4038	454	4491	13,647	18	1
PCM 20–25 21°C	3734	322	4057	12,573	26	9
PCM 21–26 21°C	3680	253	3933	12,561	28	9
PCM 22–27 21°C	3725	256	3981	12,932	27	6
No PCM 22°C	5080	1466	6547	11,435		
PCM 18–23 22°C	6284	628	6911	10,415	−6	9
PCM 19–24 22°C	5859	468	6327	9900	3	13
PCM 20–25 22°C	5332	331	5663	9642	13	16
PCM 21–26 22°C	4920	257	5177	9977	21	13
PCM 22–27 22°C	4820	259	5079	10,495	22	8
No PCM 23°C	6338	1517	7855	10,715		
PCM 18–23 23°C	8474	682	9155	8304	−17	23
PCM 19–24 23°C	8055	501	8556	8280	−9	23
PCM 20–25 23°C	7453	353	7806	8224	1	23
PCM 21–26 23°C	6774	272	7046	8635	10	19
PCM 22–27 23°C	6284	269	6553	9496	17	11
No PCM 24°C	7786	1592	9378	10,287		
PCM 18–23 24°C	10,895	799	11,694	7113	−25	31
PCM 19–24 24°C	10541	573	11,114	7116	−19	31
PCM 20–25 24°C	9935	398	10,332	7054	−10	31
PCM 21–26 24°C	9155	304	9460	7431	−1	28
PCM 22–27 24°C	8357	296	8653	8512	8	17

When looking at the heating loads, it is clear that the PCM brings little improvement and can also make it worse. The higher the heating set point, the more heating energy used and the less likely it is that the PCM integration will bring energy savings. This is explained by the fact that a heating set point situated within the phase-change temperature range of the PCM will require extra energy to overcome the added thermal mass. For instance, a poor scenario "PCM 18–23 24°C" will not

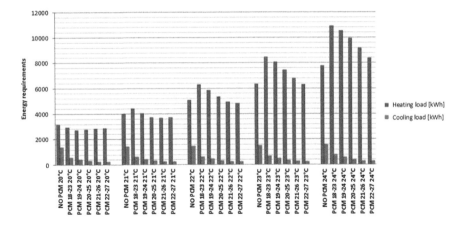

FIGURE 3.8.4   Energy requirements for the different scenarios for the house.

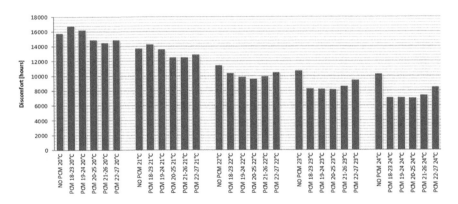

FIGURE 3.8.5   Discomfort hours for the different scenarios for the house.

save the energy needed for heating in comparison with a basic design without PCM as a lot of heat is required to go from 18°C to the heating set point of 24°C. The house is heated following a cyclic schedule, making it overcome the thermal mass every day to reheat it when people come home.

A clever scenario such as a "PCM 20–25 20°C" on the other hand appears to be a good choice. The house can reach the heating set point easily without having to go through the latent heat. Any free energy (mainly solar radiation) from this point is going to be stored in the PCM as it starts to change phase.

These observations about the energy requirements highlight the need for a good design to make the integration of PCMs economically viable. A look at Figure 3.8.5 brings more under-standing about the comfort levels.

The first observation that can be seen here is that the total of comfort hours increases when more energy is put into the HVAC system which is expected. It also shows that the addition of PCM does not bring comfort enhancement in every case.

This can be explained by the added thermal mass, which can slow down the temperature rise and therefore prevent the rooms to reach the comfort zone in winter while it is the other way around in summer. One important point is that only some PCMs seem much more likely to bring comfort enhancement. For a fixed heating set point, it is always either "PCM 20–25" or "PCM 21–26" which bring more improvement compared to the case without PCM. This can be explained by the fact that both of these types of PCM match very well with the comfort zones depicted in Figure 3.8.3 and therefore tend to keep longer the temperatures within these ranges for a longer time. One could say that the number of discomfort hours is high when assuming a total of 25,381 h as given in Table 3.8.3. However, it must be noted that the night temperature set point of 18°C is not in the comfort zone from Figure 3.8.3 even though it is a recommended temperature at night. This explains the relatively high level of discomfort from Figure 3.8.6, even though in practice comfort is achieved.

As seen in Table 3.8.4, the scenarios providing more energy savings do not necessarily provide high comfort enhancement and vice versa. The idea is then to give for each heating set point an optimal design. A trade-off has to be made between energy and comfort. This trade-off is depicted in Figure 3.8.6 and clearly shows the trend. As both outputs need to be minimised, the best designs are found on the lower part and the left of the graph in Figure 3.8.6. Then depending on the priorities of the household, the best design can be selected. For each heating set point, the chosen design is indicated with the non-solid shape symbols.

The suggested optimal scenarios are then: PCM 21–26 20°C/ PCM 21–26 21°C/ PCM 21–26 22°C/PCM 21–26 23°C/PCM 21–26 24°C. This graphical analysis exhibits the fact that the PCM changing phase within the range 21–26°C brings the best trade-off for every heating set point.

To have a better understanding of these results, 2 days were investigated in more detail. A typical day in summer and a typical day in winter. The winter analysis

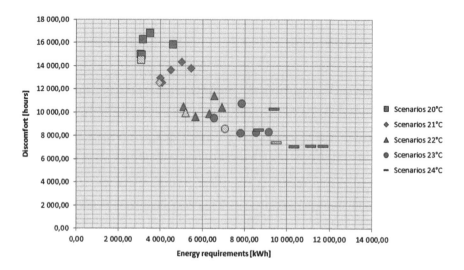

FIGURE 3.8.6 Combination of the comfort levels with the energy requirements.

depicts a scenario without PCM, and two scenarios using PCMs (good and bad). The summer analysis shows only one scenario without PCM and a good scenario with PCM as PCMs always bring improvement.

The "air temperature" in solid line is the indoor temperature targeted by the HVAC system. Therefore, some plateaus are seen when the HVAC is required as seen on the graphs. The "operative temperature" in dashed line includes the effect of radiation and is the one that determines the feeling of comfort. The HVAC load is depicted on the right to show how the different scenarios induce different heating and cooling loads.

The winter day is a typical weekday, and it is chosen to show the scenario "No PCM 20°C" compared to the scenarios "PCM 20–25 20°C" and "PCM 22–27 20°C". A constant heating set point and the analysis over the same room (Master bedroom on the first floor) give relevant results for a comparison of the influences of PCMs.

The results for this winter day are shown in Figures 3.8.7–3.8.9. As expected according to Table 3.8.4, the designs with PCM show lower heating requirements.

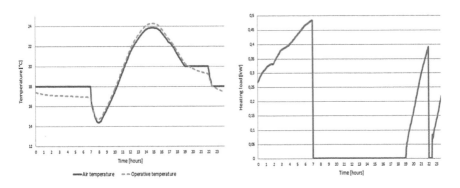

FIGURE 3.8.7 Analysis over a winter day in a bedroom for the scenario "No PCM 20°C".

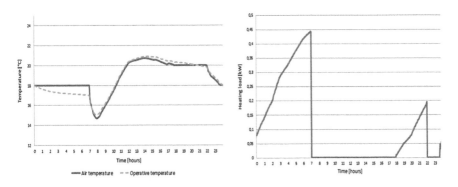

FIGURE 3.8.8 Analysis over a winter day in a bedroom for the scenario "PCM 20–25 20°C".

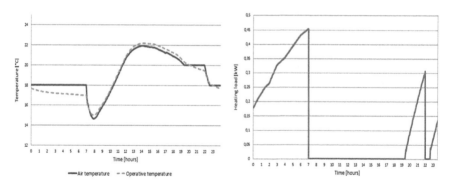

FIGURE 3.8.9 Analysis over a winter day in a bedroom for the scenario "PCM 22–27 20°C".

The one with "No PCM" requires 3.61 kWh, the one with "PCM 20–25" requires 2.40 kWh and the one with "PCM 22–27" requires 2.90 kWh. This is only for a room over a day so the difference is small, but over the whole house and the whole winter it can bring significant differences.

The goal of this comparison is to see how PCMs can save energy even in winter. This winter day is chosen because the temperature is cold outside, yet the solar radiation brings heat during the afternoon. Because on these graphs the emphasis is on the energy requirements, the "PCM 21–26" given as best trade-off energy versus comfort is not depicted here. It is indeed not the one bringing the more energy savings for a heating set point of 20°C. The room was chosen due to its orientation (North facing). It receives solar radiation in the afternoon, and the heating set point of 20°C is reached without any heating load at 5 pm when people come home. The heat is then released at a different pace in the evening, and at 10 pm, the heating set point is lowered to 18°C for the night temperature.

The scenario with "No PCM" shows a temperature rise to 24°C and then a quick drop until it reaches 20°C where the HVAC system activates. The high heating requirements come from the fact that the heat from solar radiation has not been stored in the building's envelope and therefore it needs large loads in the evening and at night.

The design "PCM 20–25" is considered optimum and shows the lowest heating requirements for this heating set point. It is because that the room can easily reach 20°C. From there, any solar gain is stored through the melting process and slows down the temperature rise, which keeps a relatively constant temperature. The stored heat is then slowly released in the evening and at night, which lowers the heating requirements.

The scenario "PCM 22–27 20°C" is an example of a non-optimum design for a set point of 20°C. It shows that the temperature rises easily to 22°C, and then the melting process barely starts. This is because that in winter, the temperature does not go high enough to melt as much PCM as in the previous scenario. Therefore, the temperature drop is slightly slowed down but less energy is stored. The consequence is that the heating requirements are higher than for the other PCM design but lower than the design without PCM.

These graphs show that designs with PCMs bring lower air temperature variations. However, what is more interesting is what happens to the operative temperature. It clearly shows that the thermal inertia is significantly increased when a PCM with a melting temperature of 20–25°C or 22–27°C is selected. Therefore, the temperatures take longer time to rise, but also to drop and hence it gives more constant temperature levels. Outside of the phase-change temperature range, the designs show relatively similar temperature variations.

The summer day is a typical weekday, and it is chosen to show the scenario "No PCM 21°C" compared to the scenario "PCM 21–26 21°C" with a constant cooling set point of 25°C. The design with PCM, according to Table 3.8.3, gives lower energy requirements and improves comfort. The results for this summer day are shown in Figures 3.8.10 and 3.8.11 for the same bedroom as for the winter study. This time, the effect of the PCM is very clear and shows much more constant temperatures over the day. It can be explained by the fact that the temperatures vary across the phase-change temperature range. The chosen design has the advantage that the PCM melts and freezes while being in the comfort zone. There are two benefits: first, it almost eliminates the need for cooling (0.03 kWh versus

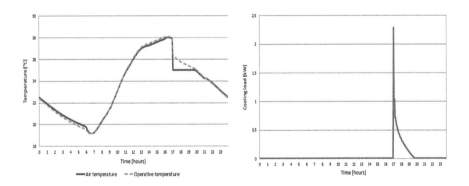

FIGURE 3.8.10 Analysis over a summer day in a bedroom for the scenario "No PCM 21°C".

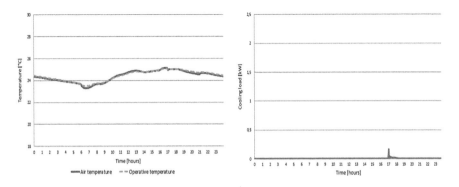

FIGURE 3.8.11 Analysis over a summer day in a bedroom for the scenario "PCM 21–26 21°C".

0.93 kWh), and second it significantly reduces the temperature swing which brings better comfort. The operative temperature varies only within a range of about 1.5°C over the whole day versus about 9°C for the design without PCM. This shows that the integration of PCMs can be effective when done properly. Not only does it lower the cooling needs, but it also brings much more stable temperature and therefore enhances comfort.

3.8.4 CONCLUSION

An approach targeting both comfort levels and energy requirements when integrating PCMs into a building was presented in this article. It was shown how the performance depends on design and operation parameters. Simulation software was used to perform year-round simulations for 30 scenarios and calculate the performance of the building. Depending on the scenario, the results showed significant variations. It is demonstrated that the definition of the comfort (set point temperature) is strongly related to the PCM material used. This indicates that the choice of the right PCM has to be done for a given comfort temperature range.

Regarding the comfort levels, the two types of PCMs that bring more improvement for any heating set point are the PCMs with a phase-change temperature range of 20°C–25°C and 21°C–26°C. For energy savings, the results vary strongly depending on the heating set point. A general observation is that most of the savings occur from cooling, and only little saving comes from heating. Combining efficiency and comfort indicators, it turns out to be that the PCM with a phase-change temperature range of 21°C–26°C is the best regardless of the heating set point.

This highlights the fact that integration of PCM brings significant benefits when the temperatures vary across the phase-change temperature range. For better results, this phase-change temperature range has to be in the comfort zone.

The integration of PCM into buildings is a complex non-linear problem, which shows significant benefits only if well designed. Being weather dependent, the specific designs chosen as optimum for New Zealand might not be the same in other countries. Especially when PCMs are used, a fine-tuning of the HVAC set points has to be performed to avoid energy losses related to inconsistent design.

ACKNOWLEDGMENTS

The authors thank the University of Auckland for funding the research and providing invaluable support and insights all along the study. The authors would like also to acknowledge the support of the Ministry of Business and Innovation of New Zealand for their support to the PCM activities in the University of Auckland.

REFERENCES

1. N. Soares, J.J. Costa, A.R. Gaspar, P. Santos, Review of passive PCM latent heat thermal energy storage systems towards buildings energy efficiency, *Energy Build.* 103 (2013) 59–82.
2. M.M. Farid, A.M. Khudhair, S.A.K. Razack, S. Al-Hallaj, A review on phase change energy storage: materials and applications, *Energy Convers. Manag.* 45 (9) (2004) 1597–1615.

3. L.F. Cabeza, A. Castell, C. Barreneche, A. De Gracia, A.I. Fernández, Materials used as PCM in thermal energy storage in buildings: a review, *Renew. Sustain. Energy Rev.* 15 (3) (2011) 1675–1695.

4. X. Jin, M.A. Medina, X. Zhang, On the importance of the location of PCMs in building walls for enhanced thermal performance, *Appl. Energy* 106 (2013) 72–78.

5. L. Shilei, Z. Neng, F. Guohui, Impact of phase change wall room on indoor thermal environment in winter, *Energy Build.* 38 (1) (2006) 18–24.

6. A. Rahman, M. E. Dickinson, M. M. Farid, Microencapsulation of a PCM through membrane emulsification and nanocompression-based determination of microcapsule strength, *Mater. Renew. Sustain. Energy* 1 (1) (2012) 1–10.

7. (a) A. Castell, M.M. Farid, Experimental validation of a methodology to assess PCM effectiveness in cooling building envelopes passively, *Energy Build.* 81 (2014) 59–71 2014;
(b) Energy Plus. Documentation [cited 2013 17 July]; Available from ⟨http:// apps1.eere. energy.gov/buildings/energyplus/energyplus_documentation.cfm⟩.

8. A. Tardieu. et al. Computer simulation and experimental measurements for an experimental PCM-impregnated office building, in: *Proceedings of the 12th Conference of International Building Performance Simulation Association*, 2011, Sydney.

9. C.L. Zhuang, A.Z. Deng, Y. Chen, S.B. Li, H.Y. Zhang, G.Z. Fan, Validation of veracity on simulating the indoor temperature in PCM light weight building by EnergyPlus, in: *Life System Modeling and Intelligent Computing*, Springer, Berlin Heidelberg (2010) 486–496.

10. P.C. Tabares-Velasco, C. Christensen, M. Bianchi, Verification and validation of Energy Plus phase change material model for opaque wall assemblies, *Build. Environ.* 54 (2012) 186–196.

11. M.M. Behzadi, M.M. Farid, Experimental and numerical investigation on the effect of using phase change materials for energy conservation in residential buildings, *HVAC Res.* 17 (3) (2011) 366–376.

12. N. Zhu, S. Wang, Z. Ma, Y. Sun, Energy performance and optimal control of airconditioned buildings with envelopes enhanced by phase change materials, *Energy Conver. Manag.* 52 (10) (2011) 3197–3205.

13. The Woolmark Company. The sleeping comfort study 2005 [cited 2013 25th Avril]; Available from ⟨http://www.thewoolcompany.co.uk/Sleep%20Better%20with%20 Wool%20Brochure-emailable.pdf⟩.

14. Farid, M.M., Private correspondence with PCM manufacturer, 2013: University of Auckland.

15. Design Builder. Comfort analysis [cited 2013 17 May]; Available from:⟨http://www. designbuilder.co.uk/programhelp/comfort_analysis.htm⟩.

3.9 Benefits of PCM Underfloor Heating with PCM Wallboards for Space Heating in Winter

Paul Devaux and Mohammed Farid
University of Auckland

3.9.1 INTRODUCTION

Moving towards more energy-efficient building has become a priority nowadays due to the climate issue that we are all familiar with. The buildings sector is the largest energy-consuming sector, accounting for over one-third of final energy consumption globally and an equally important source of carbon dioxide (CO_2) emissions. In 2013, space heating and cooling, together with water heating were estimated to account for nearly 60% of global energy consumption in buildings [1].

Within this context, performing thermal energy storage (TES) in buildings has become a priority. Energy can be mainly stored in two forms: sensible heat and latent heat [2]. Conventional building materials use sensible heat to store energy, utilising thermal mass of the construction materials. However, in some parts of the world, lightweight constructions are more popular due to their ease of construction and their architectural flexibility [3]. Such a building has a much lower thermal mass and hence its interior temperature is more influenced by the outdoor temperature. In such a building, large amount of energy is needed to keep comfortable conditions inside it.

Storing energy in an effective way can be done through the use of phase change materials (PCMs). The use of PCM, with high storage density, is more effective when used in lightweight construction [4]. Indeed, they can store between 5 and 14 times more heat per unit volume than sensible heat storage materials such as water, masonry and rock [5]. Their use has been assessed by many researchers over the past decades [1,6–12]. Unlike other materials, PCM absorbs and releases heat when changing state. This happens at a nearly constant temperature [13,14], which has two main benefits: it brings more constant temperature level inside a building and allows the PCM application to be independent of the heating and cooling systems employed [15]. Zhang et al. [16] have summarised the following effects of PCM: (a) narrow the gap between peak and off-peak loads of electricity demand; (b) save operative fees by shifting the electrical consumption from peak periods to off-peak periods; (c) use solar energy continuously, storing solar energy during the day, and releasing it at night, saving energy and improving thermal comfort; (d) store natural cooling by ventilation at night in summer and use it to prevent room temperature from rising during the day, thus reducing the cooling load of air conditioning.

Two types of PCM-based TES systems can be used in buildings: active and passive storage systems. Active systems require an additional fluid loop to charge and discharge a storage tank. On the contrary, passive systems don't require such heat exchanger to extract heat or cold from the storage. The storage is part of the building envelope, storing energy in the form of latent heat. PCM can be incorporated into various building components, such as walls [17], floors [18], and window glazing [19], using micro or macroencapsulation.

A previous work by Barzin et al. [20] presented an experimental investigation on the benefits of using PCM in underfloor heating system, by comparing two small huts of 2.63 m × 2.64 m × 2.64 m, built at Tamaki Campus of University of Auckland, New Zealand. The first hut is finished with ordinary gypsum board while the second one includes PCM boards on the floor (PCM: paraffin, melting range 27°C–29°C), and PCM DuPont sheets called Energain® on the walls and ceiling (PCM: paraffin, melting range 21.7°C).

It must be noted that there is still some lack of confidence in using computer simulation for the prediction of thermal behaviour of a building [21]. Also, the knowledge gap concerning the benefits of PCM in the building sector needs to be narrowed. These important issues will be discussed in this article. The software "EnergyPlus" is well-known as being powerful and capable of dealing with buildings incorporating PCM. EnergyPlus simulation for both common and PCM buildings have been validated previously but without the incorporation of PCM under floor system [22]. This article aims to model the thermal performance of building incorporating PCM-underfloor heating system to implement successful strategy of peak load shifting and show, for the first time, the potential saving in both energy and heating cost. The use of higher melting point PCM with the underfloor heating system will allow effective peak load shifting, while using lower melting point PCM in the walls and ceiling will provide the comfort needed in the building. The developed model will be validated against experimental measurements conducted in the huts described above... Further new performance indicators for energy and cost savings will be introduced to assess the benefit of using PCM. These indicators will be used, for the first time, to optimise both energy and cost savings through the selection of the best peak load shifting periods. The best position of the morning and evening peak periods, which will cause maximum benefit will be identified.

3.9.2 METHODOLOGY

The first objective of this article is dedicated to validate the model of Tamaki hut containing PCM in its walls, ceiling and in its underfloor heating system, which has not been done before. The second objective is to investigate the possibility of using a fixed heating schedule to control the underfloor heating system in such a way to maximise saving in energy cost.

3.9.2.1 THE INPUTS TO THE MODEL

3.9.2.1.1 The Hut Construction

Two almost identical huts of 2.63 m × 2.64 m × 2.64 m are modelled. The first hut is the control hut, finished with ordinary gypsum board on the walls, ceiling and floor. The second hut incorporates PCM on the walls, ceiling and floor as described below. The construction dimensions and materials properties are taken from Khudhair [9] (Figure 3.9.1).

On the floor of hut 2, a PCM-impregnated gypsum board (PCMGP) is used in combination with the underfloor heating system (21 wt% PCM). The table below shows the physical properties of PCM and PCM-gypsum board used with the underfloor heating system as reported by Barzin [20] (Table 3.9.1). Figure 3.9.2 shows a plot of the enthalpy of the PCM-board.

The interior walls and ceiling were lined with PCM DuPont Sheets known as ENERGAIN®. Their properties are taken form a paper published by Kuznik and Virgone [10], as shown in Table 3.9.2 and Figure 3.9.3, except for the density which was given by the manufacturer.

Field	Units	Obj1	Obj2	Obj3	Obj4	Obj5	Obj6
Name		Carpet	40 mm Wood	Gypsum Board 13m	Insulation	Particle Board	Siding
Roughness		MediumRough	MediumSmooth	MediumSmooth	Rough	MediumSmooth	MediumSmooth
Thickness	m	0.0127	0.04	0.013	0.083	0.017	0.0125
Conductivity	W/m-K	0.06	0.12	0.25	0.038	0.12	0.094
Density	kg/m3	288	510	670	32	510	640
Specific Heat	J/kg-K	1380	1380	1089	835	1380	1170
Thermal Absorptance				0.9			0.9
Solar Absorptance				0.3			0.3
Visible Absorptance				0.3			0.3

Field	Units	Obj7	Obj8	Obj9	Obj10	Obj11
Name		PCMGB	Metal 10mm	Polystyrene Foam	Gypsum Board 10m	ENERGAIN
Roughness		MediumSmooth	Smooth	MediumRough	MediumSmooth	Smooth
Thickness	m	0.01	0.01	0.06	0.01	0.0052
Conductivity	W/m-K	0.24	70	0.027	0.25	0.2
Density	kg/m3	848	7824	55	670	865
Specific Heat	J/kg-K	1280	500	1210	1089	2400
Thermal Absorptance		0.9	0.9	0.9	0.9	0.9
Solar Absorptance		0.3	0.7	0.7	0.3	0.7
Visible Absorptance		0.3	0.7	0.7	0.3	0.7

FIGURE 3.9.1   EnergyPlus input for material properties of hut 2, taken from [9].

TABLE 3.9.1
PCMGP on Floor of Hut 2, Properties Calculations

	Gypsum Board	PCM	PCMGP	Weight Ratio	0.21
Density	670	/	848	kg/m³	
Specific heat capacity	1089	2000	1280	J/kg K	
Conductivity	0.25	0.2	0.24	W/m K	
Melting range	27–29			°C	
Latent heat		120,000	25,200	J/kg	

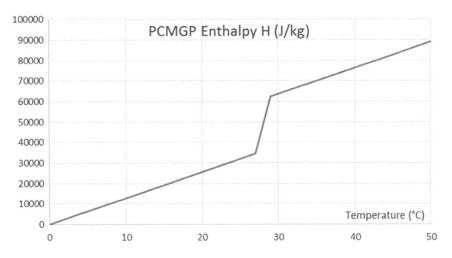

FIGURE 3.9.2 Enthalpy variation of the PCMGP on the floor of hut 2.

TABLE 3.9.2

Energain Sheet Properties

ENERGAIN Sheet

Thickness	5.2	mm
Density	865	kg/m³
Conductivity	0.2	W/m K
Specific heat	2.4	J/g K

Source: Taken from [10].

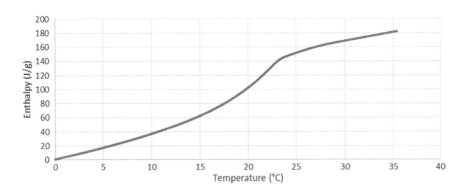

FIGURE 3.9.3 Energain sheet enthalpy variation, taken from [10].

3.9.2.1.2   Run Period and Weather Data

A 10-day period from 7/09/2013 to 16/09/2013 was used to run the simulation. This period was chosen because it is the one that could be validated against available experimental measurements. This validation is presented in the part 3 of this article.

Another input in EnergyPlus software is the file that describes all the hourly outdoor conditions: dry bulb temperature, humidity ratio, dew point, direct normal and diffused solar radiation intensity. The exhaustive list of weather data used by EnergyPlus can be found in the software documentation [23]. The dry bulb temperature was measured on site for the run period. The humidity ratio from the closest weather station (Albany, North Shore) was taken and used to calculate the dew point. The onsite measured global horizontal irradiance from the Sun was used to calculate the split between the direct normal irradiance and the diffuse horizontal irradiance. This split was done using the method proposed in *Solar Engineering of Thermal Processes* [24] (Part 2.10, Chapter 2). The approach is to correlate the fraction I_{dif}/I of the hourly radiation on a horizontal plane which is diffuse, with k_T, the hourly clearness index, which depends of the amount of cloud.

3.9.2.1.3   The Underfloor Heating System & Schedule

For this work, to compare properly the huts, the same HVAC (underfloor heating system) is implemented for both of them. The heater is simply an electric foils, with a total power of 725 W. This HVAC system is the same as the one used in the experimental work [20] and has been validated, as demonstrated below (Figure 3.9.4).

Contrary to the published experimental work [20], the heating system here is not controlled by price-based schedules. When comparing the real-time estimated price of electricity used during the experiments to control the heating system, and the actual price for this period, it appeared that they were not the same (the actual price can be found from the website http://www.emi.ea.govt.nz). This could put on hold any possibility to use price-based schedule to control HVAC system.

As a result, two other heating control methods were used here. First, for the computer validation against experimental data presented in the part 3 of this article, the heating schedule was deduced from the measured temperature variation, so that the custom schedule is not far from the experimental one. The predicted indoor temperature and peak load shifting were then compared with their corresponding measured values.

A second heating control method was created and used for the part 4 of this article, based on the assumption that the electricity price tends to peak for a few hours in the morning and evening, defining two high peak periods over the day. The length of this high peak period has been chosen to be 4 h. When the price is high, the indoor

FIGURE 3.9.4   Underfloor heating system layout, taken from [18].

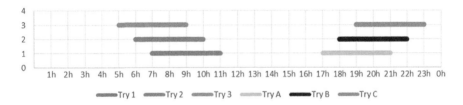

FIGURE 3.9.5   Tested positions of the high peak periods.

temperature was set in the range of 17.8°C–18.2°C. When the price is low, the temperature was set to 19.8°C–20.2°C. These settings were used to control the heater. For example, when the price is low, the heater is switched on until the indoor temperature reaches the low price upper (LPU) constraint (20.2°C), then switched off until the indoor temperature drops to the low price lower (LPL) constraint (19.8°C).

The electricity price is not used to control the heater, but it is used to calculate its cost. The 10-days period is assumed sufficient to show different electricity price variations over a day. This factor is important, as it will enable this work to show how a unique HVAC schedule for each day is sufficient to perform peak load shifting, despite the large variation in electricity price.

Further analysis is also conducted to identify the best position of the two-peak periods regarding energy consumption, energy saving, cost saving and peak load shifting. The graph below shows the three selected durations of the high two-peak periods (Figure 3.9.5).

3.9.2.2   THE OUTPUTS OF THE MODEL

EnergyPlus allows generating many outputs, as calculated minute by minute. These output are: indoor temperature of the huts, temperature of the PCM board placed on top of the underfloor heater, and the electrical power of the heater.

The electricity price available at http://www.emi.ea.govt.nz is then used to calculate the cost of electricity for both huts.

3.9.3   COMPUTER VALIDATION

This part of the article presents (a) a validation of the software EnergyPlus using the experimental measurements obtained from the hut built at Tamaki campus of University of Auckland and (b) the model ability to predict energy consumption and peak load shifting of an underfloor heating system. The results of the software are compared against the experimental work that took place during the period from 7/09/2013 to 16/09/2013.

3.9.3.1   THE RUN PERIOD

The run periods have been chosen based on the availability of experimental data, which were from September 7th to September 16th, 2013. However, the data on days September 10th and September 11th were not reliable and hence were excluded. The first day of the second period, September 12th, was one with low solar radiation. Being the first day of the simulation, EnergyPlus warm-up days will be based on this particular day, which will lead to an inaccurate results in terms of energy consumption. This is the reason why this day, September 12th, was not included in the analysis.

3.9.3.2 COMPARISON OF THE RESULTS

The graphs presented below alternate the simulation results and the experimental measurements and will be discussed below. With the simulation results, the temperature of the PCM under-floor "T pcm" (Figures 3.9.6 and 3.9.8) and the Direct Solar Radiation ("Solar Radiation") are also plotted to help understanding the thermal behaviour of the hut and the PCM.

There are 2 pairs of graphs:

- September 7–9th
- September 13–15th.

The first pair of graph shows a good agreement between the predicted indoor air temperature "T inside" (Figure 3.9.6) and the measured air temperature "Hut 2 RT" (Figure 3.9.7). They also show that the predicted periods for the on/off operation of the underfloor heater match quiet well with those implemented experimentally. However, for the second period below, the predicted periods when the heater is on (Figure 3.9.8) are not always the same as shown in the measurements (Figure 3.9.9). Looking at the evening temperatures during September 14 and September 15 may give explanation to this as will be discussed at the end of this section.

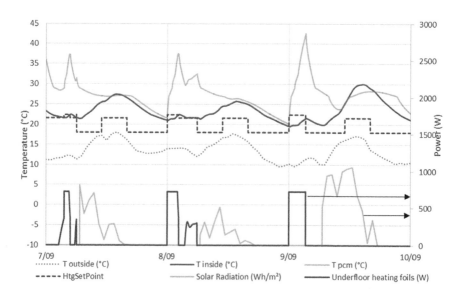

FIGURE 3.9.6 Simulation results for the Tamaki hut containing PCM wallboards and PCM underfloor heating – 7–9th September 2013.

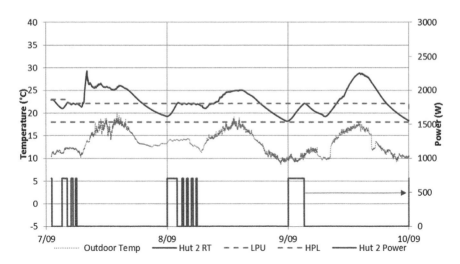

FIGURE 3.9.7 Experimental data for the Tamaki hut containing PCM wallboards and PCM underfloor heating – 7–9th September 2013.

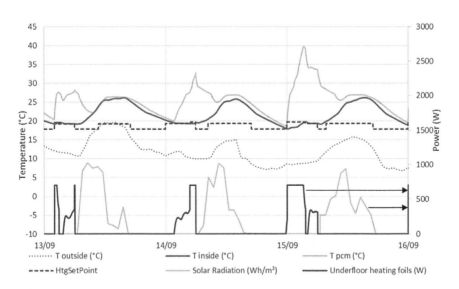

FIGURE 3.9.8 Simulation results for the Tamaki hut containing PCM wallboards and PCM underfloor heating – 13–15th September 2013.

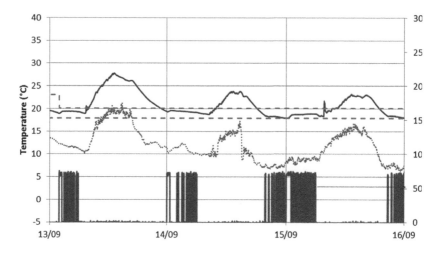

FIGURE 3.9.9 Experimental data for the Tamaki hut containing PCM wallboards and PCM underfloor heating – 13–15th September 2013.

It is interesting to notice that the calculated air temperature variation "T inside" (Figure 3.9.8) is sometimes different from the measured one "Hut 2 RT" (Figure 3.9.9). On 14th of September, the measured air temperature drops down faster than the simulation at night, showing a difference in terms of inertia between the measurements and simulation. Hence, the temperature reaches faster towards the high price lower constraint (17.8°C) at around 20:00, forcing the underfloor heating system to operate. On the contrary, in the simulation, the air temperature reaches 17.8°C only at around 24:00. This explains why the measured electricity consumption for September 14th is higher than what is predicted through the simulation. The same shifting in the electricity use occurs in September 15th.

Table 3.9.3 of the energy consumption shows a very good agreement between the measured daily energy consumptions in the hut and those predicted by EnergyPlus, with an average difference of only 1.5% over 7 days. However, looking at each specific day, EnergyPlus sometimes overestimates the consumption, while sometimes underestimates it. This can be due to a number of reasons such as: (a) the simplification of the model itself, (b) the uncertainties linked to the measurements, and (c) the creation of the EPW file that describes the outdoor conditions, in particular the calculations of the solar radiation data.

Although the error in the prediction of daily energy consumption can be high, the prediction over 7 days is very good indeed.

From this result, the use of EnergyPlus to predict the energy consumption in buildings incorporating PCM underfloor heating system can be reliable.

3.9.4   RESULTS AND DISCUSSION

The first results presented correspond to the analysis of the location of peak. The second part of the results displays the detailed energy consumption, energy saving and cost saving for the peak period position that is found to be best.

TABLE 3.9.3

Experimental and Calculated Daily Energy Consumption of the Underfloor Heating System

		Hut 2 Consumption (W h)		
		Experimental	Simulation	Difference (%)
7-Sep	Day 1	1213	1088	−10
8-Sep	Day 2	2166	2261	4
9-Sep	Day 3	2202	2380	8
13-Sep	Day 4	1115	1150	3
14-Sep	Day 5	1949	1650	−15
15-Sep	Day 6	2866	3038	6
16-Sep	Day 7	2443	2599	6
	Average	1994	2024	1.5

3.9.4.1 PEAK PERIOD POSITION ANALYSIS

To compare the effect of using different peak time periods, four criteria were considered: the average energy consumption of hut 2, the average energy saving in hut 2 (with PCM) in comparison to hut 1 (without PCM), the average cost saving, and the average peak load shifting, given by the difference between the cost saving and the energy saving:

$$\text{Peak Load Shifting} = \text{Cost Saving} - \text{Energy Saving} \qquad (3.9.1)$$

Energy Saving is calculated from the Energy Consumption (EC):

$$\text{Energy Saving} = \frac{\text{EC}_{\text{hut 1}} - \text{EC}_{\text{hut 2}}}{\text{EC}_{\text{hut 1}}} \qquad (3.9.2)$$

The cost of the consumed energy $\text{Cost}_{\text{hut}(x)}$ is calculated using the price of electricity that can be found from the website http:// www.emi.ea.govt.nz. Then, the saving is given by:

$$\text{Cost Saving} = \frac{\text{Cost}_{\text{hut 1}} - \text{Cost}_{\text{hut 2}}}{\text{Cost}_{\text{hut 1}}} \qquad (3.9.3)$$

For each of the four criteria, a performance indicator between 0 and 1 is calculated, 1 being the best performance. This will ease the visual comparison of the results. For energy consumption (EC), the best performance is the lowest of the three. The other performances are obtained by the following equation:

$$\text{Perf}(\text{EC}_i) = 2 - \frac{\text{EC}_i}{\min_i \text{EC}_i} \qquad (3.9.4)$$

For the other criteria x_i, the performance is obtained by calculating the ratio of the number to the highest of the three:

$$\text{Perf}(x_i) = \frac{x_i}{\max_i x_i} \quad\quad (3.9.5)$$

3.9.4.1.1 The Morning Peak Period

The period of the morning peak is first investigated (Table 3.9.4). The evening peak period is fixed at 17:00–21:00.

The last position (5:00–9:00) shows the best performance in terms of energy consumption, cost saving and peak load shifting. Stopping the underfloor heating system at 5:00 am so the temperature can decrease to the lower temperature set range (~18°C) seems to be the best for both lower energy consumption and best peak load shifting.

3.9.4.1.2 The Evening Peak Period

The period of the evening peak is now investigated (Table 3.9.5); the morning peak is being fixed at 5:00–9:00.

Even though the differences are not that obvious, the last criteria confirm that the best position of the evening peak period is 19:00–23:00, which means that the heating system is operating within the high temperature range until 19:00.

The best position for the peak periods are 5:00–9:00 and 19:00–23:00. The next section will detail the results of EnergyPlus using these two periods as high peak periods.

3.9.4.2 DETAILED RESULTS – GRAPHS

Figure 3.9.10 shows the results over the 10-day period. By comparing the two indoor temperatures ("HUT 1: T air" and "HUT 2: T air"), the effect of PCM in reducing peak temperature during the day is obvious. For the sunny days (see the curve "Direct Solar Radiation"), air temperature in hut 2 goes by up to 9°C less than in hut 1.

TABLE 3.9.4

Morning Peak Analysis, Results and Comparison of the 4 Criteria

Peak Period (h)	Average Daily Energy Consumption (W h)	Energy Saving	Cost Saving	Peak Load Shifting
7–11	2216	31%	34%	2.5%
6–10	2086	32%	36%	3.6%
5–9	1879	34%	38%	4.3%
	Performance indicators			
7–11	0.82	0.94	0.90	0.58
6–10	0.89	0.95	0.94	0.83
5–9	1	1	1	1

TABLE 3.9.5

Evening Peak Analysis, Results and Comparison of the Four Criteria

Peak Period (h)	Average Daily Energy Consumption (W h)	Energy Saving	Cost Saving	Peak Load Shifting
17–21	1879	34%	38%	4%
18–22	1886	33%	39%	6%
19–23	1868	32%	42%	9%
	Performance indicators			
17–21	0.99	1	0.91	0.46
18–22	0.99	0.98	0.93	0.64
19–23	1	0.96	1	1

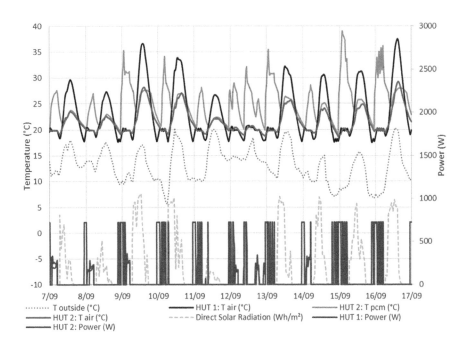

FIGURE 3.9.10 Simulation results, comparison of the two Tamaki huts with underfloor heating – The 10 days run period.

For the morning potential peak load shifting, one must look at the temperature of the PCM-impregnated gypsum board on the floor of hut 2 ("HUT 2: T pcm"). The melting range of the PCM being 27°C–29°C, makes the charging/discharging period easy to see on the graphs. The first 2 days of the simulation are the only one when the PCM doesn't have time to melt during night, hence to release heat during morning peak. Days 1 and 2 are displayed in Figure 3.9.11. The PCM board is releasing heat

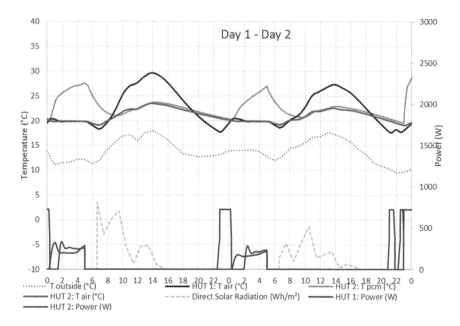

FIGURE 3.9.11 Simulation results, comparison of the two Tamaki huts with underfloor heating – Day 1–2.

in the morning of Day 3–10. For those days, the PCM on the floor is melted during the night by the underfloor heating system. However, it is not sufficient to perform a successful morning peak load shifting. Indeed, only Day 4 shows a good example of morning peak load shifting that is sought after, as hut 1 requires energy between 6 and 7 am, while hut 2 doesn't – see the curves "HUT 1: Power" and "HUT 2: Power" in Figure 3.9.12.

Most of the days, hut 1 doesn't require energy during the morning peak because after the heater is switched off at 5 am, the air temperature doesn't have enough time to drop to the lower set constraint (~18°C) before solar radiation starts to raise room temperature.

Figure 3.9.13 shows a particular day with low solar radiation. During Day 6, very little energy and cost saving was achieved, as reported in Table 3.9.6 at the end of this section. The underfloor heating had to operate both during morning and evening.

The evening peak load shifting occurs more or less every day of the 10-day period. Figure 3.9.14 presents the results for Days 7 and 8, when the evening peak load shifting is obvious, looking at the energy consumption in hut 2 in comparison to hut 1.

3.9.4.3 Detailed Results – Table

Table 3.9.6 shows the daily energy consumption, energy savings and cost savings of hut 2 in comparison to hut 1, for each day of the 10-day period.

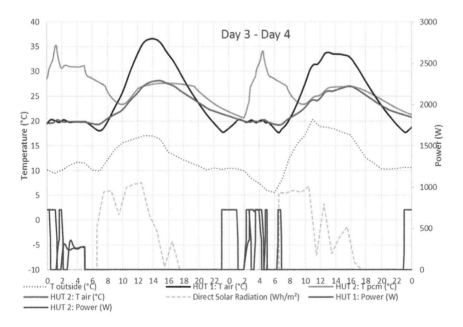

FIGURE 3.9.12 Simulation results, comparison of the two Tamaki huts with underfloor heating – Day 3–4.

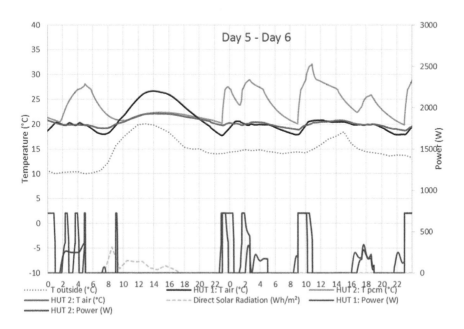

FIGURE 3.9.13 Simulation results, comparison of the two Tamaki huts with underfloor heating – Day 5–6.

TABLE 3.9.6

Detailed Results and Comparison of Both Huts with Underfloor Heating for 10 Days

	Daily Energy Consumption (W h)		Duration HVAC on (h)		Energy Cost		Energy Savings (%)	Cost Savings (%)	Peak Load Shifting?
	Hut 1	Hut 2	Hut 1	Hut 2	Hut 1	Hut 2			
Day 1	2055	1036	5.5	4.7	0.140	0.073	50	48	0
Day 2	2504	1491	6.0	5.6	0.145	0.065	40	55	1
Day 3	2078	1726	4.8	4.5	0.072	0.055	17	23	1
Day 4	3102	1486	5.3	2.7	0.135	0.041	52	70	1
Day 5	2652	1386	5.3	4.2	0.136	0.056	48	59	1
Day 6	3391	3091	10.7	6.6	0.168	0.157	9	6	0
Day 7	1903	1676	3.7	4.5	0.051	0.040	12	22	1
Day 8	3565	1691	7.5	4.6	0.171	0.068	53	60	1
Day 9	3520	3059	6.0	5.6	0.180	0.122	13	32	1
Day 10	2852	2031	4.7	3.7	0.190	0.133	29	30	1
Average	2762	1868	6.0	4.7	0.139	0.081	32	42	1

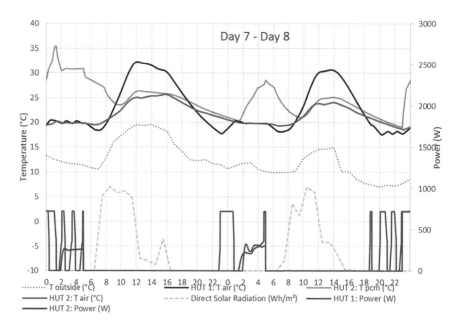

FIGURE 3.9.14 Simulation results, comparison of the two Tamaki huts with underfloor heating – Day 7–8.

The table shows the success of the PCM in creating peak load shifting in hut 2. Over the 10-day period, an overall energy saving and cost saving of 32% and 42%, respectively, have been achieved. In average, hut 2 requires heating around 1 h and 20 min less per day than hut 1.

During the day, a good amount of solar energy is absorbed by the PCM incorporated in the walls and ceiling (Energain sheets), and is released later during evening, thus creating a successful peak load shifting for the evening period, for almost all days of the week.

Concerning the potential morning peak load shifting, the graphs presented above don't always show a convincing behaviour of the underfloor heating in hut 2 (Figures 3.9.13 and 3.9.14): only a little shift in the morning peak load is performed (Figure 3.9.12). This could be due to many things, such as the heating schedule chosen here, all the uncertainties in the material properties, the simplifications of the model, which all contribute to a possible over-estimation of the thermal mass of the hut. This possibility may explain why the temperature in hut 1 doesn't drop fast enough in the morning to force its heating system operating, before the solar radiation starts to heat the room.

3.9.5   CONCLUSIONS

The two huts built at Tamaki campus of University of Auckland (New Zealand) have been modelled, for the first time, to include PCM underfloor heating system using the software EnergyPlus. The use of higher melting point PCM with the underfloor heating system allowed significant peak load shifting, while using lower melting point PCM in the walls and ceiling provided the comfort needed in the building. The results showed limited morning peak load shifting; however, the potential of this heating control has been highlighted in details. Successful evening peak load shifting with an effective control of the underfloor heating system was performed over a 10-day period, which allows cost saving of 42%, and a corresponding energy saving of 32%. The cost saving has come from both energy saving and peak load shifting.

Finally, the temperature set also has a role to play in these results. This issue was not investigated here. In further work, it would be interesting to answer the questions: what temperature ranges would be best for each day? How to predict them the day before to make the controller change the set ranges of the under-floor heating unit?

ACKNOWLEDGMENTS

The authors would like also to acknowledge the support of the Ministry of Business and Innovation of New Zealand, the Royal Society of New Zealand and the European Commission Seventh Framework Programme through the Agreement No. PIRSES-GA-2013–610692 (INNOSTORAGE) for their support to the PCM activities in the University of Auckland.

REFERENCES

1. IEA. *Transition to Sustainable Buildings – Strategies and Opportunities to 2050.* International Energy Agency (IEA); 2013.
2. Başçetinçelik A, Öztürk HH, Paksoy HÖ, Demirel Y. Energetic and exergetic efficiency of latent heat storage system for greenhouse heating. *Renew Energy* 1999;16(1–4):691–4.
3. Hoes P, Trcka M, Hensen JLM, Hoekstra Bonnema B. Investigating the potential of a novel low-energy house concept with hybrid adaptable thermal storage. *Energy Convers Manage* 2011;52(6):2442–7.
4. Chiu JNW, Martin V. Multistage latent heat cold thermal energy storage design analysis. *Appl Energy* 2013;112:1438–45.
5. Sharma A, Chen CR, Vu Lan N. Solar-energy drying systems: a review. *Renew Sustain Energy Rev* 2009;13(6–7):1185–210.
6. Kuznik F, David D, Johannes K, Roux J-J. A review on phase change materials integrated in building walls. *Renew Sustain Energy Rev* 2011;15:379–91.
7. Farid MM, Khudhair AM, Razack SAK, Al-Hallaj S. A review on phase change energy storage: materials and applications. *Energy Convers Manage* 2004;45 (9):1597–615.
8. Khudhair AM, Farid MM. A review on energy conservation in building applications with thermal storage by latent heat using phase change materials. *Energy Convers Manage* 2004;45:263–75.
9. Khudhair AM. Use of phase change materials impregnated into building materials for thermal energy storage applications. In: New Zealand: Department of Chemical & Materials Engineering, University of Auckland; 2006.
10. Kuznik F, Virgone J. Experimental investigation of wallboard containing phase change material: data for validation of numerical modelling. *Energy Build* 2009;41:561e70.
11. Zhou D et al. Review on thermal energy storage with phase change materials (PCMs) in building applications. *Appl Energy* 2012;92:593–605.
12. Oro E et al. Review on phase change materials (PCM) for cold thermal energy storage applications. *Appl Energy* 2012;99:513–33.
13. Ascione F, Bianco N, De Masi RF, de'Rossi F, Vanoli GP. Energy refurbishment of existing buildings through the use of phase change materials: energy savings and indoor comfort in the cooling season. Appl Energy 2014;113:990–1007.
14. Cabeza LF, Castell A, Barreneche C, De Gracia A, Fernández AI. Materials used as PCM in thermal energy storage in buildings: a review. *Renew Sustain Energy Rev* 2011;15(3):1675–95.
15. Soares N, Costa JJ, Gaspar AR, Santos P. Review of passive PCM latent heat thermal energy storage systems towards buildings energy efficiency. *Energy Build* 2013;59: 82–103.
16. Zhang Y, Zhou G, Lin K, Zhang Q, Di H. Application of latent heat thermal energy storage in buildings: state-of-the- art and outlook. *Build Environ* 2007;42:2197–209.
17. Castell A et al. Experimental study of using PCM in brick constructive solutions for passive cooling. *Energy Build* 2010;42(4):534–40.
18. Eicker U. Cooling strategies, summer comfort and energy performance of a rehabilitated passive standard office building. *Appl Energy* 2010;87(6):2031–9.
19. Long L et al. Performance demonstration and evaluation of the synergetic application of vanadium dioxide glazing and phase change material in passive buildings. *Appl Energy* 2014;136:89–97.
20. Barzin R, Chen JJJ, Young BR, Farid MM. Application of PCM underfloor heating in combination with PCM wallboards for space heating using price based control system. *Appl Energy* 2015;148:39–48.

21. Dutil Y et al. Modelling phage change materials behaviour in building applications: comments on material characterization and model validation. *Renew Energy* 2014;61:132–5.

22. Tabares-Velasco, PC, Christensen, C, Bianchi, M. Verification and validation of EnergyPlus phase change material model for opaque wall assemblies. Build Environ 2012;54:186–96.

23. EnergyPlus. Auxiliary programs; 2016.

24. Duffie JA, Beckman WA. *Solar Engineering of Thermal Processes*, 4th ed.; John Wiley, Madison, WI, 2013.

4 Energy-Saving, Peak Load Shifting and Price-Based Control Heating and Cooling
Active Applications

In recent work, an active air–PCM heat exchanger unit was designed and fabricated to study its potential use in space heating and cooling [1]. To this end, two experimental huts, each equipped with a solar and electric heater in winter or an air-conditioning unit in summer, were used to investigate the mentioned concept. Also, one of the huts was provided with a PCM storage unit, and the results obtained were compared with those collected from the reference hut. The results showed that the application of the PCM heat exchanger storage unit reduced indoor temperature variation between day and night by 42% in winter and 20% in summer. In addition, over 10% energy-saving in summer and 30% in winter were recorded [2].

The application of the active PCM system also played an essential role in shifting both heating and cooling loads. In winter, the PCM was charged by solar energy during the day and electric heating during off-peak period for use during peak hours. In summer, free night cooling and cheaper electricity, available at night (by running the AC unit), were used to cool the PCM during the off-peak hours to be used during the period of high-electricity price period. For such a small size facility, a daily energy-saving of up to 47% (~0.83 kWh) with a corresponding 65% cost-saving in winter and up to 23% daily energy-saving (~0.27 kWh) with a relevant 42% of cost-saving in summer was achieved [3].

The price-based control strategy used for peak load shifting raised the need for a predictive control strategy as sometimes, the energy stored during the off-peak period through the use of the electric heater or the AC unit remained unused. Hence, the model predictive control strategy was implemented. This method uses the prediction of weather conditions and electricity prices to decide on charging, storing, and discharging energy from PCM while minimizing electricity cost [4].

The performance of the designed active PCM system was then compared with that of a passive system. The results were in favour of the active system as 22% less energy was consumed to provide comfort when the same energy storage capacity was used in both systems. Also, the active system was more efficient in creating peak load shifting through better control of heat release and hence led to 32% less electricity cost [5].

REFERENCES

1. Gholamibozanjani G, Farid M. Experimental and mathematical modeling of an air-PCM heat exchanger operating under static and dynamic loads. *Energy Build* 2019;202:109354. doi:10.1016/J.ENBUILD.2019.109354.
2. Gholamibozanjani G, Farid M. Application of an active PCM storage system into a building for heating / cooling load reduction. *Energy* 2020;210:118572. doi:10.1016/j.energy.2020.118572.
3. Gholamibozanjani G, Farid M. Peak load shifting using a price-based control in PCM-enhanced buildings. *Sol Energy* 2020;211. doi:10.1016/j.solener.2020.09.016.
4. Gholamibozanjani G, Tarragona J, Gracia AD, Fernández C, Cabeza LF, Farid MM. Model predictive control strategy applied to different types of building for space heating. *Appl Energy* 2018;231. doi:10.1016/j.apenergy.2018.09.181.
5. Gholamibozanjani G, Farid M. A comparison between passive and active PCM systems applied to buildings. *Renew Energy* 2020;162:112–23. doi:10.1016/j.renene.2020.08.007.

4.1 Application of an Active PCM Storage System into a Building for Heating/Cooling Load Reduction

Gohar Gholamibozanjani and Mohammed Farid
University of Auckland

4.1.1 INTRODUCTION

Over the period between 1989 and 2016, total primary energy consumption has risen annually by 2% worldwide and 1.8% in New Zealand [1], which has contributed to a considerable amount of CO_2 emission [2]. The increasing energy consumption and environmental concerns have encouraged the utilization of cleaner and more sustainable energy resources [3]. Solar energy as a renewable, free, and green source of energy [4] is a promising candidate for space heating of buildings [5], which accounts for a significant share of the world's total energy consumption [6]. As solar energy is not available at night or during cloudy hours, the development of solar energy storage technologies is required. Thermal energy storage systems in general and phase change materials (PCMs) in particular offer a great advantage in reducing the energy consumption of buildings [7] and hence minimizing greenhouse gas emissions [8]. PCMs also provide an effective strategy to compensate for the low thermal inertia of buildings that may otherwise cause overheating and increased space cooling load requirements in buildings over summer [1]. The effectiveness of PCM depends on climatic conditions [9]. In continental climates, energy savings in air conditioning (AC), due to the assistance of PCM, have been reported to be between 4% and 95%, while in tropical climates, energy savings reported being limited to less than 17% [10].

PCM technology can be implemented into buildings, either passively [11,12] or actively [13]. In the passive application, PCM is integrated into the building construction materials. The PCM then absorbs the sunlight energy, stores it and releases it to the indoor environment on demand, through natural convection heat transfer, but with no control over the process [14]. Barzin et al. [15] studied the effect of price-based control on PCM underfloor heating in combination with PCM wallboards and hence demonstrated a successful morning peak load shifting but with only little improvement in evening peak load. The results showed 35% total energy saving with a corresponding 44.4% electricity cost saving due to peak load shifting.

In the active applications, on the other hand, forced convection is the dominant heat transfer mechanism between the storage material and indoor air for which some auxiliary electrical or mechanical devices are required [16]. PCM heat storage based on an active system has shown considerable recent research interest [17]. Some studies have been conducted on the design and characterization of an active PCM storage systems for space heating [18], cooling [19] and ventilation [20,21]. Stathopoulos et al. [22] coupled the model of an air-based active PCM storage to a building model under artificial environmental conditions. The results showed the potential of peak load shifting by using an electric heater to charge PCM during off-peak hours for use during peak hours. Osterman et al. [23] numerically investigated the effect of an active air-based PCM storage system on reducing the heating and cooling loads using solar energy and night cooling. Englmair et al. [24,25] successfully designed a solar-assisted short- and long-term energy storage system for residential hot water supply and space heating. The heat captured by solar collector was either directly sent to a water buffer storage or stored in PCM for later use. Using the stable super-cooling of sodium acetate trihydrate, a long-term heat storage was provided.

As mentioned above, previous studies were focused on the characterization of an active PCM storage, its effect on energy demand reduction under artificial weather conditions, charging the PCM using an electric heater only, or using water-based solar collector. However, to the authors' knowledge, nobody has experimentally evaluated the energy savings of a building incorporating an active PCM storage in combination with an air-based solar collector under real environmental conditions. This study investigates energy saving in both space heating and cooling at different seasons, using two identical experimental huts. One test hut was used as a reference, while the other one was used as a retrofit hut containing PCM as a storage medium. Both huts were equipped with a solar air heater and electric heater in winter, and AC unit in summer. A control strategy was developed in LabVIEW software to ensure thermal comfort in the hut based on weather conditions in the city of Auckland, New Zealand. The experimental work presented here investigates the effect of PCM thermal energy storage on (a) collecting solar energy, and excess heat available in the environment for later use in winter, (b) capturing free night cooling for later use in summer, and (c) reducing the electrical energy used for heating and cooling, through the application of an active PCM system in combination with a control strategy.

4.1.2   METHODOLOGY

4.1.2.1   EXPERIMENTAL SETUP

4.1.2.1.1   Air-PCM Heat Storage Units

Two identical air–PCM heat storage units were designed and fabricated in the University of Auckland. The heat storage units' body was made of acrylic sheets and contained 19 sets of aluminum compact storage modules (CSM) (0.45 m $\times$ 0.30 m $\times$ 0.01 m) filled with a commercial macroencapsulated PCM–RT25HC (manufactured by Rubitherm GmbH). The initial purpose of this study was to investigate the potential of an active PCM system in winter. The general comfort temperature for tenants in winter is in the range of 18°C–27°C [1]. Thus, based on Auckland weather conditions and solar radiation, a PCM with a transition temperature at 25°C could melt completely. Among the available PCMs, RT25HC had the highest heat of fusion (230 kJ/kg), which was the main reason for its selection.

The overall weight of PCM in each unit was approximately 9.5 kg RT25HC thermophysical properties can be found from references [26,27]. The melting temperatures of PCM were reported between 22°C and 26°C [26]. The specifications of PCM were then cross checked using differential scanning calorimetry (DSC) analysis and water bath experiments. Indeed, DSC was used to measure the latent heat of PCM while water bath was used to determine the melting temperature range of PCM.

Water bath: A 2 mL vial was filled with PCM and left in a water bath with a constant temperature at 30°C. The water bath experiment showed that PCM starts melting earlier than 22°C, while the majority of melting occurs over a temperature between 22°C and 26°C as shown in Figure 4.1.1.

DSC: A DSC 8000 from PerkinElmer Inc. was used in the experiment to investigate the latent heat of the PCM. Results were compatible with those provided by the manufacturer as shown in Figure 4.1.2. DSC can provide accurate values of the latent heat, but heating rate is very low, as shown in Figure 4.1.2. This problem has been reported in the literature indicating that for accurate determination of the start of melting and solidification using DSC, the heating and cooling rates must be as low as 1°C/min [28].

The heat storage units were insulated with a 20 mm layer of PVC/NBR black rubber foam (thermal conductivity of about 0.037 W/m K [29]). Figure 4.1.3 shows one of the fabricated heat storage units with and without insulation. Detailed information regarding the design and performance of the units can be found in Ref. [30] written by the authors of the current paper. The PCM heat storage units were then placed in one of the experimental huts, described in Section 4.1.2.1.3, and received heat from a vertical solar air heater installed on the exterior wall of the hut.

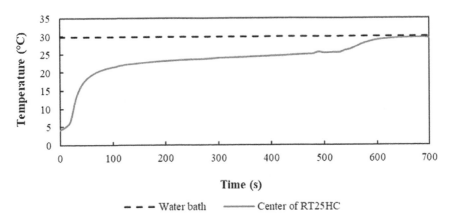

FIGURE 4.1.1 The observation of RT25HC melting temperature using a water bath.

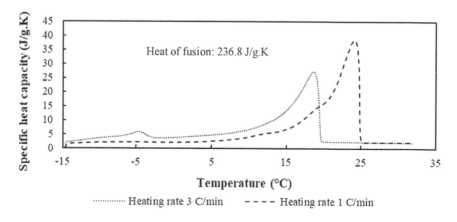

FIGURE 4.1.2 DSC analysis of RT25HC.

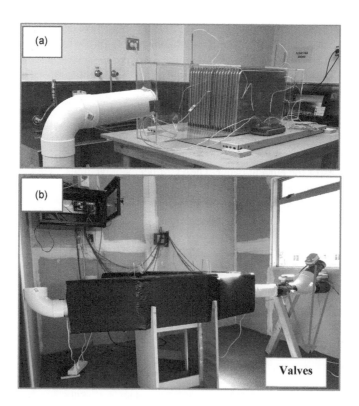

FIGURE 4.1.3 Fabricated air–PCM heat storage unit, (a) without insulation, (b) with insulation, placed in the experimental hut.

4.1.2.1.2 Solar Air Heater

Solar air heaters are typically among the most cost-effective solar technologies used to capture solar energy and convert it into thermal energy [31]. In the current study, $1\,m \times 1\,m$ solar air heater having $0.1\,m \times 1\,m$ photovoltaic (PV) module on top of it was used to provide electricity to run an air fan. The solar air heater was designed to extract air from inside the experimental hut, warm it up and then return it to the room or to the PCM heat storage units. The air circulation was stimulated through a DC fan (Model 091 CE2XB-A71GP) driven by PV module to the room or through a fan, powered by electricity, to the heat storage unit. The power consumption of the fan was only 12 W. Thus, the energy consumption by the fan was less than 5% of the total energy stored in PCM. The performance of this solar air heater is discussed with more details in Sections 4.1.2.1.3 and 4.1.2.4.

4.1.2.1.3 Experimental Huts

Experimental studies were carried out at Ardmore campus of the University of Auckland, New Zealand using two identical fully instrumented experimental huts, namely, "Hut 1" and "Hut 2", as shown in Figure 4.1.4. The huts with an interior dimension of $2.4\,m \times 2.4\,m \times 2.4\,m$ were designed to operate independently. Both huts were built using standard lightweight materials and elevated 0.4 m above the ground. They were equipped with a $1\,m \times 1\,m$ north-facing single glazed window and were insulated using a 100 mm thick pink glass wool in the walls and ceiling and a 60 mm polystyrene foam under the floor to reduce the heat losses. Regular gypsum boards were used as a surface layer on the interior walls and ceiling. The detailed information of the opaque envelope with the associated thermophysical characteristics is found in Ref. [32]. An electric heater (SUNAIR, Model OHS1000) inside the room and a solar air heater on the exterior north wall of the huts were used to provide the heating demand in winter. Otherwise, an AC unit (GREE heat pump, Model GWH09MB-K3DNA4H/O) was installed on the south wall to meet the required cooling demand for summer.

FIGURE 4.1.4 The experimental huts located at the University of Auckland, New Zealand.

Hut 1 was considered as a reference in all experiments throughout this study. But, in addition, Hut 2 was equipped with two identical air-PCM heat storage units connected in parallel. Figure 4.1.5 displays a schematic view of different possible pathways used to supply the heating and cooling demand of Hut 2. The pathways were defined using 110 mm electric polyvinyl chloride valves (*V1*, *V2*, and *V3*). Table 4.1.1 provides details on the status of valves, and the fans used to circulate air through the solar heater and PCM heat storage units. In this table, status "1" denotes the ON mode of the unit and "0" refers to the OFF mode. Section 4.1.2.4 elaborates on the conditions required to go through each pathway.

There was no internal heat gain from lighting, occupants and electric equipment as the huts were empty. The huts represent a building, which has a much larger external surface area to volume compared to a real building. Hence, they are more influenced by external environmental conditions. However, the objective of this study was to compare the thermal performance of a building with and without PCM storage. To this end, two identical huts under the same environmental conditions were used.

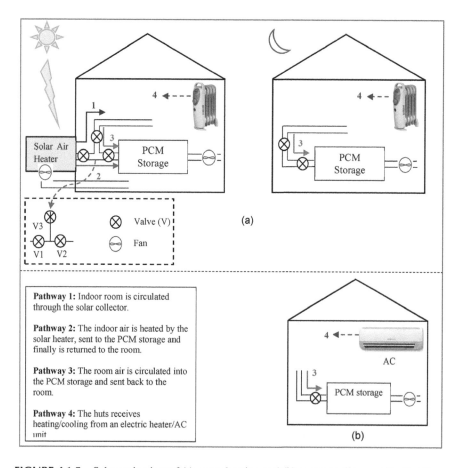

FIGURE 4.1.5　Schematic view of (a) space heating and (b) space cooling used in Hut 2.

TABLE 4.1.1

Status of Valves, Fans and Electric Heater in Different Pathways

Pathway	V1	V2	V3	Fan (Solar Heater)	Fan (PCM Storage)	Electric Heater/AC
1	1	0	1	1	0	0
2	1	1	0	0	1	0
3	0	1	1	0	1	0
4	0	0	0	0	0	1
No action	0	0	0	0	0	0

4.1.2.2 MEASUREMENT INSTRUMENTATION

T-type thermocouples were used to measure the desired temperatures at different places. The distribution of the temperature sensors (thermocouples) is shown in Figure 4.1.6 and explained in Table 4.1.2. The room temperature was taken from the average of T_9 and T_{10}. All thermocouples were calibrated against a reference thermometer (Ebro TFX430) from 0 to 50°C, which enabled a measurement sensitivity up to 0.02°C and an accuracy of ±0.45°C. In this experimental study, it was not possible to duplicate the same experiment as all parameters including the inlet temperature of PCM storage, room and outdoor temperatures were highly dependent on the environmental conditions. Nevertheless, several experiments with different inlet air temperatures to the storage system were conducted, and they showed the same trend as discussed in the study published by the authors of this paper [30]. Solar radiation was measured by using a VAEQ08E pyranometer (Figure 4.1.6, point 12). A power meter, with an accuracy of up to ±0.5%, was incorporated to measure electricity consumed by the electric heaters or the AC units. A data acquisition system was used to receive the required data and send it to a computer, placed in the control hut referred to as Hut 3 (Figure 4.1.4). Air velocity was measured manually by a digital anemometer model AM-4201, which was used to calculate air flow rate. The air velocity in the pipes, with a diameter of 10 cm, was constant at 3 m/s. Detailed information regarding the dimensions of the heat exchanger and consequently the velocity inside the chamber and between PCM modules are explained in a paper published by the authors of the current research [30].

4.1.2.3 DATA ACQUISITION

National Instruments CompactRIO (Compact Reconfigurable Input/Output) programmable controller is an advanced embedded control and data acquisition system, usually designed for applications that require high performance and reliability. CompactRIO incorporates a real-time embedded processor, a high-performance FPGA (Field Programmable Gate Array) and replaceable I/O (Input/ output) modules mounted on the frame. Each I/O module is directly connected to the FPGA [33]. On the other hand, LabVIEW software which is a laboratory virtual instrumentation tool is used as an interface to design, program, process, collect and store the required data through a user-friendly graphical programming system [34].

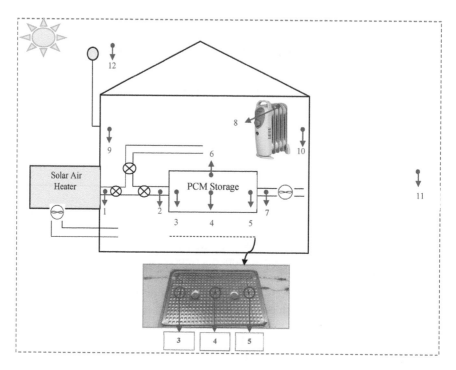

FIGURE 4.1.6   Distribution of temperature sensors in the experiments.

TABLE 4.1.2
Position of Temperature Sensors Used in this Study

Sensor	Sensing position
1	Outlet air of solar heater
2	Inlet air to PCM storage chamber
3	PCM inside the middle storage module (at the beginning)
4	PCM inside the middle storage module (in the center)
5	PCM inside the middle storage module (at the end)
6	Air in the middle of storage Chamber
7	Outlet air of PCM storage Chamber
8	Surface of heater/AC
9	Western area of the room
10	Eastern area of the room
11	Outdoor temperature (far away from the huts)

LabVIEW software includes built-in support to transfer data from I/O modules to FPGA and from FPGA to the embedded processor for real-time analysis and data logging to a dedicated computer [33]. In the case of the current study, CompactRIO system (NI cRIO-9012, National Instruments, USA) is programmed to receive the temperatures of huts, PCM, outdoor and exit of solar air heater as well as the electric heater/AC

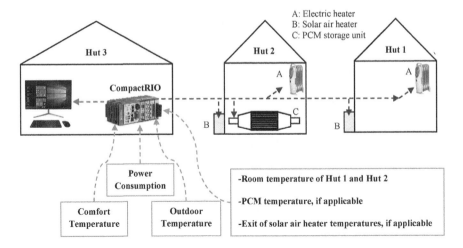

FIGURE 4.1.7 Data acquisition system in experimental huts.

unit power consumption through analog inputs. After processing the data based on the comfortable temperature level, the corrective commands are sent to the electric heater/AC unit, fans and valves (Figure 4.1.3) to initiate the operation of the electric heater/AC unit, solar air heater and PCM heat storage units as shown in Figure 4.1.7. All the required information is then stored in the host computer, placed in Hut 3.

4.1.2.4 Control System

4.1.2.4.1 Space Heating

Two different sets of experiments were carried out to investigate the benefit of active PCM storage for space heating. All experimental conditions, in both sets, were similar, except the use of electric heater in the second experimental set. In the first set, Hut 1 as a reference was heated directly by a solar air heater so long as the room temperature remained below 25°C. While Hut 2 received its required energy from solar air heater so long as the room temperature was lower than 21°C ($T_{LC} + 1$) and the surplus energy from solar heater was stored in the PCM heat storage units for later use. As soon as the room temperature in both huts reached 25°C, the solar heaters were disconnected automatically. The aim of this experiment is to highlight the effect of PCM storage on reducing the indoor temperature variation between day and night, without the use of an electric heating.

In the second experimental set, both huts were equipped with a solar air heater and an electric heater, but Hut 2 was also provided with two identical PCM heat storage units. During cold seasons, the comfort temperature was set between 20°C and 25°C, during the day, and between 19°C and 20°C, during the night [35]. In this set, the solar air heater, in Hut 1, kept circulating air to the room so long as the room temperature was within comfort level and stopped once it reached the upper bound of thermal comfort. On the other hand, when the room temperature dropped below the lower bound of the thermal comfort range (20°C during the day and 19°C at night),

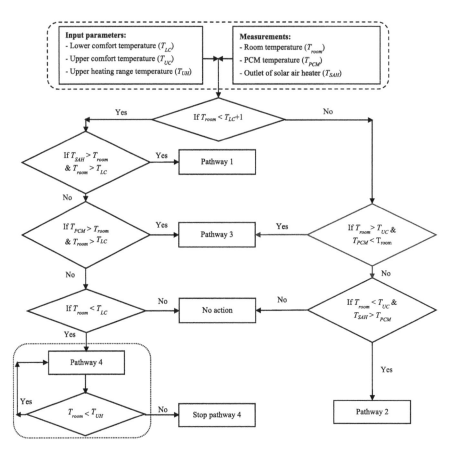

FIGURE 4.1.8   Control algorithm applied to Hut 2 for heating in winter, in the second experimental set.

the electric heater would be started. The range of operation for the electric heaters was between 20°C and 22°C during the day, and 19°C and 20°C during the night, where the lower bound of the heating range was equivalent to the lower bound of comfort temperature. Hut 2 (Figure 4.1.8) received energy from the solar heater and stored it in PCM so long as room temperature was within comfort level.

During the following cooler hours, the energy stored in PCM was used to heat the hut and hence reduced the heating demand of Hut 2. As soon as room temperature dropped below comfort, the electric heater would be started. Once the room temperature reached the upper bound of thermal comfort, all energy sources were stopped. If the room temperature exceeded the upper bound of comfort temperature, plus PCM temperature was lower than the room temperature, then the PCM heat storage units collected the excess heat from the hut through pathway 3. Figure 4.1.8 shows the flowchart of the control strategy used in Hut 2. In the second experimental set, the use of electric heater guaranteed the room temperature to stay above the lower bound of comfort level. However, the room temperature could exceed the upper bound of comfort due to incoming solar radiation through windows.

4.1.2.4.2   Space Cooling

In this section, both huts were equipped with an AC unit, while in addition, Hut 2 was provided with PCM heat storage units. When the room temperature of Hut 2 exceeded the upper bound of comfort level, and if PCM temperature was lower than room temperature, the PCM storage units removed heat from the hut while melting the PCM. During the resulting cooler hours (mainly at night), the collected heat was automatically released to the room. Hence, PCM was solidified for use in the following daytime. If the coolness stored in the PCM was insufficient to maintain the indoor comfort temperature and room temperature exceeded the upper bound of comfort level, the AC unit would start automatically. In Hut 1, the AC unit started as soon as the room temperature reached the upper bound of comfort level. Once the AC units were switched ON, they continued until the room temperature reached the lower bound of comfort level (Figure 4.1.9). In the flowcharts, "no action" denotes the "OFF" mode of both cooling sources (AC and PCM).

4.1.3   RESULTS AND DISCUSSION

4.1.3.1   Space Heating

4.1.3.1.1   Application of Air-Based PCM System for Heating in Combination with Solar Heater

In the first experimental set explained in Section 4.1.2.4, the PCM heat storage units were placed in Hut 2 and connected to the solar heater, while Hut 1 was heated

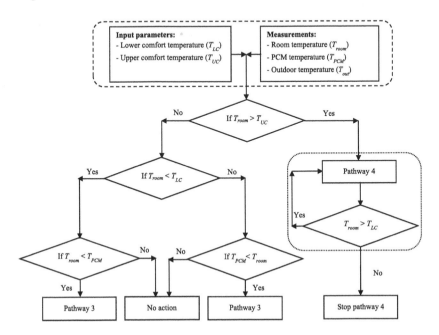

FIGURE 4.1.9   Control algorithm applied to Hut 2 for air-conditioning in the second set of experiments.

directly by a similar solar air heater. In this experiment, no electric heater was used. As displayed in Figure 4.1.10, over a 3-day period from midday 24 July until midday 27 July, the results showed that the PCM heat storage units improved the thermal efficiency of Hut 2 in comparison to Hut 1. This improvement was reflected in the period when indoor room temperature remained within the comfort zone as well as the indoor minimum and maximum temperatures. Hut 2 remained within the comfort zone 54% longer period than Hut 1. In fact, utilizing the high latent heat of PCM, solar energy was stored during sunny hours to be used during colder hours, in the evening. This storage ability helped Hut 2 to go through a lower temperature rise during sunny hours, and a lower temperature decrease during cooler hours, respectively. On the other hand, Hut 1 experienced a sharp temperature rise and drop during the same periods, as shown by square symbols in Figure 4.1.10. As a result, it can be observed that unlike Hut 2, Hut 1 exceeded the upper bound of comfort temperature (by 2°C–4°C). In addition, the minimum temperature in Hut 2 was always about 3°C higher than that in Hut 1 (shown by a double arrow vector in Figure 4.1.10), due to PCM-assisted heating.

4.1.3.1.2 Application of Air-Based PCM in Combination with Solar and Electric Heaters

In this section, both huts were heated using solar and electric heaters, but in addition, Hut 2 was provided with two identical heat storage units. The objective of this experiment was to maintain the indoor room temperature of both huts within the comfort level and prevent the temperature from falling below the lower bound of comfort temperature (19°C at night and 20°C during the day). Besides, if room temperature exceeded the upper bound of comfort temperature, the excess heat was absorbed by using PCM (pathway 3, shown in upper right rhombus box in Figure 4.1.8), cooling the air in Hut 2. Figure 4.1.10, in Section 4.1.3.1.1, displays the effect of PCM on capturing the excess heat from Hut 2, which occurred due to incoming solar radiation through windows (dashed oval).

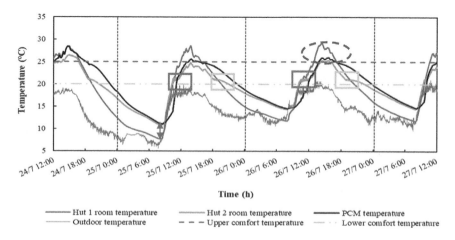

FIGURE 4.1.10 Thermal performance of huts in the first experimental set, over a 3-day period.

Based on the control strategies discussed in Section 4.1.2.4.1, the thermal performances of both huts are shown in Figure 4.1.11, over a 28-day period between 26 June (midday) and 24 July (midday) 2019. It can be seen that the indoor temperature of both huts was kept above the lower bound of the comfort level (19°C at night and 20°C during the day) by the electric heaters. However, the indoor temperature of both huts sometimes exceeded the upper bound of the comfort temperature (25°C) during sunny hours because of the high surface area per volume of huts, which is very different from real buildings. This overheating can be observed over the period between 26 June and 2 July. Hence, to prevent the undesired overheating, the windows of both huts were shaded by a curtain, which improved the indoor temperature of the huts reasonably, from 3 July until 24 July.

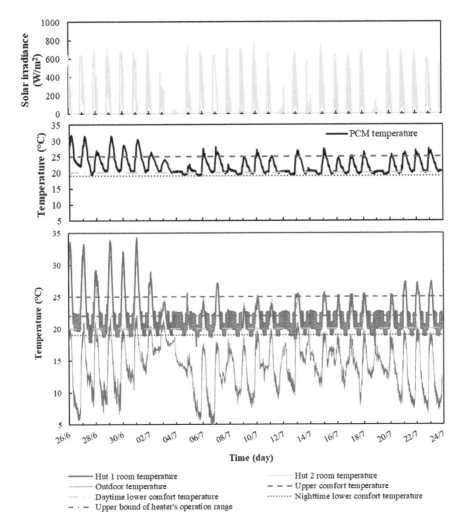

FIGURE 4.1.11 Thermal performance of Hut 2 in the presence of heat storage units and an electric heater, compared to the Hut 1 with electric heater only, over a 28-day period.

In these experiments, the PCM heat storage units absorbed the daytime sunlight energy (pathway 2) and released it during the cooler hours (pathway 3), and hence reduced the heating demand. This load reduction through the PCM storage can be observed more clearly through the thermal behavior of the huts during two consecutive winter days (midday 15 July until midday 17 July), shown in Figure 4.1.12. The associated environmental conditions, such as solar radiation, outdoor temperature, and electricity price, are given.

Figure 4.1.12 also shows the status of the electric heaters in both huts. As shown in this figure, the comfort level was set between 20°C and 25°C during the day (from 5 am until 11 pm) and 19°C–20°C during the night (from 11 pm until 5 am in the following day). In addition, the operating temperature range for the electric heaters was between 19°C and 20°C during the night and between 20°C and 22°C during the day. Results showed that from midnight until 8:30 am on 15 July, the electric heaters were in operation in both huts. In fact, from midnight until 5 am, the electric heaters started as soon as the room temperature dropped below the lower bound of comfort temperature at night (19°C) and continued until the room temperature reached 20°C and then stopped. Over the period between 5 am and 8:30 am, the range of the

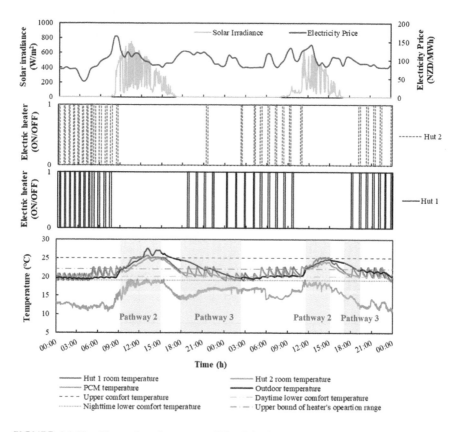

FIGURE 4.1.12 Thermal performance of Hut 2 in the presence of heat storage units and electric heater compared to the Hut 1 with an electric heater only (15–17 July).

electric heater's operation was between 20°C and 22°C (daytime setting). After the electric heaters were automatically switched OFF, the room temperature in both huts decreased, and once the room temperature reached the lower bound of the comfort temperature, heaters were automatically switched ON again. These processes were repeated and hence created the fluctuations seen in Figure 4.1.12, from midnight until 8:30 am. Then, solar radiation heated the huts until the room temperature reached the upper bound of the comfort temperature (25°C). In Hut 1, the captured energy through solar heater directly heated the room, while, in Hut 2, the collected solar energy was stored in the PCM heat storage (Pathway 2, shown in Figure 4.1.8) and then used to heat the room. Thus, the room temperature of Hut 2 increased at a slower rate compared to Hut 1. After staying within the comfort level for a couple of hours, the electric heater in Hut 1 was automatically switched ON at 6:30 pm and operated continuously until next morning; However, in Hut 2, it started at around 9 pm for only 20 min, followed by using energy stored in PCM until 2:30 am next day. The electric heater then continued operating until the next morning. Hence, the assistance of PCM heat storage units resulted in 28% electrical energy saving.

It is also worth noting that based on the Auckland weather condition, the hours between 6 pm and 9:30 pm are considered peak hours, during which the electricity price was higher (around 130 NZD/MWh compared to 90 NZD/MWh during low demand hours [36]). Therefore, utilizing PCM heat storage not only saved energy but also electricity cost through peak-load reduction.

The purpose of the abovementioned experimental measurements carried out over a 28-day period during June and July 2019 (cold winter) was to determine the effect of PCM heat storage on the energy consumption of the huts in cold seasons. Figure 4.1.13 shows that the implementation of PCM storage had a consistently positive effect on the thermal performance of Hut 2, leading to lower electrical energy consumption and cost compared to Hut 1.

The effectiveness of PCM was influenced by outdoor temperature and proper charge and discharge processes. In other words, if PCM melts completely during a day, it can provide maximum utilization for the following night. On the other hand, the cooler a night is, the faster the PCM will be discharged; hence, it can satisfy the

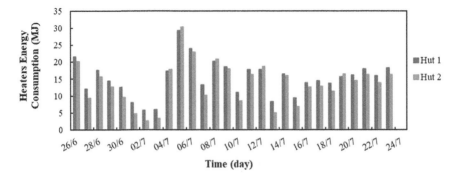

FIGURE 4.1.13 Energy consumption of electric heaters in the huts over a 28-day period in June and July 2019.

heating demand for a shorter period. For instance, during the day of 2 July, PCM was melted completely and during the following night, the outdoor temperature did not drop much (remained at around 16°C). Thus, as shown in Figure 4.1.14, Hut 2 did not require any electrical heating until around 5 am on 3 July, while Hut 1 used electric heating from around 8:30 pm on 2 July. This load reduction in Hut 2 resulted in about 54% energy saving compared to Hut 1. In contrast, if the PCM fails to go through an appropriate charge and discharge processes such as the day on 14 July (Figure 4.1.15a) (which was not charged completely) or releases its energy very fast due to the very cold outdoor temperature, such as the day on 6 July in Figure 4.1.15b, the majority of the heating demand will be provided by the electric heater. As a result, only 2.2% and 4.5% energy savings were achieved on 14 July and 6 July, respectively (Figure 4.1.13).

From Figure 4.1.13, it is also obvious that sometimes the energy saving was negative in Hut 2, such as the day of 12 July. The reason is that PCM was solidified during the nighttime of 12 July (midnight until 6 am). On the other hand, during the daytime of 12 July, the weather was cloudy and PCM was not fully charged as shown in Figure 4.1.16. Hence, electric heating was not only used to ensure comfort temperature but also used to melt the PCM.

Figure 4.1.17 illustrates the daily energy saving in Hut 2 compared to Hut 1, over a 28-day period during the months of June and July as well as a 19-day period in May 2019 (mild season). Results showed that during the months of June and July which are the coldest seasons of Auckland city, the daily energy saving reached up to 54% with a corresponding 10.3% accumulative energy saving, due to the assistance of PCM heat storage units and solar air heater. During the month of May, which is autumn in Auckland and the weather is more moderate, between 10% and 100% daily energy saving with a corresponding 40% accumulative energy saving was achieved.

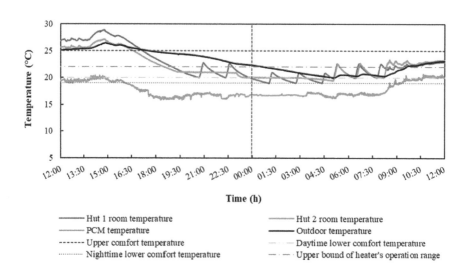

FIGURE 4.1.14 Thermal performance of the huts over the period between midday 2 July and midday 3 July.

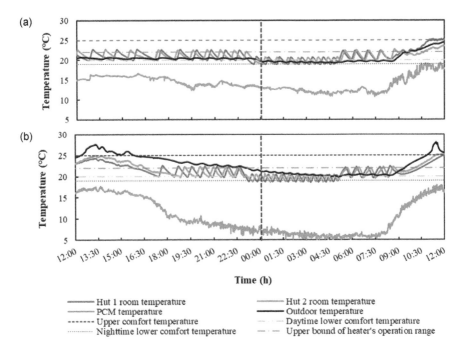

FIGURE 4.1.15   Thermal performance of the huts over the period between (a) midday 14 July and midday 15 July, (b) midday 6 July and midday 7 July.

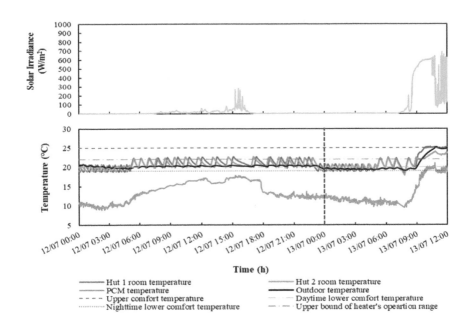

FIGURE 4.1.16   Thermal performance of the huts over the period between midnight 12 July and midday 13 July.

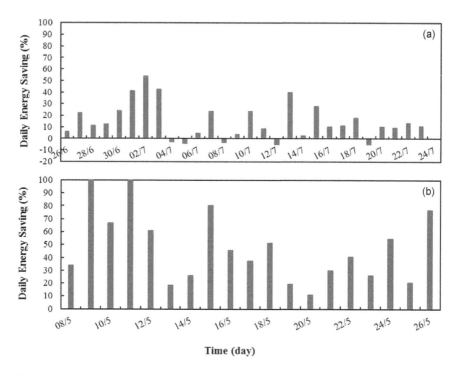

FIGURE 4.1.17 Daily energy saving for Hut 2 compared to Hut 1 over (a) a 28-day period in June-July and (b) a 19-day period in May 2019.

It is surprising that although the weather on 3 July was cloudy and there was no direct solar radiation (Figure 4.1.11), diffuse solar radiation was captured through the solar air heater and used to charge the PCM in the heat storage units. As a result, about 40% energy saving per day was achieved.

4.1.3.2 SPACE COOLING

The PCM storage units were used in combination with an AC unit in Hut 2, while a similar AC unit was used in Hut 1. The control systems, discussed in Section 4.1.2.4.2, were applied to Huts 1 and 2, respectively. In this set of experiments, the comfort level was set between 22°C and 25°C [35]. Figure 4.1.18 illustrates the thermal behavior of both huts based on the new design over a one-day experiment on March 30, 2019. The relevant environmental conditions, such as solar radiation and outdoor temperature, and electricity prices, are also shown. As mentioned before, the operation of the AC units in the huts is shown, where the value "1" denotes the ON mode of the AC unit, and "0" refers to the OFF mode.

Based on this experiment, at around 11:30 am the room temperature of both huts exceeded the lower bound of comfort level (22°C). The room temperature of Hut 1 kept increasing until it reached the upper bound of comfort level at 25°C, and hence, the AC unit started automatically (at 12:40 pm). The AC unit continued operating

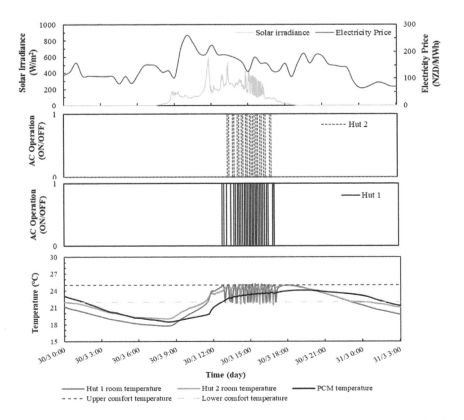

FIGURE 4.1.18   Thermal performance of Hut 2 in the presence of PCM storage and AC unit compared to the Hut 1 with an AC unit only, under low solar radiation.

until the room temperature dropped to 22°C. However, the presence of PCM storage units in Hut 2 (pathway 1 in Figure 4.1.9) allowed the AC unit to start about 30 min later. After the AC units were automatically switched OFF, the room temperature in both huts started to increase. However, the room temperature of Hut 2 increased at a lower rate compared to Hut 1 due to the additional cooling induced by PCM. Once the room temperature reached the upper bound of the comfort level, AC units were automatically switched ON again. These processes were repeated and hence created the fluctuations seen in Figure 4.1.18 from 12:40 pm until 5 pm. During this period, the AC unit in Hut 2 operated 21% less than the one in Hut 1.

At around 11 pm, PCM was solidified by circulating the night time indoor air into the PCM storage for use in the following day. The PCM used in this study begins freezing at a relatively high temperature (26°C) compared to its melting, which starts at 22°C [26], making this PCM a suitable type for building applications.

It is also worth noting that the electricity price in Auckland during the cooling demand period was significantly higher (around 200 NZD/MWh compared to 100 NZD/MWh during low demand hours). Hence, utilizing PCM storage would not only save energy but also reduce electricity costs through peak-load shaving.

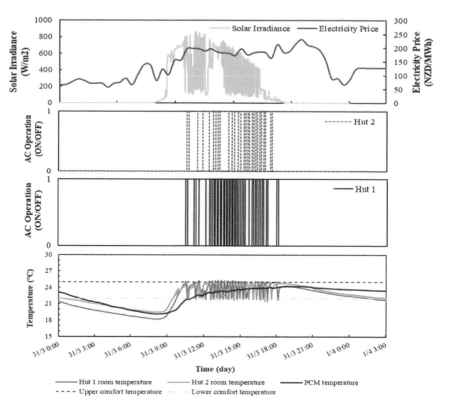

FIGURE 4.1.19 Thermal performance under medium solar radiation of Hut 2 in the presence of PCM storage and AC unit compared to the Hut 1 with AC unit only.

Based on the thermal performance of the huts on a summer day with low solar radiation (Figure 4.1.18), about 30% of energy saving was achieved. In fact, because of the low solar radiation, the room temperature of the huts did not get very high; hence, less energy was required to cool down the rooms during the daytime. During a day with medium solar radiation, Figure 4.1.19 shows a daily energy saving of approximately 10%.

The same set of experiments was carried out over 20 days of warm season (spanning from the end of March to the second weeks of April 2019) and 9 days of the real summer (January 2020) to determine the energy consumption of the AC units in both Huts 1 and 2. Figures 4.1.20 and 4.1.21 show the daily energy consumption of the AC units in both huts used to ensure comfort and daily energy savings achieved due to the implementation of PCM in combination with the control strategy. These figures show that the incorporation of PCM storage units had a consistently positive effect on the thermal performance of Hut 2, leading to lower usage of the AC unit compared to Hut 1. The level of energy savings estimated for the huts confirms this observation. Daily energy saving varied from 10% to 60% in warm season, and 2%–26% during the hot summer. The corresponding accumulative energy saving over the warm and summer periods was about 30.4% and 10%,

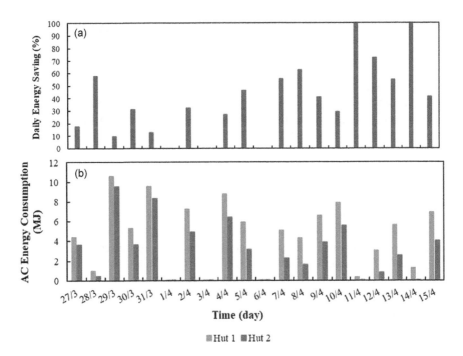

FIGURE 4.1.20 (a) Daily energy saving and (b) daily energy consumption of the AC unit of the huts over 20 days in March and April 2019.

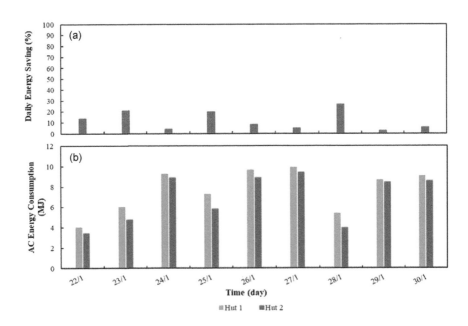

FIGURE 4.1.21 (a) Daily energy saving and (b) daily energy consumption of the AC unit of the huts over 9 days in January 2020.

respectively. In general, the lowest energy savings occurred on the hottest days and vice versa. The hotter the day, the more energy required to cool the room, and the lower the proportion of this energy could be provided through the PCM storage. For instance, the energy saving for the day of March 29, 2019 was only about 10%. In contrast, the day of 28 March was less hot, requiring less amount of energy; hence, it showed about 57% energy saving. However, on some days, almost no energy saving was achieved. The reason was the cool weather, such as the day on 1 April, or very hot weather outside, such as the day on 27 January (Figure 4.1.22). The former did not have any cooling demand during the daytime, while the latter required a considerable amount of cooling under high solar radiation. Also, the outdoor temperature on the preceding night did not drop enough to solidify the PCM and hence did not assist in AC.

Energy saving can also be affected by the nighttime outdoor temperature as well. As shown in Figure 4.1.23, from 27 March until April 2, 2019, the outdoor temperature in Auckland dropped only to 15°C at night. However, in the second period from 2 to 15 April, the temperature reached as low as 10°C at night. The low temperature in the second period allowed PCM to solidify entirely and increase energy saving. For example, the energy requirement during the daytime for Hut 1 on 27 March and 8 April was similar. However, in Hut 2, the AC unit operated on 8 April less frequently than the day on 27 March (Figure 4.1.24) due to the low outdoor temperature at night (Figure 4.1.23), which enabled PCM to capture the nighttime coolness effectively and hence more energy saving was achieved (62.8% versus 17.5%). The

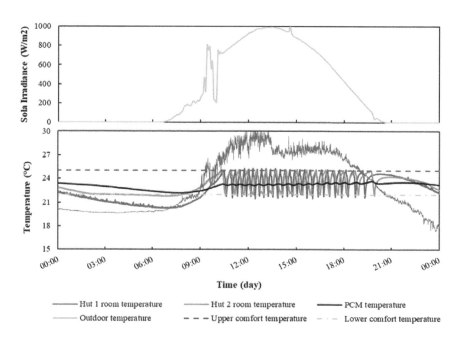

FIGURE 4.1.22 Thermal performance of both huts under high solar radiation of 27 January.

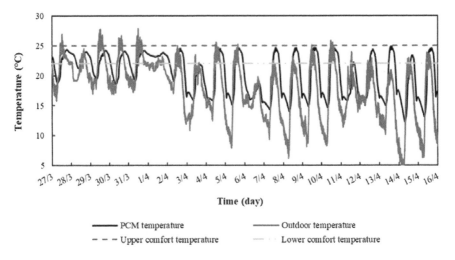

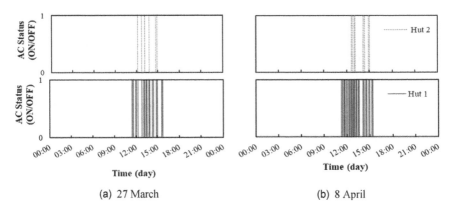

FIGURE 4.1.23   PCM performance based on the application of PCM storage units in combination with AC unit, over a 20-day period.

FIGURE 4.1.24   AC operation mode of the huts on 8 April with a cooler night compared to 27 March.

heating demand and outdoor nighttime temperature, during the summer period, were also high. Thus, the maximum energy saving was about 26%.

4.1.3.3   FURTHER COMMENTS AND DISCUSSION

PCM melting/solidification temperature is one of the main criteria that must be taken into account in the selection of PCM. The desired PCM should undergo full charge and discharge to maximize the storage benefit of PCM through its latent heat [37]. RT25HC, with a melting temperature between 22°C and 26°C and high heat of fusion, was a right choice for space heating. During sunny hours, PCM could easily melt through the available solar air heater. At night, the lower bound of comfort level was

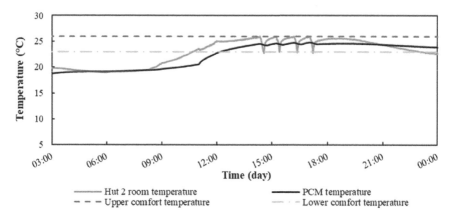

FIGURE 4.1.25   Effect of setpoint temperature range on PCM heat removal ability in summer.

reduced by 1°C to ensure that PCM could solidify at the temperature range between 19°C and 20°C. Figure 4.1.11 confirms the complete melting and solidification of the PCM (black curve) on most of the days. The degree of flatness of the curve is an indication of whether PCM has gone through a complete solidification or not.

However, RT25HC was not the best option for space cooling application under Auckland weather conditions. The nighttime low temperature could solidify the PCM as shown in Figure 4.1.23. However, setting the upper bound of comfort temperature at 25°C did not allow the entire capacity of PCM to be used for heat removal during the following day; PCM failed to melt completely as its temperature reached up to 24°C. Increasing the upper bound of setpoint temperature to 26°C resulted in a better heat removal during the day as the PCM melted up to 25°C (Figure 4.1.25). However, the indoor temperature at 26°C would not be pleasant to occupants. Therefore, based on the Auckland weather conditions, a PCM with a lower melting temperature is required.

Furthermore, the full capacity of the solar air heater, in the space heating experiments, was not fully utilized due to the melting of PCM. Hence, the amount of PCM, PCM melting temperature and the size of collector should be optimized, which is the focus of our future work.

The effect of natural convection in the melt layer of PCM depends on its thickness [38]. In the study published by the authors of this paper [30], using the same 5 mm thick contained, it was shown that the effect of natural convection was negligible.

4.1.4   CONCLUSIONS

Two identical experimental huts, located at the University of Auckland, were used to evaluate the potential of an active PCM storage system for heating and cooling load reduction in different seasons of a year. A solar air heater and an electric heater in winter or an AC unit in summer were used to ensure comfort condition, while in addition one of the huts was provided with two identical PCM heat storage units. The experiments were conducted under weather conditions of Auckland, New Zealand. Conclusions based on the results are:

- PCM heat storage showed a good performance in storing the daytime solar radiation in winter and free night cooling in summer for later use, reducing the indoor daily temperature fluctuations and heating/cooling loads.
- The use of PCM storage unit for space heating (in the absence of the electric heater) allowed Hut 2 to remain within the comfort zone 54% longer than Hut 1.
- A daily energy saving of up to 54% with a corresponding 10.3% accumulative electrical energy saving was achieved over 28 days of winter (months of June and July). These daily savings reached up to 100% and 40% over 19 days of mild winter (month of May).
- Daily energy saving varied from 10% to 60% during warm months of March and April. Accumulative energy savings of about 30% and 10% were achieved, over 20 days in warm period and 9 days in the hot time of the year, respectively.
- Maximum energy saving in summer was made when the nighttime temperature fell well below the melting temperature of the PCM while very limited saving was observed on extremely hot days.

AUTHOR STATEMENT

Gohar Gholamibozanjani: Paper writing, Designing and conducting the experiments, Analyzing the results. Mohammed Farid: Supervision, providing experimental equipment, advising on the experimental study and manuscript revision.

DECLARATION OF COMPETING INTEREST

The authors declare that they have no known competing financial interests or personal relationships that could have appeared to influence the work reported in this article.

ACKNOWLEDGMENT

The first author thanks the University of Auckland for providing her with an International Doctoral Research Scholarship.

NOMENCLATURE

AC	Air conditioner
CompactRIO	Compact Reconfigurable Input/Output
CSM	Compact storage module
DSC	Differential scanning calorimetry
FPGA	Field Programmable Gate Array
I/O	Input/Output
PCM	Phase change material
SAH	Solar air heater
T	Temperature (°C)

T_{LC}	Lower bound of comfort temperature (°C)
T_{out}	Outdoor temperature (°C)
T_{PCM}	Phase change material temperature (°C)
T_{room}	Room temperature (°C)
T_{UC}	Upper bound of comfort temperature (°C)
T_{UH}	Upper bound of heater operation range temperature (°C)
V	Valve

REFERENCES

1. U.S. Energy Information Administration (EIA). International energy statistics n.d. https://www.eia.gov/international/data/world (accessed June 14, 2019).
2. Vallati A, Ocłon' P, Colucci C, Mauri L, de Lieto Vollaro R, Taler J. Energy analysis of a thermal system composed by a heat pump coupled with a PVT solar collector. *Energy* 2019;174:91–6.
3. Yang W, Xu R, Yang B, Yang J. Experimental and numerical investigations on the thermal performance of a borehole ground heat exchanger with PCM backfill. *Energy* 2019;174:216–35.
4. Stevović I., Mirjanić D., Stevović S. Possibilities for wider investment in solar energy implementation. *Energy* 2019;180:495–510.
5. Herrando M, Pantaleo AM, Wang K, Markides CN. Solar combined cooling, heating and power systems based on hybrid PVT, PV or solar-thermal collectors for building applications. *Renew Energy* 2019;143:637–47.
6. Najjar M, Figueiredo K, Hammad AWA, Haddad A. Integrated optimization with building information modeling and life cycle assessment for generating energy efficient buildings. *Appl Energy* 2019;250:1366–82. doi:10.1016/j.apenergy.2019.05.101.
7. Gholamibozanjani G, Tarragona J, Gracia AD, Ferna'ndez C, Cabeza LF, Farid MM. Model predictive control strategy applied to different types of building for space heating. *Appl Energy* 2018;231. doi:10.1016/ j.apenergy.2018.09.181.
8. Lizana J, Chacartegui R, Barrios-Padura A, Valverde JM. Advances in thermal energy storage materials and their applications towards zero energy buildings: a critical review. *Appl Energy* 2017;203:219–39.
9. Ascione F, Bianco N, De Masi RF, de' Rossi F, Vanoli GP. Energy refurbishment of existing buildings through the use of phase change materials: energy savings and indoor comfort in the cooling season. *Appl Energy* 2014;113:990–1007. doi:10.1016/J. APENERGY.2013.08.045.
10. Saffari M, de Gracia A, Fernández C, Cabeza LF. Simulation-based optimization of PCM melting temperature to improve the energy performance in buildings. *Appl Energy* 2017;202:420–34. doi:10.1016/j.apenergy.2017.05.107.
11. Beyhan B, Cellat K, Konuklu Y, Gungor C, Karahan O, Dundar C, et al. Robust micro-encapsulated phase change materials in concrete mixes for sustainable buildings. *Int J Energy Res* 2017;41:113–26.
12. Cellat K, Tezcan F, Kardacs G, Paksoy H. Comprehensive investigation of butyl stearate as a multifunctional smart concrete additive for energy-efficient buildings. *Int J Energy Res* 2019;43:7146–58.
13. Kozak Y, Abramzon B, Ziskind G. Experimental and numerical investigation of a hybrid PCMeair heat sink. *Appl Therm Eng* 2013;59:142–52. doi:10.1016/J. APPLTHERMALENG.2013.05.021.
14. Kuznik F, Virgone J. Experimental assessment of a phase change material for wall building use. *Appl Energy* 2009;86:2038–46. doi:10.1016/J.APENERGY.2009.01.004.

15. Barzin R, Chen JJJ, Young BR, Farid MM. Application of PCM underfloor heating in combination with PCM wallboards for space heating using price-based control system. *Appl Energy* 2015;148:39–48. doi:10.1016/j.apenergy.2015.03.027.

16. Chiu JNW, Gravoille P, Martin V. Active free cooling optimization with thermal energy storage in Stockholm. *Appl Energy* 2013;109:523–9.

17. Lazaro A, Dolado P, Marín JM, Zalba B. PCM-air heat exchangers for free-cooling applications in buildings: experimental results of two real-scale prototypes. *Energy Convers Manag* 2009;50:439–43. doi:10.1016/j.enconman.2008.11.002.

18. Halawa E, Saman W. Thermal performance analysis of a phase change thermal storage unit for space heating. *Renew Energy* 2011;36:259–64. doi:10.1016/j.renene.2010.06.029.

19. Madyira DM. Experimental study of the performance of phase change mate- rial air cooling rig. *Procedia Manuf* 2016;7:420–6. doi:10.1016/j.promfg.2016.12.023.

20. Labat M, Virgone J, David D, Kuznik F. Experimental assessment of a PCM to air heat exchanger storage system for building ventilation application. *Appl Therm Eng* 2014;66:375–82. doi:10.1016/j.applthermaleng.2014.02.025.

21. Promoppatum P, Yao S, Hultz T, Agee D. Experimental and numerical investigation of the cross-flow PCM heat exchanger for the energy saving of building HVAC. *Energy Build* 2017;138:468–78. doi:10.1016/j.enbuild.2016.12.043.

22. Stathopoulos N, El Mankibi M, Issoglio R, Michel P, Haghighat F. Air-PCM heat exchanger for peak load management: experimental and simulation. *Sol Energy* 2016;132:453–66. doi:10.1016/j.solener.2016.03.030.

23. Osterman E, Butala V, Stritih U. PCM thermal storage system for "free" heating and cooling of buildings. *Energy Build* 2015;106:125–33. doi:10.1016/j.enbuild.2015.04.012.

24. Englmair G, Moser C, Furbo S, Dannemand M, Fan J. Design and functionality of a segmented heat-storage prototype utilizing stable supercooling of sodium acetate trihydrate in a solar heating system. *Appl Energy* 2018;221:522–34. doi:10.1016/J.APENERGY.2018.03.124.

25. Englmair G, Kong W, Brinkø Berg J, Furbo S, Fan J. Demonstration of a solar combi-system utilizing stable supercooling of sodium acetate trihydrate for heat storage. *Appl Therm Eng* 2020;166:114647. doi:10.1016/j.applthermaleng.2019.114647.

26. Rubitherm Technologies GmbH. Macroencapsulation - CSM n.d. https://www.rubitherm.eu/en/index.php/productcategory/organische-pcm-rt (accessed August 1, 2019).

27. Khanna S, Reddy KS, Mallick TK. Performance analysis of tilted photovoltaic system integrated with phase change material under varying operating conditions. *Energy* 2017;133:887–99. doi:10.1016/j.energy.2017.05.150.

28. Riga AT, Patterson GH. Oxidative behavior of materials by thermal analytical techniques. 1997.

29. Hebei Kingflex Insulation Co. Ltd. Rubber foam n.d. http://kingflex.coowor.com/shop/product-detail/20160104163923M1NS.htm (accessed August 1, 2019).

30. Gholamibozanjani G, Farid M. Experimental and mathematical modeling of an air-PCM heat exchanger operating under static and dynamic loads. *Energy Build* 2019;202:109354. doi:10.1016/J.ENBUILD.2019.109354.

31. Kraemer D, Poudel B, Feng H-P, Caylor JC, Yu B, Yan X, et al. High-performance flat-panel solar thermoelectric generators with high thermal concentration. *Nat Mater* 2011;10:532.

32. Devaux P, Farid MM. Benefits of PCM underfloor heating with PCM wallboards for space heating in winter. *Appl Energy* 2017;191:593–602. doi:10.1016/J.APENERGY.2017.01.060.

33. Das AD, Mahapatra KK. Real-Time Implementation of fast Fourier transform (FFT) and finding the power spectrum using LabVIEW and CompactRIO. *Int Conf Commun Syst Netw Technol.* 2013:169–73.

34. Zou X, Li B, Zhai Y, Liu H. Performance monitoring and test system for grid- connected photovoltaic systems. *Asia-Pacific Power Energy Eng Conf*; Shanghai, China, 2012. pp. 1–4.

35. CEN (European Committee for Standardization). Indoor environmental input parameters for design and assessment of energy performance of buildings addressing indoor air quality, thermal environment, lighting and acoustics. 2007. doi:10.1520/E2019-03R13.

36. Electricity Authority. NZX Electricity Price Index n.d. https://www1.electricityinfo. co.nz/price_indexes. [Accessed 10 July 2019].

37. Khudhair AM, Farid MM. A review on energy conservation in building applications with thermal storage by latent heat using phase change materials. *Energy Convers Manag* 2004;45:263–75. doi:10.1016/S0196-8904(03)00131-6.

38. Farid M, Husian R. An electrical storage heater using phase change method of heat storage. *Energy Convers Manag* 1990;30:219–30. doi:10.1016/0196-8904(90)90003-H.

4.2 Peak Load Shifting Using a Price-Based Control in PCM-Enhanced Buildings

Gohar Gholamibozanjani and Mohammed Farid
University of Auckland

4.2.1 INTRODUCTION

Peak demand or peak load is referred to as the maximum demand over a specific billing period. Peak demand usually varies for different types of buildings (Seem, 1995). For instance, peak demand for a commercial building lasts for a short period; however, it accounts for half of the overall electricity bill. Hence, peak load shifting does not only reduce the peak demand, but it also saves substantially on energy costs (Sun et al., 2013). A control system and utility management can reduce the electricity demand in a building by switching off unnecessary electrical facilities such as nonessential lights or water heaters. Demand management can result in a 15%–20% reduction in electricity cost of industrial and commercial buildings (Seem, 1995). Studies show that peak load management can lead to a 10–15 billion dollar cost-saving in the US market annually (Sadineni and Boehm, 2012). Out of the commonly adopted methods for peak demand management, peak load shifting is most widely used (Sun et al., 2013). Load shifting takes advantage of electricity rate differences between different periods by shifting on-peak loads to off-peak hours (Piette, 1990).

In industrialized countries, heating, cooling, and air-conditioning account for almost 30% of the total electrical energy used (Lizana et al., 2018). Hence, the use of thermal energy storage could contribute significantly to peak load shifting. Thermal energy storage systems that store high- or low-temperature energy for later use (Lu et al., 2020) are among the most effective means of load shifting (Lee and Jones, 1996). Phase change materials (PCMs), as the media of thermal energy storage, are promising because of their high latent heat and thermal storage capacity per unit volume, and they are able to absorb and release energy at almost constant temperature (Esakkimuthu et al., 2013). PCMs can store daytime solar energy to be used for space heating during early evening peak hours during cold seasons (Gholamibozanjani et al., 2018). On the other hand, storing free cooling available in summer at night through PCM does not only shift the cooling load to off-peak hours, but it also reduces electricity consumption (Barzin et al., 2015a). If PCM cannot provide thermal storage of free energy, then it can be used to shift the electrical load used in conventional heating and cooling of buildings (Barzin et al., 2016).

The integration of PCMs into buildings entails some extra costs for equipment (Treado and Yan Chen, 2013). Thus, the cost of this equipment must be weighed against the utility benefits that could result from the widespread use of such a system (Markarian and Fazelpour, 2019). Hence, the establishment of a control strategy has become essential to optimize the use of PCMs in buildings and to lower investment costs (Fiorentini et al., 2019). Some factors, such as peak power reduction, thermal comfort, and indoor air quality, should be taken into account in the development of an advanced control strategy.

Some control strategies have been applied to buildings equipped with a passive or active PCM system. Solgi et al. (2017) used a night purge ventilation technique to reduce the cooling demand of an office building in summer. The night ventilation was scheduled to automatically start once the outdoor temperature dropped below 30°C and to stop at 7 am, thus solidifying the PCM using the coolness available at night and then melting it to absorb heat during the day. Chen et al. (2019) implemented a control strategy into a simulation-based study to reduce the energy consumed for space cooling in a PCM-enhanced office building. To this end, fresh air was provided to the building during office hours (8 am to 6 pm), while a PCM storage came to operation if room temperature exceeded the upper bound of comfort level by 1°C. The results showed that between 16.9% and 50.8% of electricity energy-saving was achieved. Barzin et al. (2015b) studied the potential of PCM for shifting heating peak loads. Using DuPont wallboards as passive thermal storage in combination with a price-based control, led to a successful peak load shifting. Two temperature ranges were considered for the operation of the electric heater. During off-peak hours, the controller kept the room temperature between 21°C and 23°C (to heat the room and charge the PCM) while, during the peak hours, the room temperature was kept between 17°C and 19°C (to discharge PCM). The results showed up to 62.4% daily cost-saving, depending on the price of electricity.

An active PCM system is preferred when more precise control of a system or greater heat transfer performance is required. As an example, the integration of PCM into heat pumps enhances their coefficient of performance (Kelly et al., 2014). Stathopoulos et al. (2016) showed the benefit of a PCM storage for peak load shifting. In their experimental work in France, a heat exchanger unit composed of a set of plates containing paraffin with a melting point of 37°C was integrated into a ventilation system in a test cell. Through a rough estimation of the test cell's energy requirement and the energy provided from PCM storage, the capability of the system for peak load shifting was demonstrated. Some studies have integrated "Model Predictive Control" strategy, numerically, in an objective of minimizing energy (Serale et al., 2018), energy cost (Touretzky and Baldea, 2016), and PCM performance (Papachristou et al., 2018), which would increase the performance of the system by 10% (Gholamibozanjani et al., 2018). However, ON/OFF controller is the least complicated control system which needs replacement only if it does not deliver desired benefits at the right cost.

This article aims to evaluate the performance of an office-size building in the presence of active air-PCM heat storage in combination with a price-based control (using ON/OFF controller) for shifting both heating and cooling loads, which has not been previously tested by others. To this end, a control strategy was implemented, into two experimental huts, to prioritize the use of different sources of energy based on hourly electricity price. The experiments were conducted under real environmental conditions of Auckland city in New Zealand.

4.2.2 METHODOLOGY

4.2.2.1 EXPERIMENTAL SETUP

4.2.2.1.1 PCM Storage Unit

Two identical air-PCM heat exchanger units were designed and fabricated in the University of Auckland. Figure 4.2.1 shows one of the fabricated storage units, together with its schematic plan view. The body of the storage units was made of acrylic sheets and contained 19 sets of aluminum storage modules (0.45 m × 0.30 m × 0.01 m) filled with a commercial PCM-RT25HC (manufactured by Rubitherm GmbH) having a melting point of 22°C–26°C. The 19 modules were stacked vertically with 5 mm gaps between them, while the space between the modules and wall was set at 2.5 mm. The overall weight of PCM in each unit was approximately 9.5 kg. RT25HC thermophysical properties can be found in references (Rubitherm Technologies GmbH, n.d.) and (Khanna et al., 2017).

PCM melting/solidification temperature is one of the most important features that must be considered to utilize the entire potential of PCM. The selected PCM should undergo full charge and discharge so as to use the maximum capacity of its latent

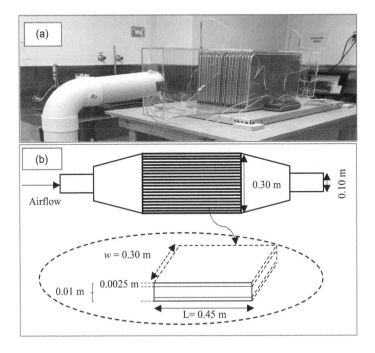

FIGURE 4.2.1   (a) The fabricated PCM storage unit and (b) schematic top-view of the unit.

heat (Khudhair and Farid, 2004). PCM material should be chosen to match its parameters to the corresponding application. In this study, PCM was selected based on the available data on solar energy, comfortable temperature, and weather conditions in the city of Auckland. Indeed, the initial selection criteria of PCM was to be used for space heating. In winter, the comfort temperature in buildings is between 18°C and 27°C (CEN European Committee for Standardization, 2007). The presence of solar air heater, which could heat up the air (passing at 3.5 m/s through a 10 cm pipe) up to 50°C, in conjunction with an active PCM storage unit enabled the use of a PCM with higher melting temperature (peak at 25°C). RT25HC that has a relatively high latent heat was a good choice. However, in passive applications, PCM should be selected based on the desired comfort condition of the people as there is no control over the heat transfer process and PCM performs based on its melting/solidification temperature (Barzin et al., 2015b), while it is not the case in active system. It is important to bear in mind that the melting/solidification range of PCM depends on its purity. Some PCMs experience a very narrow temperature range (less than 1°C), while others melt/solidify within a wider range (4°C, for example) (Hasan et al., 2016).

In terms of PCM mass, EnergyPlus software was first used to estimate the energy demand of the huts based on previous studies done by the corresponding author of the paper (Tardieu et al., 2011), (Devaux and Farid, 2017). Next, through a numerical simulation, different configurations and orientations of an air-PCM heat storage was investigated to find the capability of the unit for creating a 2-h peak load shifting. Finally, the current design was confirmed and validated by some experiments (Gholamibozanjani and Farid, 2019). Based on the heating demand of the huts, it was decided to use two of the designed storage units. The latent heat thermal storage capacity of each unit was 2185 kJ (0.6 kWh). The stored heat was released at a flow rate of 0.036 kg/s with a maximum power of 0.9 kW.

From the safety point of view, the PCM storage unit can be placed inside the buildings as they are encapsulated, and no leakage is expected, but it can equally be placed outside the buildings to save indoor space.

One T-type thermocouple was inserted into the center of the PCM inside one of the modules to measure PCM temperature. Airflow through the gaps between the modules exchanges heat with the PCM. The whole assembly was insulated with a 20 mm layer of PVC/NBR black rubber foam (thermal conductivity of about 0.037 W/m·K (Hebei Kingflex Insulation Co. Ltd., n.d.)). Detailed information can be found in reference (Gholamibozanjani and Farid, 2019) by the authors of this paper.

4.2.2.1.2 Experimental Huts

Experimental studies were carried out at Ardmore campus of the University of Auckland using two identical fully instrumented experimental huts, namely "Hut 1" and "Hut 2", as shown in Figure 4.2.2. The huts that had interior dimensions of 2.4 m × 2.4 m × 2.4 m were designed to operate independently. Both huts were built using standard lightweight materials and were elevated 0.4 m above the ground. They were equipped with a 1 m × 1 m north-facing single glazed window. Both huts received their required energy for space heating from an electric heater (SUNAIR, Model OHS1000) and a 1 m × 1 m solar air heater, which was used to capture solar energy and convert it into thermal energy. For space cooling, an air conditioner (AC)

FIGURE 4.2.2 The experimental huts located at the University of Auckland, New Zealand.

unit (GREE heat pump, Model GWH09MB-K3DNA4H/O) was installed to provide the extra cooling required. One of the huts (Hut 2) was provided with two identical PCM storage units connected in parallel to provide sufficient thermal storage, while Hut 1 was considered as a reference in all experiments in this study.

Figure 4.2.3 displays a schematic view of the different energy sources and the possible flow pathways used to supply heating and cooling to the huts. The pathways were controlled using electric polyvinyl chloride (PVC) valves ($V1$, $V2$, $V3$). A fan (Model AD0912XB-A71GP), driven by a photovoltaic module, was used to suck the indoor air into the solar collector, warm it up and return it to the room. Another inline fan, powered by electricity, was also used to assist in circulating air into the PCM storage unit. The power consumption of the fan was only 12 W, which accounted for less than 5% of the total energy stored in PCM. Table 4.2.1 provides details on the status of valves, and the fans used to circulate air through the solar heater and PCM heat storage units. In this table, status "1" denotes the ON mode of the unit, and "0" refers to the OFF mode.

The huts represent a building with much larger external surface area per volume compared to a real building, which increases the energy demand and makes it more influenced by the external environmental conditions. However, the objective of this study was to investigate the potential of an active PCM system in combination with a price-based control for peak load shifting, using two identical huts. The benefit of using passive PCM technology in real buildings has been verified by the corresponding author through computational simulation using EnergyPlus (Vautherot et al., 2015). Further simulation work should be conducted to incorporate the active storage unit as a heat source/sink within EnergyPlus.

4.2.2.2 MEASUREMENT INSTRUMENTATION

A T-type thermocouple was inserted into the central metal container of the thermal storage unit to measure the PCM temperature. A thermocouple was also installed on a sheltered area of the western wall of the huts to measure the indoor room temperatures. Thermocouples were calibrated against a reference thermometer, Ebro

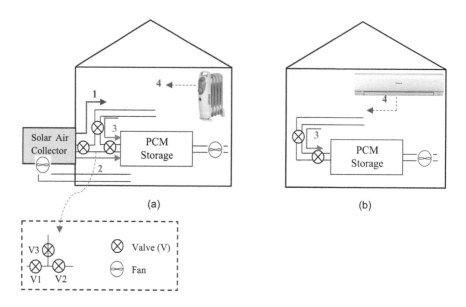

FIGURE 4.2.3   Different possible pathways used for (a) space heating and (b) space cooling.

TABLE 4.2.1

The Status of Valves, Fans and Electric Heater in the Different Pathways

Pathway	Valve 1 (V1)	Valve 2 (V2)	Valve 3 (V3)	Fan (Solar Heater)	Fan (PCM Storage)	Electric Heater
1	1	0	1	1	0	0
2	1	1	0	0	1	0
3	0	1	1	0	1	0
4	0	0	0	0	0	1
No action	0	0	0	0	0	0

TFX430, with a stated accuracy of ±0.5°C in the temperature range of 0°C–45°C. A kWh meter, with accuracy±0.5%, was installed to measure the electricity consumed by the electric heater or the AC unit (Barzin et al., 2015c). A data acquisition system received all data and recorded it to a computer placed in the control hut referred to in Figure 4.2.2 as Hut 3. The airflow rate was measured manually using a digital anemometer model AM-4201. The weather conditions, including ambient temperature, solar radiation, and wind speed and direction, were monitored using HOBOlink, model RX3000 Station - CELL-3G, connected to the CompactRio control unit.

4.2.2.3 Data Acquisition

Data acquisition is the process of measuring physical conditions and converting them into digital numeric values that can be manipulated by computers. A real-time embedded controller can be used to communicate with a field-programmable gate array (FPGA) module, digital input, digital output and analog output of compact reconfigurable technology (cRIO) mounted on an Ethernet chassis (Postolache et al., 2006). In this study, a National Instruments CompactRio (NI cRIO-9012, National Instruments, USA) was used to receive the temperature of the huts, PCM, and outdoors as well as to record electricity prices. Corrective signals would then be sent to control the operation of the PCM storage units, AC unit (in summer), solar and electric heaters (in winter) (Figure 4.2.4). A laboratory virtual instrumentation tool, LabVIEW software, was used as an interface to design, program, process, collect, and store the required data through a user-friendly graphical programming system (Zou et al., 2012). LabVIEW software includes built-in support to transfer data from input/output modules to FPGA and from FPGA to the embedded processor for real-time analysis and data logging to a dedicated computer (Das and Mahapatra, 2013).

4.2.2.4 CONTROL SYSTEM

Designing effective load scheduling, energy, and comfort management techniques is a promising and sustainable solution for the building sector to fulfill their energy demands at a reduced cost while also reducing CO_2 emission (Ahmad and Khan, 2019). In the current research, control strategies were applied to provide peak load shifting in buildings for different seasons. The control systems were implemented based on space heating demand for the cold seasons and space cooling demand for the warm seasons. For peak load shifting, the use of electricity during high wholesale market price periods was postponed and used during cheaper periods. To this end, PCM storage units stored solar energy or cheaper electrical energy available at low peak load to be utilized for space heating during peak hours. During warmer seasons, the PCM unit also stored free cooling (available at night) or lower-price electricity to be used for space cooling during peak periods. Electricity price was calculated using the average of electricity prices for the months of June and July for cold seasons, and that of January and February for the warm seasons, which is shown in the figures of Sections 4.2.3.1 and 4.2.3.2.

Throughout this study, the control system operated the PCM storage units, electric and solar heaters for space heating based on the flowchart shown in Figure 4.2.5. The profile of the electricity price in Auckland indicates that there are two main electricity

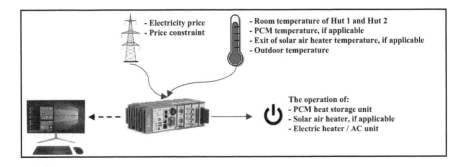

FIGURE 4.2.4   Data acquisition system in experimental huts.

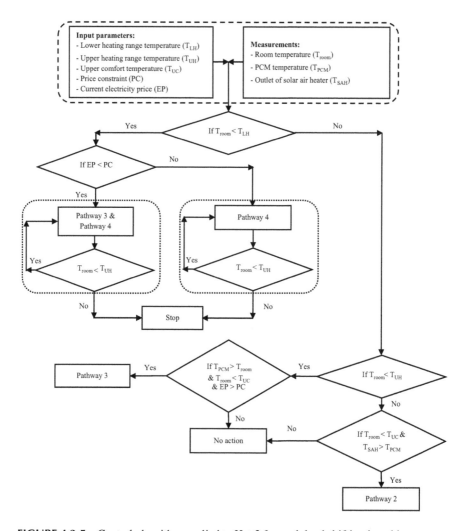

FIGURE 4.2.5   Control algorithm applied to Hut 2 for peak load shifting in cold seasons.

peaks during cold seasons: morning and evening (Barzin et al., 2015c). The control algorithm allowed PCM to absorb daytime solar energy by melting and release it during the early evening peak hours. Alternatively, cheaper electricity available at night was utilized to melt the PCM using the electric heater. The stored latent heat of solidification was then introduced to the room to shift the following morning peak period.

In cold seasons, the indoor comfort temperature was assumed between 19°C and 25°C, while the heater's operation range was set between 19°C and 20°C. When room temperature fell below the upper bound of heater's operation range (20°C), and if electricity price was high (electricity price greater than price constraint), the energy stored in PCM was used to maintain comfort. If the room temperature dropped below the lower bound of comfort level (19°C) and the price of electricity was low, the electric heater would start to heat the room and melt the PCM. During expensive hours, the heater was used to heat the room only, and hence the PCM was not charged. In the flowcharts, "no action" denotes the "OFF" mode of all heating or cooling sources (AC, solar and electric heater and PCM storage unit). When room temperature was above the upper bound of the heating range, but still within the comfort level, the solar heater would operate to store energy in PCM for later use.

In contrast, Hut 1 was heated by an identical solar heater (so long as room temperature remained within the comfort level) or by the electric heater (once the room temperature dropped below the lower bound of comfort temperature).

In warm seasons, the control strategy followed the algorithm shown in Figure 4.2.6. The comfort level was set between 19°C and 20°C during the nighttime (lower-rate electricity hours), while it was set between 22°C and 25°C during daytime. Once the room temperature exceeded the upper bound of comfort level, the AC unit would

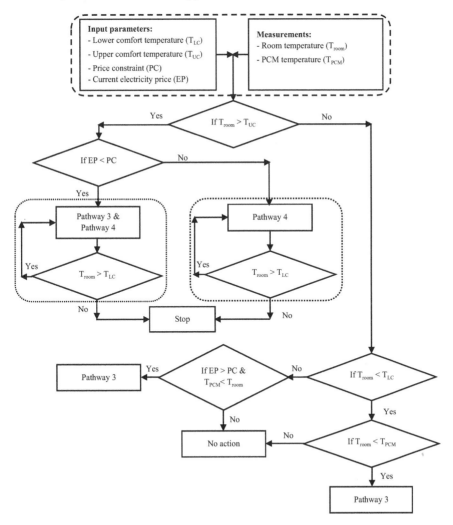

FIGURE 4.2.6 Control algorithm applied to Hut 2 for peak load shifting in warm seasons.

hence start to sustain comfort until it reached the lower bound of comfort level. During the off-peak period, the AC unit was used not only to cool down the room but also to solidify the PCM for later use. During peak hours, the PCM unit in Hut 2 assisted in absorbing the excess heat from the room and hence shifted the cooling peak load while Hut 1 used the AC unit only to ensure comfort.

4.2.3   RESULTS AND DISCUSSION

4.2.3.1   HEATING PEAK LOAD SHIFTING

To test the benefit of PCM in combination with a price-based control for space heating, the algorithm shown in Figure 4.2.5, was followed. To this end, an experiment was conducted over the period between 7 Sep and 9 Sep 2019, as shown in Figure 4.2.7. The indoor room temperature was set between 19°C and 25°C to ensure thermal comfort, while the heater's operation range was set between 19°C and 20°C.

As discussed in Section 4.2.2.4, in a cold season, there were two periods of high electricity price (higher than 140 NZD/MWh); one in the morning and the other was in the evening, based on the Auckland electricity rates. PCM helped to provide the heating demand of the hut during morning peak from 6 am to 9 am, while the period from 9 am to 1 pm, often remained within the comfort level through solar heating.

As shown in Figure 4.2.7, starting from midday on 7 Sep, PCM began to melt by the solar energy available during the sunshine hours. At around 6 pm, the room temperature of Hut 2 dropped and reached the upper bound of the heater's operation temperature (20°C). As 6 pm was considered evening electricity peak hour, the PCM storage units, which were charged by solar energy, were automatically started to heat the room (Pathway 3). By using the PCM latent heat of solidification, the room temperature

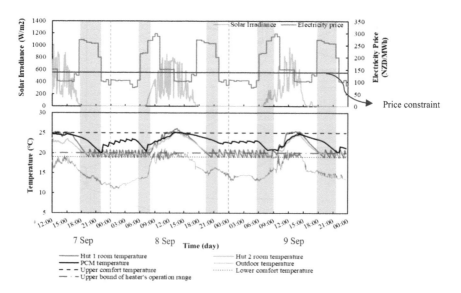

FIGURE 4.2.7   The thermal performance of the huts over three days of September 2019.

remained within the comfort level and hence shifted the evening peak load for about 4h. Once the room temperature dropped below the comfort level or lower bound of the heater's operation range (19°C), the electric heater was started to sustain comfort. It continued until the room temperature reached the upper bound of heater's operation temperature (20°C). That time was an off-peak period, so the electric heater kept satisfying the heating demand until 6 am the next day, causing fluctuations as shown in Figure 4.2.7. During this off-peak period, with a relatively cheap electricity rate, the room and also the PCM were heated for later use. Thus, an increase in the PCM temperature was observed. The stored energy in PCM was then utilized for the following morning peak hours, from 6 am on 8 Sep. Hence, PCM storage also contributed to the leveling of the morning electricity peak load. The periods of PCM discharge are highlighted in gray in Figure 4.2.7. The contribution of PCM storage to peak loads shifting can also be seen during the days of 8 Sep and 9 Sep.

In contrast to the electric heater of the Hut 2, which was OFF during electricity peak hours, the heater of the Hut 1 operated continuously from 7 pm on 7 Sep until 8 am on 8 Sep.

4.2.3.1.1   Energy-Savings

The control algorithm, shown in Figure 4.2.5, was used to continue the peak load shifting experiment over the period between 7 Sep and 1 Oct 2019 (Figure 4.2.8). Figure 4.2.8 illustrates the amount of energy consumed by the electric heaters to ensure comfort in both huts as well as the cumulative energy-saving achieved over the desired period. The results showed that the implementation of PCM storage in combination with price-based control had a consistently positive effect on the thermal performance of Hut 2, leading to reduced energy requirements compared to Hut 1. However, on some days, PCM did not show any reduction in energy (Figure 4.2.8). For example:

- A negative energy-saving was achieved on the day of 11 Sep and 24 Sep (Figure 4.2.9). The low-rate electricity at night was utilized to store energy in PCM; the heat was then extracted to shift the subsequent morning peak load. For the evening peak, however, no solar energy was available during that day to charge the PCM (Figure 4.2.8). Hence, the electric heater had to provide the heating demand of the room in the evening. A little amount of sensible heat was also absorbed by the PCM, thus, more energy was consumed in Hut 2.
- PCM also did not lead to an energy-saving on the days of 12, 20, and 29 Sep, despite they were sunny days. The reason was that the electrically generated heat, which was stored in PCM, remained unused during the morning peak period as the room temperature was already within the comfort level.
- The day of 26 Sep was relatively cold despite being sunny and had the highest heating demand compared to other days. A considerable amount of energy was required to warm up the room and PCM; hence, there was no energy-saving.
- These results raised the need for a predictive control strategy that manages the charge/discharge of the PCM based on future weather conditions.

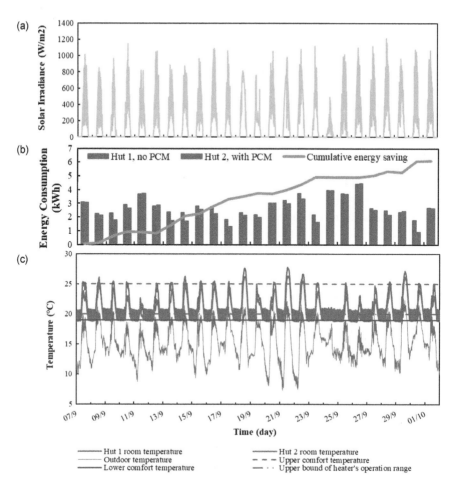

FIGURE 4.2.8   The thermal performance of the huts over the period between 7 Sep and 1 Oct, under the price-based control strategy.

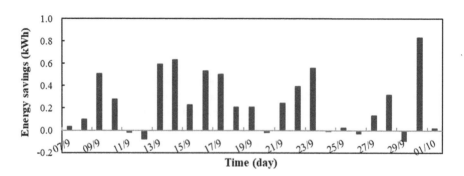

FIGURE 4.2.9   The energy savings of Hut 2 (with PCM) compared to Hut 1 (with no PCM) under the price-based control strategy for space heating.

4.2.3.1.2 Cost Savings

The PCM storage was successful in shifting the heating loads to off-peak periods and saving electricity costs (Figure 4.2.10), even when no energy-saving was achieved. For example, as explained in Section 4.2.3.1.1, on the day of 24 Sep, which was a cloudy day, PCM was not charged by solar energy for use during the evening peak period; thus, no energy-saving was achieved (Figure 4.2.9). However, the storage of low-rate thermal energy during the preceding night resulted in the shifting of morning peak load, and consequently some cost-saving.

As can be seen from Figure 4.2.10, the minimum cost-saving was achieved for the day on 26 Sep. During the morning peak hours of the day on 26 Sep, PCM released its stored energy, which was charged electrically during the low-electricity price period, to sustain comfort (Figure 4.2.11). The discharge of PCM took about 4h, from 9 am to 1 pm, as shown by a gray box in Figure 4.2.11. However, this discharge occurred when Hut 1 did not require any heating due to the available solar energy captured by the solar heater, despite being within the expensive hours (Section 4.2.3.1). In addition, the low outdoor temperature increased the need for electric heating to warm up

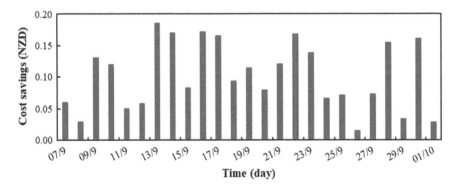

FIGURE 4.2.10 The electricity cost savings of Hut 2 (with PCM) compared to Hut 1 (with no PCM) under the price-based control strategy for space heating.

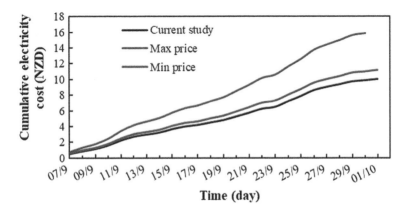

FIGURE 4.2.11 The thermal performance of the huts on 26 Sep.

the room and PCM (sensibly). Thus, only a small cost-saving was achieved for Hut 2 compared to Hut 1.

In contrast, maximum cost-saving was achieved on the day of 13 Sep (about 0.18 NZD/day), as shown in Figure 4.2.10. The reason was that the outdoor temperature did not drop significantly at the preceding night. Hence, less energy was required to ensure comfort and melt the PCM (pathways 3 and 4) for use during the morning peak period. On the other hand, that day was a sunny day with a moderate outdoor temperature, and hence lower heat was required to maintain comfort. Thus, by utilizing the high latent heat of PCM, the surplus solar energy was stored to be used during cooler hours, in the evening.

To emphasize the efficiency of the price-based control method, the lowest and highest electricity price for a day have been chosen and their cumulative curves were plotted over the period between 7 Sep and 1 Oct 2019 (Figure 4.2.12). Figure 4.2.12 shows that the experimental cumulative curve of price-based control method approaches the cumulative curve produced under the lowest electricity prices, which provides an attractive illustration of the current method.

4.2.3.2 COOLING PEAK LOAD SHIFTING

The price-based control was implemented for the PCM-enhanced hut (Hut 2) following the algorithm shown in Figure 4.2.6 to investigate peak load shifting in warm seasons. The electricity rate schedule for warm seasons was different from that of the cooling seasons. Typically, the electricity price in Auckland during the daytime (from 6 am to 5:30 pm) is higher than for the other hours, while during the period

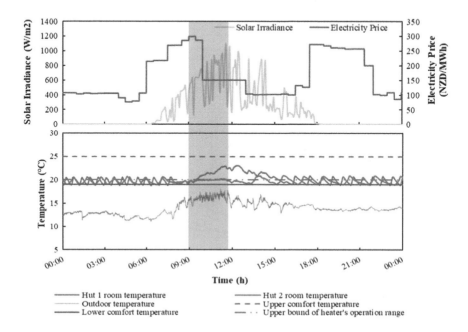

FIGURE 4.2.12   Cumulative cost of energy in Hut 2, with active storage unit.

between 2 pm and 5:30 pm, it reaches its maximum value, as highlighted in gray in Figure 4.2.13. The objective here has been to store coolness in the PCM during off-peak hours, from 12 am to 6 am, and release the latent heat during peak hours, from 2 pm to 5:30 pm.

Unlike space heating, which utilized solar energy to store heat for the following cooler hours, PCM in space cooling did not always have any high-grade free cooling sources to store coolness for the subsequent warmer hours. Therefore, the setpoint temperatures for the day and night were adjusted differently to utilize more cooling energy during the cheaper tariff hours. Indeed, the operating temperature range of the AC unit during the daytime (from 6 am to midnight) was from 22°C to 25°C, while it was between 19°C and 20°C, at night (from midnight to 6 am). During the off-peak period, the PCM was solidified through free cooling available at night (pathway 3) if the indoor room temperature was low enough (Figure 4.2.13a). If not, the AC unit provided the cooling demand of the hut through pathways 3 and 4 (Figure 4.13b).

Figure 4.13b shows that at the midnight of 5 Feb, the room temperature was above the upper bound of comfort level (20°C). However, the electricity rate was at its lowest value. Hence, the AC unit was started to cool down both the room and PCM through pathways 3 and 4 in Hut 2, while in Hut 1, the AC unit operated to cool down only the room (pathway 4). The AC was then stopped when the room temperature reached the lower bound of the comfort level at night (19°C). This process was repeated until 5 am. At 6 am, the setpoint temperature was automatically adjusted between 22°C and 25°C. As the room temperature was already lower than 25°C, there was no need for cooling until 11:30 am. Then, the AC units of both huts were automatically started to ensure comfort (pathway 4). The operation of the AC continued until 2:30 pm. As 2:30 pm was considered a peak hour, the PCM storage automatically operated so long as the indoor temperature of Hut 2 remained within comfort level; otherwise, the AC helped to sustain comfort. However, Hut 1 used the AC unit only during the entire demand period. During this peak period, the AC units of Hut 2 operated less frequently than that in Hut 1. At around 6 pm, the outdoor temperature dropped, and hence no cooling was required for the huts. After midnight, the set temperature

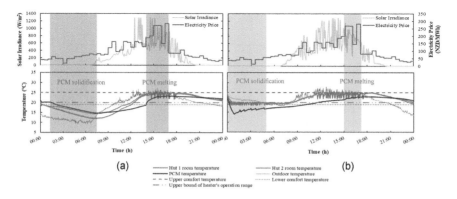

FIGURE 4.2.13 Implementation of the price-based control for cooling peak load shifting on (a) 6 Feb and (b) 5 Feb 2019.

was lowered by only 1°C to charge the PCM during that low electricity price period, which required the AC to run.

The second set of experiments was performed over the periods between 16 Jan and 19 Jan, and 2 Feb and 8 Feb (Figure 4.2.14). The results confirmed that the implementation of price-based control always led to electricity cost-saving, even when only a small percentage of energy-saving was achieved, such as the day on 2 Feb (Figure 4.2.15). On 2 Feb, the outdoor temperature during the preceding night dropped to only 19°C. Consequently, the indoor temperature did not fall enough to solidify the PCM. Instead, PCM stored coolness (generated by the AC unit) and utilized it during the following peak hours. The energy-saving was 3% with a corresponding 6% cost-saving.

By contrast, the maximum cost-saving occurred on 6 Feb, a relatively cold day. At night the outdoor and indoor temperatures dropped to as low as 10°C and 14°C, respectively; hence, PCM was solidified without the need for the AC to operate. On the other hand, the outdoor temperature on the following daytime was not very high; thus, less energy was required to sustain comfort. Also, the cooling demand period overlapped the peak hours, and hence PCM storage operated to maintain comfort. As a result, 0.27 kWh daily energy-saving (23%) with a corresponding 0.05 NZD of cost-saving (42%) was achieved.

In terms of energy consumption, February was warmer than January. As shown in Figure 4.2.14, the nighttime outdoor temperature during the days of January reached as low as 12°C, while the minimum outdoor temperature in February (except 6 Feb) was as high as 19°C. Therefore, less free cooling energy was available at theses nights of February, and hence more electrical energy was needed to run the AC unit.

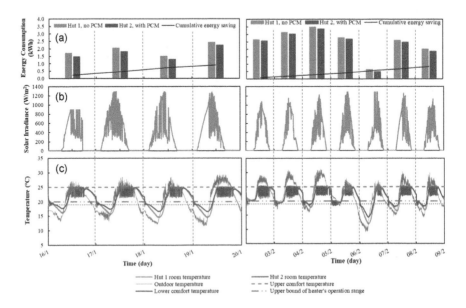

FIGURE 4.2.14 (a) Energy consumption of the huts, (b) solar radiation and (c) indoor and outdoor temperatures of the huts.

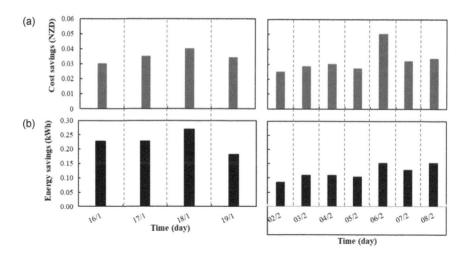

FIGURE 4.2.15 (a) The cost-saving and (b) energy-saving for Hut 2 (with PCM) compared to Hut 1 (with no PCM) under the price-based control strategy for space cooling.

4.2.3.3 FURTHER COMMENTS AND DISCUSSION

RT25HC, with a melting temperature between 22°C and 26°C, was the right choice for space heating applications. Indeed, a solar air heater with a peak temperature of 50°C (under Auckland weather conditions), could easily melt the chosen PCM. Also, the nighttime temperature of 14°C would easily solidify it. The results of space heating, achieved in Section 4.3.3.1, confirmed the proper selection of PCM, showing up to 47% of energy-saving (~0.83 kWh/day) with a corresponding 65% cost-saving. In contrast, RT25HC was not the best option for space cooling applications under Auckland weather conditions. PCM was solidified using a nighttime low temperature or through the cooling generated by the AC unit. However, PCM was not fully melted during the warm period of daytime as the maximum indoor room temperature was limited to 25°C. Thus, it did not use its entire latent heat capacity for removing the indoor heat. Nevertheless, up to 23% of daily energy-saving (~0.27 kWh) with a corresponding 42% of cost-saving was achieved.

The setpoint temperature is also another critical parameter in improving energy and cost-saving. In the current study, the summer daytime comfort level was between 22°C and 25°C, compared to 24°C and 26°C suggested by Barzin et al. (2015a). Having higher setpoint temperatures provides the opportunity to melt the PCM more completely. Thus, the same experiment described in Section 4.2.3.2 was conducted but with a setpoint temperature between 23°C and 26°C. The experiment was performed on 13 Feb, under weather conditions similar to those of 19 Jan. The results showed that lower energy was needed to cool down the room; in addition, more excess heat was absorbed by the PCM. Hence, more energy-saving was achieved: 0.22 kWh/day on 13 Feb (Figure 4.2.16) versus 0.18 kWh/day on 19 Jan (Figure 4.2.15). It is worth noting that the regular price-based control strategy may fail to use the entire stored energy in PCM as the algorithm does not predict the future heating/cooling demand.

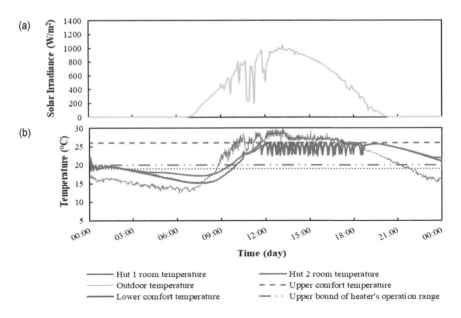

FIGURE 4.2.16 (a) Solar radiation and (b) the thermal performance of the huts on the day of 13 Feb.

Hence, the implementation of price-based control in conjunction with weather forecast data may result in a better performance.

The price of PCM is still relatively high, with the lowest price being 7.5 NZ$/kg. PCM has been implemented in a number of applications in passive buildings, in which 10 wt% PCM in the wall frame provides a maximum benefit (Zhang et al., 2005). However, due to better heat extraction and storage rate, active PCM systems are more efficient in peak load shifting compared to passive systems (Gholamibozanjani and Farid, 2020). Hence, the application of active system could reduce the quantity of PCM and improve the economic potential of PCM application in buildings, consequently.

4.2.4 CONCLUSION AND FUTURE STUDIES

The use of active PCM storage in combination with a price-based control was successful in creating heating and cooling peak load shifting in the studied buildings. The results showed up to 47% of daily energy-saving in winter (~0.83 kWh/day) with a corresponding 65% of electricity cost-saving (~0.185 NZD/day), and up to 23% daily energy-saving (~0.27 kWh/day) with a corresponding 42% of cost-saving (~0.05 NZD/day), in summer. The selected PCM, RT25HC, was a suitable PCM for heating purpose. However, it was not the best option for cooling in summer, since its melting temperature (22°C–26°C) is higher than the upper bound of the comfort level (25°C). Hence, the latent heat storage capacity of the PCM was not fully captured. The authors believe that based on Auckland weather conditions, a PCM with a melting temperature of 23°C, could work for both space heating and cooling. PCM amount and temperature should be optimized based on the desired application and environmental weather conditions, which is the focus of authors' future studies. Moreover, using predictive control strategies, which rely on weather forecast, would enhance peak load shifting.

DECLARATION OF COMPETING INTEREST

The authors declare that they have no known competing financial interests or personal relationships that could have appeared to influence the work reported in this article.

ACKNOWLEDGMENT

The first author thanks the University of Auckland for providing her with an International Doctoral Research Scholarship.

NOMENCLATURE

AC	Air conditioner
CompactRIO	Compact Reconfigurable Input/Output
EP	Current electricity price
FPGA	Field Programmable Gate Array
I/O	Input/Output
PC	Price constraint
PCM	Phase change material
PVC	Polyvinyl Chloride
SAH	Solar air heater
T	Temperature (°C)
T_{LH}	Lower bound of heating range temperature (°C)
T_{out}	Outdoor temperature (°C)
T_{PCM}	Phase change material temperature (°C)
T_{room}	Room temperature (°C)
T_{UC}	Upper bound of comfort temperature (°C)
T_{UH}	Upper bound of heater operation range temperature (°C)
V	Valve

REFERENCES

Ahmad, A., Khan, J.Y., 2019. Real-Time Load Scheduling, Energy Storage Control and Comfort Management for Grid-Connected Solar Integrated Smart Buildings. *Appl. Energy* 259, 114208. doi:10.1016/j.apenergy.2019.114208.

Barzin, R., Chen, J.J.J., Young, B.R., Farid, M.M., 2015a. Application of PCM energy storage in combination with night ventilation for space cooling. *Appl. Energy* 158, 412–421. doi:10.1016/j.apenergy.2015.08.088.

Barzin, R., Chen, J.J.J., Young, B.R., Farid, M.M., 2015b. Peak load shifting with energy storage and price-based control system. *Energy* 92, 505–514. doi:10.1016/j.energy.2015.05.144.

Barzin, R., Chen, J.J.J., Young, B.R., Farid, M.M., 2015c. Application of PCM underfloor heating in combination with PCM wallboards for space heating using price-based control system. *Appl. Energy* 148, 39–48. doi:10.1016/j. apenergy.2015.03.027.

Barzin, R., Chen, J.J.J., Young, B.R., Farid, M.M., 2016. Application of weather forecast in conjunction with price-based method for PCM solar passive buildings - An experimental study. *Appl. Energy* 163, 9–18. doi:10.1016/j.apenergy.2015.11.016.

CEN (European Committee for Standardization), 2007. *Indoor Environmental Input Parameters for Design and Assessment of Energy Performance of Buildings Addressing Indoor Air Quality, Thermal Environment, Lighting and Acoustics*, Geneva: International Standard Organization. doi:10.1520/E2019-03R13.

Chen, X., Zhang, Q., Zhai, Z.J., Ma, X., 2019. Potential of ventilation systems with thermal energy storage using PCMs applied to air conditioned buildings. *Renew. Energy* 138, 39–53. doi:10.1016/j.renene.2019.01.026.

Das, A.D., Mahapatra, K.K., 2013. Real-Time Implementation of fast Fourier transform (FFT) and finding the power spectrum using LabVIEW and CompactRIO. In: *2013 International Conference on Communication Systems and Network Technologies*, NW Washington, DC, USA, pp. 169–173.

Devaux, P., Farid, M.M., 2017. Benefits of PCM underfloor heating with PCM wallboards for space heating in winter. *Appl. Energy* 191, 593–602. doi:10.1016/J. APENERGY.2017.01.060.

Esakkimuthu, S., Hassabou, A.H., Palaniappan, C., Spinnler, M., Blumenberg, J., Velraj, R., 2013. Experimental investigation on phase change material based thermal storage system for solar air heating applications. *Sol. Energy* 88, 144–153. doi:10.1016/J. SOLENER.2012.11.006.

Fiorentini, M., Serale, G., Kokogiannakis, G., Capozzoli, A., Cooper, P., 2019. Development and evaluation of a comfort-oriented control strategy for thermal management of mixed-mode ventilated buildings. *Energy Build.* 109347. doi:10.1016/J.ENBUILD.2019.109347.

Gholamibozanjani, G., Farid, M., 2020. A comparison between passive and active PCM systems applied to buildings. *Renew. Energy* 162, 112–123. doi:10.1016/j.renene.2020.08.007.

Gholamibozanjani, G., Farid, M., 2019. Experimental and mathematical modeling of an air-PCM heat exchanger operating under static and dynamic loads. *Energy Build.* 202, 109354 doi:10.1016/J.ENBUILD.2019.109354.

Gholamibozanjani, G., Tarragona, J., Gracia, A.D., Fernández, C., Cabeza, L.F., Farid, M.M., 2018. Model predictive control strategy applied to different types of building for space heating. *Appl. Energy* 231. doi:10.1016/j.apenergy.2018.09.181.

Hasan, A., Hejase, H., Abdelbaqi, S., Assi, A., Hamdan, M.O., 2016. Comparative effectiveness of different phase change materials to improve cooling performance of heat sinks for electronic devices. *Appl. Sci.* 6. doi:10.3390/app6090226.

Hebei Kingflex Insulation Co. Ltd., n.d. Rubber foam [WWW Document]. http://kingflex. coowor.com/shop/product-detail/20160104163923M1NS.htm (accessed 8.1.19).

Kelly, N.J., Tuohy, P.G., Hawkes, A.D., 2014. Performance assessment of tariff-based air source heat pump load shifting in a UK detached dwelling featuring phase change-enhanced buffering. *Appl. Therm. Eng.* 71, 809–820. doi:10.1016/J.APPLTHERMALENG.2013.12.019.

Khanna, S., Reddy, K.S., Mallick, T.K., 2017. Performance analysis of tilted photovoltaic system integrated with phase change material under varying operating conditions. *Energy* 133, 887–899. doi:10.1016/j.energy.2017.05.150.

Khudhair, A.M., Farid, M.M., 2004. A review on energy conservation in building applications with thermal storage by latent heat using phase change materials. *Energy Convers Manag.* 45, 263–275. doi:10.1016/S0196-8904(03)00131-6.

Lee, A.H.W., Jones, J.W., 1996. Laboratory performance of an ice-on-coil, thermal- energy storage system for residential and light commercial applications. *Energy* 21, 115–130. doi:10.1016/0360-5442(95)00095-X.

Lizana, J., Friedrich, D., Renaldi, R., Chacartegui, R., 2018. Energy flexible building through smart demand-side management and latent heat storage. *Appl. Energy* 230, 471–485. doi:10.1016/J.APENERGY.2018.08.065.

Lu, W., Liu, G., Xiong, Z., Wu, Z., Zhang, G., 2020. An experimental investigation of composite phase change materials of ternary nitrate and expanded graphite for medium-temperature thermal energy storage. *Sol. Energy* 195, 573–580. doi:10.1016/J. SOLENER.2019.11.102.

Markarian, E., Fazelpour, F., 2019. Multi-objective optimization of energy performance of a building considering different configurations and types of PCM. *Sol. Energy* 191, 481–496. doi:10.1016/J.SOLENER.2019.09.003.

Papachristou, A.C., Vallianos, C.A., Dermardiros, V., Athienitis, A.K., Candanedo, J.A., 2018. A numerical and experimental study of a simple model-based predictive control strategy in a perimeter zone with phase change material. *Sci. Technol. Built Environ.* 4731. doi:10.1080/23744731.2018.1438011.

Piette, M.A., 1990. Analysis of a commercial ice-storage system: Design principles and measured performance. *Energy Build.* 14, 337–350. doi:10.1016/0378-7788(90)90096-2.

Postolache, O., Pereira, J.M.D., Girão, P.S., 2006. Real-time sensing channel modelling based on an FPGA and real-time controller. *Conf. Rec. - IEEE Instrum. Meas. Technol. Conf.* pp. 557–562 doi:10.1109/IMTC.2006.235182.

Rubitherm Technologies GmbH, n.d. Macroencapsulation - CSM [WWW Document]. https:// www.rubitherm.eu/en/index.php/productcategory/organische-pcm-rt (accessed 8.1.19).

Sadineni, S.B., Boehm, R.F., 2012. Measurements and simulations for peak electrical load reduction in cooling dominated climate. *Energy* 37, 689–697.

Seem, J.E., 1995. Adaptive demand limiting control using load shedding. *HVAC&R Res.* 1, 21–34.

Serale, G., Fiorentini, M., Capozzoli, A., Cooper, P., Perino, M., 2018. Formulation of a model predictive control algorithm to enhance the performance of a latent heat solar thermal system. *Energy Convers. Manag.* 173, 438–449. doi:10.1016/j. enconman.2018.07.099.

Solgi, E., Kari, B.M., Fayaz, R., Taheri, H., 2017. The impact of phase change materials assisted night purge ventilation on the indoor thermal conditions of office buildings in hot-arid climates. *Energy Build.* 150, 488–497. doi:10.1016/J.ENBUILD.2017.06.035.

Stathopoulos, N., El Mankibi, M., Issoglio, R., Michel, P., Haghighat, F., 2016. Air-PCM heat exchanger for peak load management: Experimental and simulation. *Sol. Energy* 132, 453–466. doi:10.1016/j.solener.2016.03.030.

Sun, Y., Wang, S., Xiao, F., Gao, D., 2013. Peak load shifting control using different cold thermal energy storage facilities in commercial buildings: A review. *Energy Convers. Manag.* 71, 101–114. doi:10.1016/j.enconman.2013.03.026.

Tardieu, A., Behzadi, S., Chen, J.J.J., Farid, M.M., 2011. Computer simulation and experimental measurments for an experimental PCM-impregnated office building. In: *Proceedings of the Building Simulation 2011:12th Conference of International Building Performance Simulation Association*, Sydney, Australia, pp. 56–63.

Touretzky, C.R., Baldea, M., 2016. A hierarchical scheduling and control strategy for thermal energy storage systems. *Energy Build.* 110, 94–107. doi:10.1016/j.enbuild.2015.09.049.

Treado, S., Yan Chen, Y.C., 2013. Saving building energy through advanced control strategies. *Energies* 6, 4769–4785. doi:10.3390/en6094769.

Vautherot, M., Maréchal, F., Farid, M.M., 2015. Analysis of energy requirements versus comfort levels for the integration of phase change materials in buildings. *J. Build. Eng.* 1, 53–62. doi:10.1016/j.jobe.2015.03.003.

Zhang, M., Medina, M.A., King, J.B., 2005. Development of a thermally enhanced frame wall with phase-change materials for on-peak air conditioning demand reduction and energy savings in residential buildings. *Int. J. energy Res.* 29, 795–809.

Zou, X., Li, B., Zhai, Y., Liu, H., 2012. Performance monitoring and test system for grid-connected photovoltaic systems, in: *2012 Asia-Pacific Power and Energy Engineering Conference*. Shanghai, China, pp. 1–4.

4.3 Model Predictive Control Strategy Applied to Different Types of Building for Space Heating

Gohar Gholamibozanjani
University of Auckland

Joan Tarragona
University of Lleida
University of Perugia

Alvaro de Gracia
University of Lleida
University of Perugia

Cèsar Fernández and Luisa F. Cabeza
University of Lleida

Mohammed Farid
University of Auckland

4.3.1 INTRODUCTION

About 36% of global energy used worldwide is attributed to buildings [1], which also contribute to about 17% of total direct energy-related CO_2 emissions to the environment [2]. Heating, ventilation, and air conditioning (HVAC) make a major contribution to energy consumption in buildings [3].

Design professionals, especially architects and engineers, are experiencing an unprecedented level of demand to apply novel approaches to buildings to improve their thermal performance. The integration of thermal energy storage (TES) systems into buildings can satisfy their growing demand for energy as well as reduce environmental pollution caused by the excessive use of energy. Among different energy storage systems, latent heat energy storage (LHES) using phase change materials (PCMs) can greatly enhance the energy efficiency of buildings owing to their large energy storage capacity, which is available within a narrow temperature range [4]. However, the incorporation of TES in buildings to minimize energy consumption and energy costs, while maintaining a comfortable thermal environment, requires comprehensive pre-analysis and thorough mathematical study.

The development of computer technologies and modeling techniques has enabled the prediction of energy consumption levels in buildings [5]. By means of design control methods using dynamic models, prediction of the thermal performance of building systems is now more cost-effective and less time-consuming. Indeed, dynamic models have become crucial for the development of control programs to optimize energy consumption and provide a comfort zone for the occupants of buildings [6]. In this regard, smart control of TES would maximize its energy and economic benefits, hence justifying its initial high investment costs.

Model predictive control (MPC) through the well-established strategy of classical control has attracted research attention in the area of energy-efficient buildings. Although MPC strategies have been used in process control for several decades, they have not been applied to building automation until recently. Basic criteria that MPC strategies need to meet are simplicity, well-estimated system dynamics, steady-state properties, and suitable prediction properties [7]. For instance, Ebrahimpour and Santro [8] used the moving horizon estimation of lumped load and occupancy to improve the accuracy of the dynamic model and MPC performance, subsequently. The advantage of the MPC strategy over conventional building control methods is that it considers the future prediction of ambient temperature, solar radiation and occupancy, as well as system operating constraints, in the design of the control system [9]. However, in conventional methods, the control system is based on occupancy status of the building only, so the heating system is switched off if there is no one in the building. Further, TES is not used to cut down the operating cost of the building [10].

By taking into account internal gains, equipment, weather, and cost, an MPC can provide the required level of thermal comfort [11]. Ma et al. [12] conducted a numerical study to control the cooling system of a building. The building was equipped with a water tank and a series of chillers to provide the cold water. A cost saving of about 24% was achieved through the implementation of an MPC strategy and using weather profile prediction. Morosan et al. [13] also studied thermal regulation using an MPC strategy and weather profile prediction. The control design in their study was based on available control strategies, which have centralized and decentralized structures. In the centralized structure, a single controller is used to provide a comfortable indoor temperature for a multizone building. However, in the decentralized structure, each zone has its own controller. As the centralized structure has computational complexity, and the decentralized one ignores heat transfer between zones, they proposed a distributed control strategy to take advantage of both control structures. Their findings showed that by implementing the distributed structure, in which case the local controllers of different zones share their future behavior, the performance of system was improved.

MPC strategy is being used in HVAC systems for optimal heating and cooling [14], and reduction of peak energy demand in buildings [15]. In the study of energy efficient heating, Siroky et al. [16] carried out an experimental analysis of an MPC strategy using weather prediction approach. Over a 2-month experiment modeled on a building in Prague, the Czech Republic, an energy saving of about 15%–28%

was achieved. Differences found in energy saving were due to the effect of various parameters, such as insulation level and variation in outside temperature. The results revealed a good consistency with the results of a large-scale simulation carried out in another study [17]. It is clear that MPC not only minimizes energy consumption but also contributes to reduction in peak energy demand, which in turn can lower the operating costs of a building. Ma et al. [18] studied the effect of MPC strategy on reduction of peak electricity demand for cooling in a commercial building. Owing to the automatic off-peak pre-cooling effect and shifting of energy demand from peak to off-peak hours, the analysis using MPC resulted in a significant cost saving.

Other research has studied the role of MPC strategies in buildings using TES. For example, Zhao et al. [19] conducted an economic MPC-based study to optimize the energy demand of a Hong Kong zero-carbon building. A stratified chilled water storage tank was integrated into the model as TES. The results showed reductions of 6%–22% in energy consumption, 23%–29% in operating costs, and 12%–48% in CO_2 emissions, depending on the connection to grid and season of the year.

A considerable number of studies have applied MPC strategies to the HVAC systems of buildings to make them more energy-efficient [20]. The majority of this work has taken advantage of sensible TES [21] to further improve energy savings. Much less work has been done on the incorporation of LHES into systems. In one example, Papachristou et al. [22] incorporated PCM into the fabrics of a building in Canada. The PCM was charged through forced air circulation in room. Their objective was to develop a low-order thermal network model for the design of MPC strategy as well as optimization of PCM performance. Finally, the comparison of the modeling results with experimental data showed a great match in predicting the peak power demand and room temperature profile. Touretzky and Baldea [23] embedded LHES into a chilled water tank energy storage system. Proposing a hierarchical control strategy, they tried to manage the cooling demand schema of building and enhance the operation of its electric grid. They investigated cost savings for different load leveling, which is a way to distribute power requirement more evenly during the day, to change the electricity demand patterns of building. As a result, a cost saving of about 88% was achieved for the higher load. It is worth noting that energy storage was the main element of load-leveling strategy. In this case, electricity was used to charge the water tank as well as the PCM with heat. However, utilizing solar energy would have been more advantageous. As the most abundant and free source of energy, solar energy has an enormous potential for heating and cooling of domestic and commercial buildings [24]. To take advantage of solar energy, Fiorentini et al. [25] performed an experimental study of a hybrid MPC strategy to mainly control the cooling process of a residential building in Australia. The building was modeled through a simple R-C model. For the sake of thermal energy management, they considered two hierarchical control modes for HVAC system; one with a 24-h prediction horizon and a 1-h control step, followed by a 1-h prediction horizon and a 5-min control step. Using an air-based photovoltaic thermal collector and a PCM storage, the controller was able to satisfy the cooling demand of the room with a higher heat pump average coefficient of performance than the reference standard air conditioner.

To the authors' knowledge, economic MPC strategy has not been applied to buildings incorporating active PCM storage system charged by solar energy. In this content, a solar-assisted active HVAC system was controlled to minimize heating costs while providing the required comfortable temperature. This strategy was applied to domestic and service buildings and offices in winter, based on weather conditions of Auckland city in New Zealand. Considering the weather condition, the heating demand of buildings was calculated through EnergyPlus simulation rather than using a simplified model. This data was saved in an excel interface, which then was called in Python to apply the MPC strategy. Figure 4.3.1 gives an overview of the performance of MPC in the buildings of current study. In addition, the effect of some parameters such as receding horizon, decision time step, and mass capacity of PCM on energy costs, was studied.

4.3.2 METHODOLOGY

4.3.2.1 DESCRIPTION OF THE SYSTEM

4.3.2.1.1 General Overview

Building automation systems are designed to control heating processes in service and domestic buildings as well as offices. In this study, the building system comprised a solar air collector, a heat exchanger filled with PCM, a backup heater, a fan to drive air from the heat exchanger to a standard basic building model (see more details in Section 4.3.2.2), and energy to keep it at specific temperatures at specified times. The energy required to maintain thermal comfort of a standard basic building is termed *'demand'*. The heating demand of the standard room in this study could be supplied directly from the solar collector, the stored solar energy in the heat exchanger, and also from backup heater. There are three different operation modes for these energy sources, as described in the following paragraphs. By choosing the correct sequence for the operational modes, MPC can automatically provide a comfortable room temperature and reduce electricity cost.

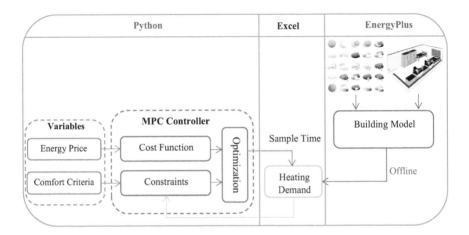

FIGURE 4.3.1   Schematic view of the MPC controller performance in the building of current study.

Solar air collector mode: The solar collector mode is active when demand is coincident with the presence of sunlight. If solar energy is greater than demand, surplus thermal energy will be stored in the heat exchanger by charging PCM in it. In the event that the collected solar energy is greater than both demand and the energy required to charge PCM, the excess will be discharged into environment.

Heat exchanger discharge mode: Energy from PCM will be discharged if demand is greater than available energy from the solar collector. However, if at a given time the electricity rate is cheap, the optimizer tends to use backup heater and leave the PCM energy for hours of high electricity price.

Backup heater mode: If the heat exchanger and solar collector are not able to satisfy demand, the backup heater starts working. On the other hand, when the system is operating during hours of low-cost electricity, even though solar energy is available or the heat exchanger is charged, MPC can decide to use the backup heater based on weather prediction.

In the design of system, the backup heater and heat exchanger were located inside the building. Thus, the energy loss from both may be considered negligible. Figure 4.3.2 shows an overview of automation system used in the current study.

The aim of the automation system was to take advantage of low-cost nighttime electricity rates and diurnal solar energy, which in turn would culminate in the reduction of global and peak energy demand, as well as energy costs. Being aware that the heating demand and energy consumption of a building can influence operational costs [26], a simulation-based optimization process was carried out to satisfy the MPC strategy (Section 4.3.2.4). Specifically, the heating demand of the building was evaluated through the simulation software (Section 4.3.2.2). Then, demand along with other input data was sent to the optimizer (Section 4.3.2.3) to minimize the heating cost of the building.

4.3.2.1.2 Solar Air Collector

Solar energy is the most available source of energy of all renewable and fossil-based energy resources [27]. Moreover, due to its clean, environmentally-friendly, and sustainable features, the application of solar energy has gained considerable attention [28]. Solar energy can be utilized using various technologies such as solar water heaters, solar cookers, solar dryers, solar ponds, solar architecture, solar air conditioning,

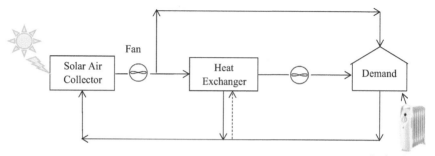

FIGURE 4.3.2   Overview of the automation system.

and solar chimneys [29]. Hence, a fruitful and cost-effective approach is required to extract solar energy, convert it into thermal energy and then store it [30].

The focus of this study was on space heating of buildings. Therefore, a non-concentrating, flat plate solar collector was assumed as the source of energy. Flat plate collectors usually contain glass or plastic glazed covers, dark-colored absorber plates, insulation, tubes filled with a heat transfer fluid, and other ancillaries [31]. The performance of a solar collector is evaluated by its efficiency, which is defined as the ratio of the useful thermal energy collected to the total amount of radiation hitting the surface of the collector over a specific period of time [32]. Assuming a constant value for some parameters, such as transmission coefficient of glazing, absorption coefficient of plate, collector heat removal factor, and collector overall heat loss coefficient, the efficiency will be a linear function of solar radiation intensity and the difference between inlet airflow of the collector and ambient temperature (Equation 4.3.1) [33]. This assumption is used to simplify working equations and the encoding process to accelerate the speed of the calculations:

$$\eta_{SAC} = 0.74 - 8.22\left(\frac{T_{in,SAC} - T_{amb}}{I}\right) \tag{4.3.1}$$

where η is the efficiency, T the temperature, and I the intensity of solar radiation. The subscript SAC is the acronym for the solar air collector. Subscripts in and amb represent the inlet and ambient temperature of the solar collector, respectively.

The collector in this study was a 1 m × 1 m solar air collector, incorporating a 0.1 m × 1 m photovoltaic (PV) panel on top of it (Figure 4.3.3) to drive the fan needed to circulate the air through the collector. The efficiency of the solar collector was calculated using Equation (4.3.1).

4.3.2.1.3 Heat Exchanger

The heat exchanger unit in this study (Figure 4.3.4) was made of acrylic materials and a set of 19 thin aluminum containers (0.45 m × 0.30 m × 0.01 m) holding a commercial PCM – RT25HC. The plates were positioned parallel to the airflow with 5 mm gap between them. The heat exchanger unit was coupled with fan of the solar collector or the fan used to drive air from heat exchanger to room allowing the circulation of air through the plates. The amount of PCM in the heat exchanger unit was approximately 9.5 kg. Table 4.3.1 details the properties of PCM. The whole assembly was insulated using a 0.040 m thick layer of polystyrene and mineral wool (i.e. thermal resistance of about 1 m²·K/W).

To evaluate the thermal behavior of the PCM, the specific heat capacity of the mushy phase was calculated using Equation (4.3.2) [35]:

$$C_{p,PCM} = \begin{cases} C_{p,s} & T_{PCM} < T_m - \dfrac{\Delta T_m}{2} \\ C_{p,s} + \dfrac{LH_{PCM}}{\Delta T_m} & T_m - \dfrac{\Delta T_m}{2} < T_{PCM} < T_m + \dfrac{\Delta T_m}{2} \\ C_{p,l} & T_{PCM} > T_m + \dfrac{\Delta T_m}{2} \end{cases} \tag{4.3.2}$$

FIGURE 4.3.3 Integrated solar collector and photovoltaic module.

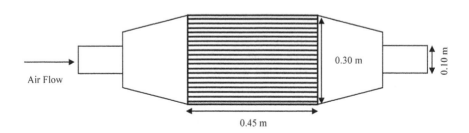

FIGURE 4.3.4 Schematic top view of the heat exchanger.

where C_p is specific heat capacity, LH is heat of fusion, and ΔT_m is an arbitrary small value representing the range of phase change temperature. Subscripts s, l, and m represent solid phase, liquid phase, and melting condition of PCM, respectively.

The findings rest on the assumption that all the PCM had the same temperature [36]. Indeed, based on this isothermal model, de Gracia et al. [37] confirmed that discrepancies between the simplified model and the experimental data are negligible enough for engineers and architects to predict the performance of the LHES without the need for complicated computational resources. On the other hand, as the charging and discharging process of PCM are mainly driven by force convection, heat losses and gains were only considered during the storage period. On this basis, the

TABLE 4.3.1

Physical and Thermal Properties of the PCM -RT 25 HC [34]

Parameter	Unit	Value
PCM type	–	Organic
Melting temperature range	°C	22–26
Heat of fusion	kJ/kg	230
Specific heat	kJ/kg·K	2 (s)
		2 (l)
Thermal conductivity	W/m·K	0.2
Density	kg/m³	880 (s)
		770 (l)
Volume expansion	%	12.5
Max operating temperature	°C	65

temperature of PCM at different time steps could be obtained from Equation (4.3.3), meaning that PCM temperature at each time step was equal to the PCM temperature at previous time step, plus the energy earning from solar collector, minus the energy leaving the heat exchanger to the room:

$$T_{PCM(k+1)} = T_{PCM(k)} + \frac{\dot{Q}_{SAC-HE(k)} - \dot{Q}_{HE-Room\ (k)}}{M_{PCM} \times C_{p,PCM}} \times \Delta t \qquad (4.3.3)$$

$$\Delta t = t_{k+1} - t_k \qquad (4.3.4)$$

where $\dot{Q}$ is the thermal power of device, k is the time step, Δt is the decision time step, and M is the PCM mass. Subscripts SAC-HE and HE-Room represent energy from solar air collector to heat exchanger, and heat exchanger to room, respectively.

To facilitate the calculations, it was assumed that the outlet temperature of the heat exchanger was the linear function of PCM temperature (Equation 4.3.5) [38]:

$$\varepsilon_{HE(k)} = \frac{T_{out,HE(k)} - T_{Room}}{T_{PCM(k)} - T_{Room}} \qquad (4.3.5)$$

where subscript "out, HE" represents the outlet condition of the heat exchanger.

4.3.2.1.4 Backup Heater

Electric resistance heating is 100% efficient, as it converts nearly all the electrical energy to thermal energy. Hence, the electric power of heater is equal to the thermal power used to provide the heating demand [39]. To estimate the power of fan, Hastings [40] proposed Equation (4.3.6):

$$\dot{Q}_f = \frac{\Delta p}{\eta_f} Q_{air} \qquad (4.3.6)$$

where Δp is the system overall pressure drop and Q is the air volume flow rate. In this study, $\dfrac{\Delta p}{\eta_f}$ was considered to be constant to simplify the problem and implement MPC strategy more simply.

4.3.2.2 Heating Demand Simulation

The simulation part of the study was performed through EnergyPlus v8.1. EnergyPlus is a powerful building energy simulation software, which is used widely by designers and engineers all over the world. This tool is able to estimate building energy demand according to the envelope design, weather conditions, occupancy status, and HVAC system design and control [41]. EnergyPlus takes advantage of the features and capabilities of BLAST and DOE-2 programs, which have been supported by the US government, and builds new features such as variable time steps for HVAC simulation and user-configurable systems [42]. It is able to assess heat balance loads at fixed time steps, as well as the response of HVAC, plant and electrical systems at variable time steps. This integration has led to more precise and realistic temperature prediction, estimation of adsorption, desorption, radiant heating and cooling systems, advanced infiltration, and multizone airflow calculations [43].

In this study, ASHRAE standard 140 Case 600 was selected as the reference building to simulate the heating demand of three types of building through EnergyPlus software. According to this standard, the basic building model is a rectangular single zone with dimensions 8 m wide, 6 m long and 2.7 m high, and without any interior partitions. The building has two 12 m² windows, facing south (Figure 4.3.5) [44]. However, as the current simulation was based on conditions in Auckland, New Zealand, a slight modification was carried on the test model to change the windows to north facing. The design of the standard building is based on lightweight construction materials. More information regarding the building specifications, such as envelope components and their properties, infiltration, internal loads and mechanical systems can be found in the publication cited in Ref. [44]. As ASHRAE standard was

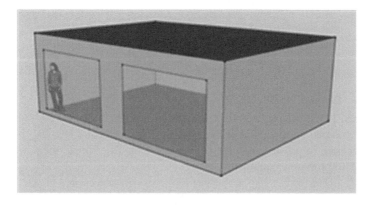

FIGURE 4.3.5 ASHRAE Standard 140 Case 600.

TABLE 4.3.2

Schedules of the Service, Office and Domestic Buildings [46]

Building	Comfort Temperature Range (°C)	Schedule Time
Service	20–25	24 h
Office	20–24	8 am–4 pm
Domestic	20–24	6 pm–12 am

used to implement the reference building in EnergyPlus, no validation was done for simulation to calculate heating demand [45].

This study investigated the heating demand for different types of buildings, namely offices, and domestic and service buildings, each of which entailed its own schedule. The schedule represented the time of day that HVAC system needed to operate to maintain the space temperature at a specific range. Table 4.3.2 depicts the details of the schedules applied in simulation.

4.3.2.3 NUMERICAL OPTIMIZATION

Solving Constraint Integer Programs (SCIP) Optimization Suite is a software toolbox, which can generate and solve algebraic optimization problems. Its modeling language is ZIMPL, the linear programming solver is SoPlex, and the constraint integer programming and branch-cut-and-price framework is SCIP. SCIP is able to quickly solve both mixed-integer linear and nonlinear programming, MIP and MINLP. The other features of SCIP optimization Suite are UG frameworks used to parallelize branch-and-bound-based solvers, and GCG framework, which performs as a generic branch-cut-and-price solver [47].

As a well-defined and high-level programming interface, Python can be used to write SCIP codes [48]. Python is an outstanding tool, offering open-source, object-oriented, user-friendly, versatile, portable, extensible, and customizable software [49]. In addition, it has a standard library, which allows users to have access to a large number of useful modules, inter-process communication, and operating and file systems [50].

In this study, SCIP 4.0.0 as the optimizer and Python 2.7.12 as the programming interface, both of which were installed on an Ubuntu Linux operative system, were used to produce the dynamic model of system.

4.3.2.4 MPC STRATEGY

An MPC tool is a successful optimization-based strategy, by which the behavior of a controlled system can be explicitly predicted over a receding horizon [51]. The main elements of MPC are objective function, prediction horizon, decision time step, manipulated variables, optimization algorithm, and feedback signals [52]. Along with its dynamic modeling of the process, MPC acts in a way to optimize the objective function, subject to some constraints [53]. The basic structure shown in Figure 4.3.6 is

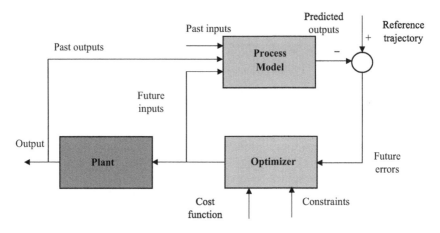

FIGURE 4.3.6   Basic structure of MPC strategy [54].

used to implement the MPC strategy. In this strategy, a model is utilized to predict the outputs based on the past inputs and outputs and current inputs as well as the proposed optimal future control signals. These signals are calculated through the optimizer considering the objective function, constraints and future errors, for a determined horizon. The predicted output, then, is compared to reference trajectory and an error is calculated. The cycling process is continued until a minimal error is obtained [54].

In the MPC strategy, Equation (4.3.7) is used to introduce the objective function and follows constraints required to satisfy the demand [16]:

$$\min_{u_0,\dots,u_{N-1}} \sum_{k=0}^{N-1} l_k(x_k,u_k) \quad \text{Objective function} \tag{4.37}$$

subject to

$$x_{k=0} = x_0 \quad \text{Current state} \tag{4.3.8}$$

$$x_{k+1} = f(x_k,u_k) \quad \text{Dynamics} - \text{state update} \tag{4.3.9}$$

$$(x_k,u_k) \in X_k \times U_k \quad \text{Constraints} \tag{4.3.10}$$

where l is the cost function, N is the prediction horizon, $x_k \in R^n$ is the state and $u_k \in R^m$ is the control input. X_k and U_k define the constraints of state and inputs values, respectively. In general, objective function should satisfy stability and target performance parameters. To prove its stability, the objective function should be able to follow Lyapunov-like functions [55]. In terms of the target performance, the objective function usually needs to minimize one behavior and maximize another one [16]. For example, the objective function of the current study tends to minimize the cost and maximize the inside thermal comfort.

The objective function used in this study (Equation 4.3.11) aimed to minimize the total energy cost of the heating, including the cost of backup heater and the fan applied to the system to drive air from TES to room at on-peak hours:

$$\text{Objective function :} \quad \underset{\substack{\dot{Q}_{BH(0)}\cdots\dot{Q}_{BH(N-1)} \\ \dot{Q}_{f(0)}\cdots\dot{Q}_{f(N-1)}}}{\overset{\infty}{min}} \sum_{k=0}^{N-1}\left(\dot{Q}_{BH(k)} + \dot{Q}_{f(k)}\right) \times \Delta t \times \text{electricity price} \tag{4.3.11}$$

$$\text{subjected to:} T_{PCM \blacklozenge k=0} = T_{PCM0} \quad \text{Current state} \tag{4.3.12}$$

$$T_{PCM(k+1)} = T_{PCM(k)} + \frac{\dot{Q}_{SAC-HE(k)} - \dot{Q}_{HE-Room(k)}}{M_{PCM} \times C_{p,PCM}} \times \Delta t \quad \text{Dynamics} \tag{4.3.13}$$

$$\text{Demand}_{(k)} = \dot{Q}_{SAC-Room(k)} + \dot{Q}_{HE-Room(k)} + \dot{Q}_{BH(k)} \quad \text{Constraint} \tag{4.3.14}$$

$$\dot{Q}_{SAC(k)} \geq \dot{Q}_{SAC-HE(k)} + \dot{Q}_{SAC-Room(k)} \quad \text{Constraint} \tag{4.3.15}$$

$$\dot{m}_{SAC(k)} \geq \dot{m}_{SAC-HE(k)} + \dot{m}_{SAC-Room(k)} \quad \text{Constraint} \tag{4.3.16}$$

$$\dot{m}_{SAC-Room(k)} = \frac{\dot{Q}_{SAC-Room(k)}}{C_{p,air}\left(T_{out,SAC} - T_{Room}\right)} \quad \text{Constrain} \tag{4.3.17}$$

$$\dot{m}_{SAC-HE(k)} = \frac{\dot{Q}_{SAC-HE(k)}}{C_{p,\,air}\left(T_{out,SAC} - T_{out,HE(k)}\right)} \quad \text{Constraint} \tag{4.3.18}$$

$$\dot{m}_{HE-Room(k)} = \frac{\dot{Q}_{HE-Room(k)}}{C_{p,air}\left(T_{out,HE(k)} - T_{Room}\right)} \quad \text{Constraint} \tag{4.3.19}$$

where $\dot{m}$ is the mass flow rate and subscripts f and SAC-Room represent the fan and solar collector to room, respectively. The subscript PCM0 represents the initial condition of PCM. The operating range of variables is defined as below:

$$0 \leq \dot{m}_{SAC(k)} \,\&\, \dot{m}_{HE-Room(k)} \leq 0.2 \tag{4.3.20}$$

$$20 \leq T_{PCM(k)} \leq 60 \tag{4.3.21}$$

$$0 \leq \dot{Q}_{HE(k)} \,\&\, \dot{Q}_{BH(k)} \leq \text{Maximum demand} \tag{4.3.22}$$

$$0 \leq \dot{Q}_{SAC(k)} \leq \text{Maximum energy from SAC} \tag{4.3.23}$$

The energy cost obtained through MPC strategy was then compared with 'simple control' method to estimate the cost savings for electrical energy. The basic idea of the simple control method is to compare the system output with the determined setpoint and minimize error by tuning process control inputs, i.e. repeating a measurement and computation procedure for each time step. The simple control method fails to predict outputs as the horizon is ignored [56].

4.3.3   RESULTS AND DISCUSSION

4.3.3.1   EFFECT OF RECEDING HORIZON

One significant parameter in the evaluation of MPC performance is horizon [57]. The period of time in which the objective function is being optimized is called prediction horizon. In this section, results for the effect of MPC horizon on the heating costs of the aforementioned building types are discussed. Heating demand of a service building in a typical day of winter in Auckland was obtained through EnergyPlus simulation (Figure 4.3.7). Figure 4.3.8 displays the impact of receding horizon on electricity cost savings for service schedule profile for 1 and 7 running days, beginning with the first day of winter in Auckland in 2017. The cost savings achieved by the current strategy are compared with those with the simple control method. The decision time step of the figure is 15 min. Figure 4.3.8 shows that increasing the horizon decreased the electricity consumption, which in turn resulted in a reduction in energy cost. Greater horizons enabled the model to cover a wider range of weather conditions as well as the electricity cost data, based on which the model decided whether to store or release the PCM energy to satisfy heating demand. In addition, it can be seen from the growth rate shown in the graphs that energy savings were more prominent when the duration of simulation increased from 1 to 7 days. This is because the model had more flexibility in relation to the charging and discharging time of the PCM. To clarify, in the 1-day (i.e. 24 h) simulation, heating demand resulting from the EnergyPlus simulation of the service schedule profile was for midnight until 9 am, and then from 8:30 pm to midnight (Figure 4.3.7). Accordingly, PCM was only able to be charged during the day and discharged from 8:30 pm until midnight without causing any effect on the first period of demand. However, in the 7-day simulation, this period

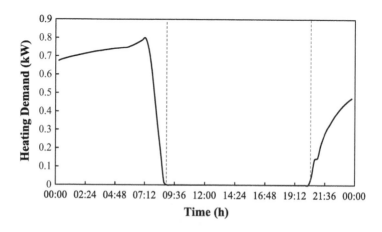

FIGURE 4.3.7   The building heating demand in a 1-day simulation.

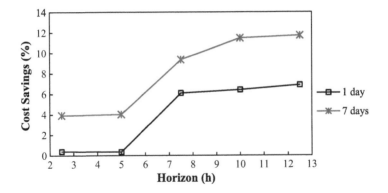

FIGURE 4.3.8   Electricity cost savings at different horizons for 1 and 7 running days.

was extended, giving PCM the ability to release its energy after second midnight and
so save more energy.

In Figure 4.3.8, the sharp increase shown in cost saving from 5 to 7.5 h is explained
by Figures 4.3.9 and 4.3.10, which exhibit the performance of MPC system for horizons 5 and 7.5 h, respectively. In these figures, electricity cost was extracted from
Ref. [58], and solar radiation and demand were obtained from the simulation as input
data. The PCM temperature and the sources of energies utilized to provide demand
are the output information. Based on the first few winter days in Auckland, the most
expensive time of the day was warm period when there was no demand. At around
8 pm, demand started to grow, in which case for a system setting based on 5 h prediction (Figure 4.3.9), it could use only a small proportion of solar energy (about
0.1 kW/m²) stored in PCM. Hence, at 8:30 pm, a limited amount of energy could
be released from the heat exchanger (vertical arrows). However, based on the 7.5 h
prediction (Figure 4.3.10), the system was able to take advantage of a larger amount
of solar energy and release it at the time of need. Therefore, more energy savings and
cost reductions were achieved.

The system decided to charge PCM based not only on the presence of the Sun but
also on the conditions forecast. It can be observed that with the horizon of 5 h, the
maximum temperature of PCM in heat exchanger was 23°C. However, for a horizon of 7.5 h, it reached about 37°C. With higher horizons, the system was able to
anticipate a wider range of heating demand hours and store greater quantities of
the solar energy in PCM. The results for 7 running days also confirm this finding
(Figure 4.3.11). Figure 4.3.11 shows the behavior of the PCM's temperature at different horizons of 5, 7.5, and 10 h for 7 days, for given solar radiation, electricity cost,
and building heating demand. The trends show that with horizons ascending from 5
to 7.5 and 10 h, the maximum temperature of PCM, as well as retention time at that
temperature, rose. Indeed, according to the prediction for 5 h, the system was exposed
to demand when solar energy was not available. Thus, PCM remained unloaded. In
contrast, applying the 10 h prediction allowed the PCM to be charged at a higher
temperature and for a longer time.

Further, the MPC strategy performed in such a way that the electricity and energy
stored in the heat exchanger were consumed during the corresponding cheap and

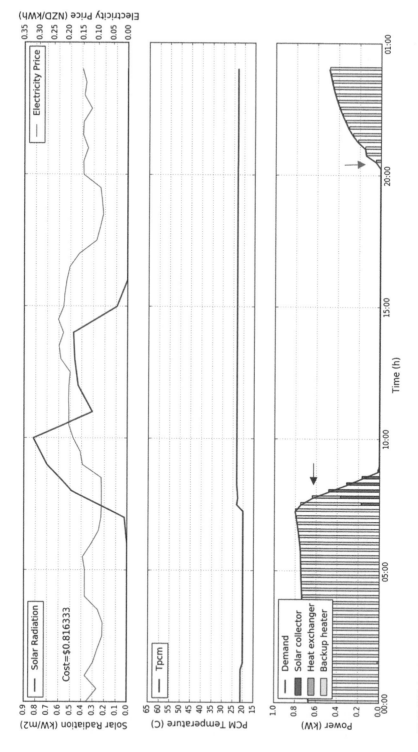

FIGURE 4.3.9   The performance of the MPC with the horizon of 5 h for 1 day.

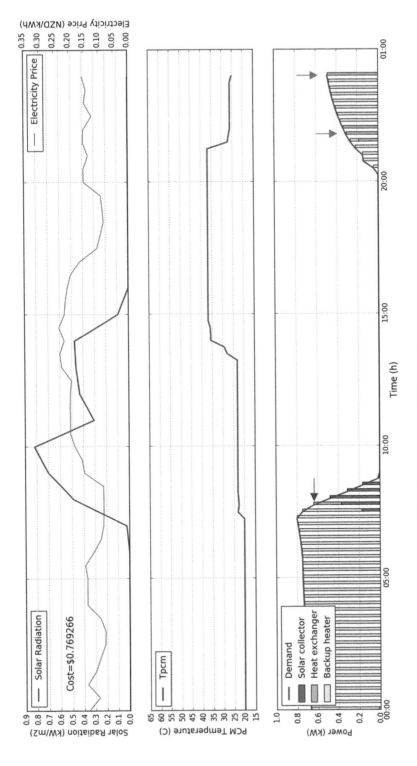

FIGURE 4.3.10 The performance of the MPC with the horizon of 7.5 h for 1 day.

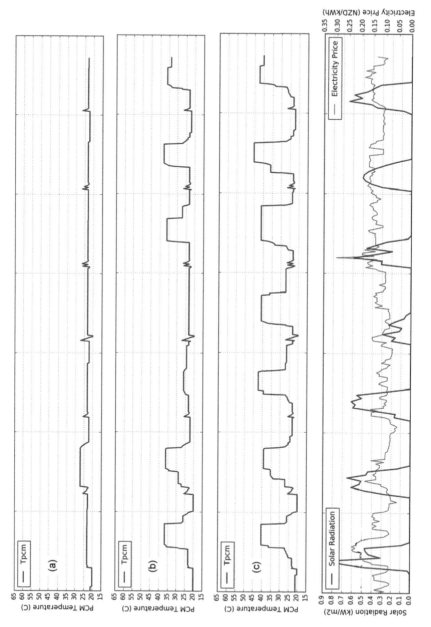

FIGURE 4.3.11   The PCM temperature profile at horizons of (a) 5 h, (b) 7.5 h, and (c) 10 h for 7 days.

costly hours, respectively. According to Figures 4.3.9 and 4.3.10, at around 8 am, there was a PCM discharge, even though the electricity cost was quite cheap (horizontal arrow). This is because at that time, with horizons of 5 and 7.5 h, no demand was foreseen. Thus, the MPC system attempted to extract the energy from heat exchanger during cheap hours, to avoid using backup heater energy and minimize the cost.

4.3.3.2 EFFECT OF DECISION TIME STEP

The duration between each optimization process is known as decision time step. Some researchers have studied the impact of decision time step on the performance of MPC strategies [59]. Table 4.3.3 presents the electricity cost of heating as a function of decision time step for the first 1 and 7 days of winter in Auckland. As shown in the table, for a specific number of horizons (i.e. 24 in this study), the longest decision time step appeared to be more efficient. Indeed, according to the relation between the horizon and decision time step (Equation 4.3.24), lengthening the decision time step expanded the prediction hours, which in turn culminated in reducing the energy consumption:

$$\text{Receding horizon } (h) = N \times \Delta t(h) \tag{4.3.24}$$

However, the accuracy of system is another critical parameter that should be taken into account. While less complicated and time-consuming, large decision time steps may lead to some important data being missed, hence deteriorating the precision of the outcome. In this case, the reduction rate of the electricity cost for 7 days was less than for 1 day (Figure 4.3.12). Based on the discussion in Section 4.3.3.1, an opposite scenario was expected. In the 7-day simulation, almost the same cost savings were achieved for both 30 and 60 min decision time steps, even though the prediction time of the latter is two times that of the former. Due to skipping some essential information such as demand, weather condition, electricity cost, and PCM temperature, the system failed to provide an authentic estimation of energy and cost requirement. In other words, for the decision time step of 60 min, PCM reached its maximum point quickly so that no further improvement was attained. This conclusion is verified by PCM temperature profile in Figure 4.3.13, which indicates that PCM temperature oscillated sharply between the boundaries temperatures of the program (20°C and 60°C) with 60 min decision time step. The temperature variation was more gradual with the 30 min time step, indicating it is better to make the time step as small as possible. However, as discussed, decreasing the time steps increases the time and cost of computations. On the other hand, longer time steps make the system more sensitive to sharp changes, meaning a contingency plan has to be developed to avoid malfunctioning of the system, such as not meeting demand and overheating PCM in the heat exchanger unit. Hence, a balance between the cost and accuracy of the program needs to be established.

4.3.3.3 EFFECT OF MASS CAPACITY OF PCM

As a latent heat storage medium, PCM plays a significant role in the heating process of a space. To exploit the maximum efficiency of PCM, the investigation of effective

TABLE 4.3.3

Energy Cost of Heating for 1- and 7-Day Simulations at Different Decision Time Steps

Decision Time Step (min)	Horizon (h)	Energy Cost (NZD) for 1 Day	Energy Cost (NZD) for 7 Days
5	2	0.802	7.205
15	6	0.790	6.350
30	12	0.750	5.795
60	24	0.660	5.760

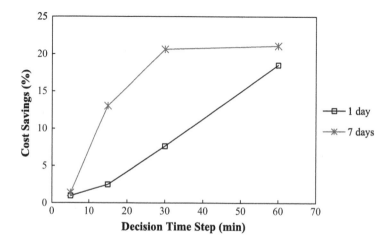

FIGURE 4.3.12 Electricity cost savings at different decision time steps for 1 and 7 running days.

parameters of PCM performance was seen as beneficial. Accordingly, the objective in this part of the study was to demonstrate the impact of the amount of PCM in the heat exchanger on energy saving of the described service building, for 1 and 7 running days. The decision time step and horizon were considered to be 0.5 and 15h, respectively. As expected, Figure 4.3.14 shows that increasing the PCM mass helped the system to store more solar energy. Accordingly, the heat exchanger was able to release a greater amount of energy, leading to more electrical energy being saved. Applying 28.5 kg of PCM (three times the initial PCM) into the system saved about 27% in electricity costs in 7 days, which would provide remarkable savings over the whole winter, as well as over a couple of years in a row. However, it is not a good idea to raise the amount of PCM as much as possible. Instead, the PCM mass implemented in the heat exchanger should be optimized according to the efficiency of unit as well as the capital investment of PCM.

In contrast to the 7-day simulation, increasing the amount of PCM from 19 to 28.5 kg in the1-day simulation did not improve the efficiency of the system. The

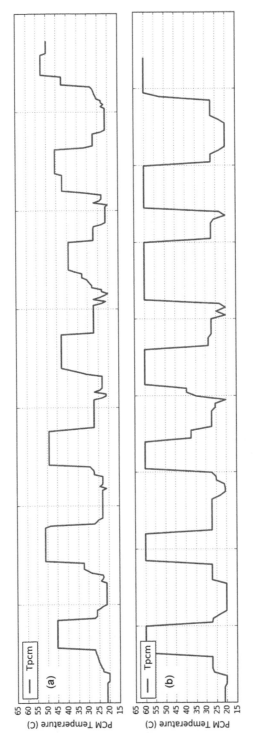

FIGURE 4.3.13 PCM temperature profile for decision time steps of (a) 30 min and (b) 60 min, at given solar radiation and electricity cost.

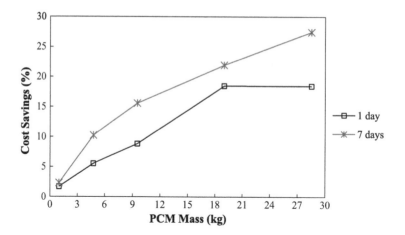

FIGURE 4.3.14   Effect of PCM mass on the electricity cost savings for 1 and 7 running days.

reason is that at the given prediction horizon and decision time step, almost all the heating demand of the period between 8:30 pm and midnight was satisfied by the stored energy of the PCM (Figure 4.3.15). Hence, further addition of PCM was not effective.

4.3.3.4   MPC PERFORMANCE IN DIFFERENT BUILDINGS

MPC strategies can be applied to different types of buildings with different heating demands. Figure 4.3.16 displays the heating demand for service, domestic, and office building scenarios, as well as energy sources used to supply the demand. In this simulation, decision time step and horizon are 0.25 and 10 h, respectively. The results of the simulation are compatible with the schedules for the buildings outlined in Table 4.3.2. The plan for service building was to maintain comfort zone over a 24-h period. Hence, demand was zero only when solar energy was available to warm up the space. In this case, according to the available energy sources, the heat exchanger provided a small proportion of demand, while backup heater supplied a large proportion. Domestic buildings only need to be kept within specified range from 6 pm until midnight. Therefore, the energy stored in PCM during the day can largely meet demand. For the case of office building in this study, demand appeared particularly high during the day and could be mostly met by solar energy, or energy stored in the heat exchanger. Thus, the MPC strategy performed more effectively for office and domestic buildings than for service building in this study. Confirming the results, the 7-day simulation for different building types (Figure 4.3.17) also showed the highest cost saving (56%) for domestic building, where a large amount of the heating demand was satisfied by discharging the PCM. The service building showed the lowest cost saving, as they should provide the comfortable indoor temperature for 24 h.

     Furthermore, when demand was compared in the 7-day simulation (Figure 4.3.18), the intensity of demand in office building was larger than for the domestic building, which in turn showed greater intensity of demand than the service building. The reason for this is associated with the period of demand for each building. In the office

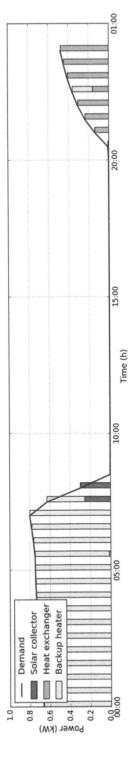

FIGURE 4.3.15 The performance of the MPC system with 19 kg of PCM and horizon of 15 h.

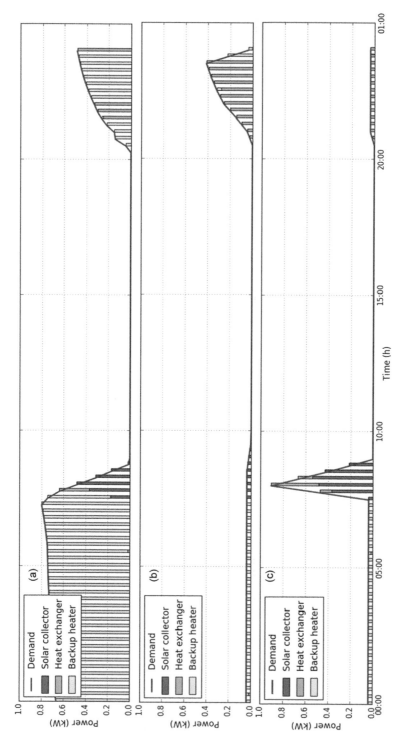

FIGURE 4.3.16   The profile demand and the sources of energy used to supply the demand for (a) service, (b) domestic and (c) office buildings.

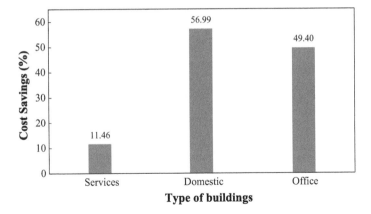

FIGURE 4.3.17 Cost savings achieved by different type of building for the 7-day simulation.

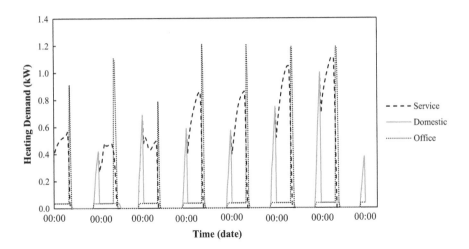

FIGURE 4.3.18 Heating demand associated with different buildings.

building, demand rose after the low temperature at night, which made the building very cold, meaning it needed more energy than the other buildings to compensate. In contrast, the demand intensity for the domestic buildings was the lowest as it appeared at around 8 pm, after taking advantage of daytime sunlight and warm temperatures. Finally, the demand intensity for the service building was between that of the two buildings. It needed to be considered all the time, so the building's temperature did not get too low.

The summary of the investigated parameters, operating conditions and schedules as well as the findings of the current study are shown in Table 4.3.4. The effect of horizon, decision time step, PCM mass, and operating schedule on cost savings of heating process of a building in New Zealand were studied.

TABLE 4.3.4

Summary of the Operating Conditions and Findings

Parameter	Horizon Numbers	Decision Time Step (min)	PCM Mass (kg)	Simulation Time (Day)	Schedule	Cost Savings (%)
Horizon numbers	10, 20, 30	15	9.50	1	Service	0.34–6.80
				7		4–11.70
Decision time step	24	5, 15, 30, 60	9.50	1	Service	0.97–18.50
				7		1.37–21.10
PCM mass	30	30	1, 4.75, 9.50, 19, 28.50	1	Service	1.69–18.45
				7		2.29–27.45
Schedule	40	15	9.50	7	Service	11.46
					Domestic	56.99
					Office	49.40

4.3.4 CONCLUSIONS

A numerical study was carried out of an MPC strategy to control the heating process for three versions of a standard building, all equipped with a heat exchanger containing PCM and a solar air collector to capture the solar energy and direct it to the heat exchanger. In this simulation-based optimization investigation, EnergyPlus, Python, and SCIP software packages were used as the simulator, interface, and optimizer, respectively. The objective function of the MPC was to select the appropriate schedule of operation for the whole system to minimize the electricity cost and meet the heating demand of the service, domestic, and office buildings used in the simulation. The results confirmed that implementation of the smart MPC strategy was more beneficial for the domestic building, followed by the office and then service buildings. The effect of various parameters on the performance of the MPC strategy was investigated. Results showed that the greater the prediction horizon, the higher the cost saving achieved. Moreover, increasing the time step enhanced the horizon hours on the one hand but ignored some significant input information on the other hand. Thus, smaller time steps were better in terms of the accuracy of the simulation. However, using smaller time steps requires more processing time and more powerful instruments, which increases the cost of computations. In addition, with regard to the PCM content of the heat exchanger, a higher amount was more beneficial in terms of cost savings for electricity. However, increases in the capital cost of PCM would necessitate deciding on an optimized amount of PCM for the heat exchanger. Finally, the results show that more cost saving was achieved in the 7-day simulation of service building than for the 1-day simulation.

ACKNOWLEDGEMENTS

The study was partially funded by the Spanish Government (ENE2015-64117-C5-1-R (MINECO/FEDER) and ENE2015-64117-C5-3-R (MINECO/FEDER)). The authors at the University of Lleida would like to thank the Catalan Government for the quality accreditation given to their research group (2017 SGR 1537). GREiA is the certified agent for TECNIO in the category technology developers for the Government of Catalonia. The research leading to these results also received funding from the European Union's Seventh Framework Programme (FP7/2007–2013) under grant agreement no. PIRSES-GA-2013-610692 (INNOSTORAGE).

NOMENCLATURE

Δp	overall pressure drop (kPa)
Δt	decision time step (s)
ΔT	temperature difference (°C)
C_p	specific heat capacity (kJ/kg·K)
I	intensity of solar radiation (W/m^2)
k	simulation time step
l	cost function (NZD)
M	mass (kg)
$\dot{m}$	mass flow rate
N	number of prediction horizon
Q	volume flow rate (m^3/s)
$\dot{Q}$	thermal power (kJ/s)
T	temperature (°C)
u	control input
U	Constraint of input value
x	control input
x_0	initial state of the control input
X	Constraint of input value

GREEK SYMBOLS

η	Efficiency
ε	conversion coefficient to estimate the outlet temperature of heat exchanger as a function of PCM temperature

SUBSCRIPTS

f	fan
amb	ambient
HE-Room	directing from heat exchanger to room
in, SAC	inlet of solar air collector
l	liquid
m	melting point
out, SAC	outlet of solar air collector
out, HE	outlet of heat exchanger
s	solid
SAC-HE	directing from solar air collector to heat exchanger
SAC-Room	directing from solar air collector to room

REFERENCES

1. Barzin R, Chen JJJ, Young BR, Farid MM. Application of PCM energy storage in combination with night ventilation for space cooling. *Appl Energy* 2015;158:412–21. doi:10.1016/j.apenergy.2015.08.088.

2. Nejat P, Jomehzadeh F, Taheri MM, Gohari M, Muhd MZ. A global review of energy consumption, CO2 emissions and policy in the residential sector (with an overview of the top ten CO2 emitting countries). *Renew Sustain Energy Rev* 2015;43:843–62. doi:10.1016/j.rser.2014.11.066.

3. Kirubakaran V, Sahu C, Radhakrishnan TK, Sivakumaran N. Energy efficient model based algorithm for control of building HVAC systems. *Ecotoxicol Environ Saf* 2015;121:236–43. doi:10.1016/j.ecoenv.2015.03.027.

4. Zhou D, Zhao CY, Tian Y. Review on thermal energy storage with phase change materials (PCMs) in building applications. *Appl Energy* 2012;92:593–605. doi:10.1016/j.apenergy.2011.08.025.

5. Hien WN, Poh LK, Feriadi H. Computer-based performance simulation for building design and evaluation: the Singapore perspective. *Simul Gaming* 2003;34:457–77. doi:10.1177/1046878103255917.

6. Couenne F, Hamroun B, Jallut C. Experimental investigation of the dynamic behavior of a large-scale refrigeration e PCM energy storage system. Validation of a complete model. *Energy* 2016;116:32–42. doi:10.1016/j.energy.2016.09.098.

7. Prívara S, Cigler J, Váňa Z, Oldewurtel F, Sagerschnig C, Žáčeková E. Building modeling as a crucial part for building predictive control. *Energy Build* 2013;56:8–22. doi:10.1016/j.enbuild.2012.10.024.

8. Ebrahimpour M, Santoro BF. Moving horizon estimation of lumped load and occupancy in smart buildings. *IEEE Conf Control Appl* 2016;2016:468–73. doi:10.1109/CCA.2016.7587874.

9. Killian M, Kozek M. Ten questions concerning model predictive control for energy efficient buildings. *Build Environ* 2016;105:403–12. doi:10.1016/j.buildenv.2016.05.034.

10. Braun JE. Load control using building thermal mass. *J Sol Energy Eng* 2003;125:292–301. doi:10.1115/1.1592184.

11. Carrascal-Lekunberri E, Garrido I, Van Der Heijde B, Garrido AJ, Sala JM, Helsen L. Energy conservation in an office building using an enhanced blind system control. *Energies* 2017;10. doi:10.3390/en10020196.

12. Ma Y, Borrelli F, Hencey B, Packard A, Bortoff S. Model predictive control of thermal energy storage in building cooling systems. In: *Proc 48h IEEE Conf Decis Control Held Jointly with 2009 28th Chinese Control Conf 2009*: pp. 392–7. doi:10.1109/CDC.2009.5400677.

13. Moroşan P-D, Bourdais R, Dumur D, Buisson J. Building temperature regulation using a distributed model predictive control. *Energy Build* 2010;42:1445–52. doi:10.1016/j.enbuild.2010.03.014.

14. Yudong M, Matusko J, Borrelli F. Stochastic model predictive control for building HVAC systems: complexity and conservatism. *Control Syst Technol IEEE Trans* 2015;23:101–16. doi:10.1109/TCST.2014.2313736.

15. Oldewurtel F, Ulbig A, Parisio A, Andersson G, Morari M, IEEE. Reducing peak electricity demand in building climate control using real-time pricing and model predictive control. In: *49th IEEE Conf Decis Control*; 2010. pp. 1927–32. doi:10.1109/cdc.2010.5717458.

16. Široký J, Oldewurtel F, Cigler J, Prívara S. Experimental analysis of model predictive control for an energy efficient building heating system. *Appl Energy* 2011;88:3079–87. doi:10.1016/j.apenergy.2011.03.009.

17. Gyalistras D, Fischlin A, Zurich E, Morari M, Jones CN, Oldewurtel F, et al. Use of weather and occupancy forecasts for optimal building climate control facts of the project. 2010; 41.

18. Ma J, Qin J, Salsbury T, Xu P. Demand reduction in building energy systems based on economic model predictive control. *Chem Eng Sci* 2012;67:92–100. doi:10.1016/j. ces.2011.07.052.

19. Zhao Y, Lu Y, Yan C, Wang S. MPC-based optimal scheduling of grid-connected low energy buildings with thermal energy storages. *Energy Build* 2015;86:415–26. doi:10.1016/j.enbuild.2014.10.019.

20. Knudsen MD, Petersen S. Model predictive control for demand response of domestic hot water preparation in ultra-low temperature district heating systems. *Energy Build* 2017;146:55–64. doi:10.1016/j.enbuild.2017.04.023.

21. Labidi M, Eynard J, Faugeroux O, Grieu S. A new strategy based on power demand forecasting to the management of multi-energy district boilers equipped with hot water tanks. *Appl Therm Eng* 2017;113:1366–80. doi:10.1016/j. applthermaleng. 2016.11.151.

22. Papachristou AC, Vallianos CA, Dermardiros V, Athienitis AK, Candanedo JA. A numerical and experimental study of a simple model-based predictive control strategy in a perimeter zone with phase change material. *Sci Technol Built Environ* 2018;4731. doi:10.1080/23744731.2018.1438011.

23. Touretzky CR, Baldea M. A hierarchical scheduling and control strategy for thermal energy storage systems. *Energy Build* 2016;110:94–107. doi:10.1016/j. enbuild.2015.09.049.

24. Hawlader MNA, Uddin MS, Khin MM. Microencapsulated PCM thermal-energy storage system. *Appl Energy* 2003;74:195–202. doi:10.1016/S0306-2619(02)00146-0.

25. Fiorentini M, Wall J, Ma Z, Braslavsky JH, Cooper P. Hybrid model predictive control of a residential HVAC system with on-site thermal energy generation and storage. *Appl Energy* 2017;187:465–79. doi:10.1016/j.apenergy.2016.11.041.

26. Fumo N, Mago P, Luck R. Methodology to estimate building energy consumption using EnergyPlus Benchmark Models. *Energy Build* 2010;42:2331–7. doi:10.1016/j. enbuild.2010.07.027.

27. Lewis NS. Toward cost-effective solar energy use. *Science (80-)* 2007;315:798–802.

28. Panwar NL, Kaushik SC, Kothari S. Role of renewable energy sources in environmental protection: a review. *Renew Sustain Energy Rev* 2011;15:1513–24. doi:10.1016/j. rser.2010.11.037.

29. Thirugnanasambandam M, Iniyan S, Goic R. A review of solar thermal technologies. *Renew Sustain Energy Rev* 2010;14:312–22. doi:10.1016/j.rser.2009.07.014.

30. Mekhilef S, Saidur R, Safari A. A review on solar energy use in industries. *Renew Sustain Energy Rev* 2011;15:1777–90. doi:10.1016/j.rser.2010.12.018.

31. Tian Y, Zhao CY. A review of solar collectors and thermal energy storage in solar thermal applications. *Appl Energy* 2013;104:538–53. doi:10.1016/j. apenergy.2012.11.051.

32. Chabane F, Moummi N, Benramache S, Bensahal D, Belahssen O. Collector efficiency by single pass of solar air heaters with and without using fins. *Eng J* 2013;17:43–55. doi:10.4186/ej.2013.17.3.43.

33. Struckmann F. Analysis of a flat-plate solar. Collector 2008. doi:10. 3846/mla.2011.108.

34. Rubitherm Technologies GmbH. Macroencapsulation - CSM n.d. https://www.rubith erm.eu/en/index.php/productcategory/makroverkaspelung-csm [accessed 01.03.18].

35. Promoppatum P, Yao S, Hultz T, Agee D. Experimental and numerical investigation of the cross-flow PCM heat exchanger for the energy saving of building HVAC. *Energy Build* 2017;138:468–78. doi:10.1016/j.enbuild.2016.12.043.

36. Hed G, Bellander R. Mathematical modelling of PCM air heat exchanger. *Energy Build* 2006;38:82–9. doi:10.1016/j.enbuild.2005.04.002.

37. De Gracia A, Castell A, Fernández C, Cabeza LF. A simple model to predict the thermal performance of a ventilated facade with phase change materials. *Energy Build* 2015;93:137–42. doi:10.1016/j.enbuild.2015.01.069.

38. Cengel Ya. *HEAT TRANSFER: A Practical Approach.* Second. New York: McGraw-Hill; 2003.

39. Energy Saver. Electric Resistance Heating. US Dep Energy 2016. https://energy.gov/energysaver/electric-resistance-heating [accessed 01.03.18].

40. Hastings R. *Solar Air Systems: A Design Handbook.* New York: Routledge; 2013.

41. Stadler M, Firestone R, Curtil D, Marnay C, Berkeley L. On-site generation simulation with energyplus for commercial buildings. *Buildings* 2006:242–54.

42. Crawley DB, Lawrie LK, Winkelmann FC, Buhl WF, Huang YJ, Pedersen CO, et al. EnergyPlus: creating a new-generation building energy simulation program. *Energy Build* 2001;33:319–31. doi:10.1016/S0378-7788(00)00114-6.

43. Crawley DB, Hand JW, Kummert M, Griffith BT. Contrasting the capabilities of building energy performance simulation programs. *Build Environ* 2008;43:661–73. doi:10.1016/J.BUILDENV.2006.10.027.

44. Henninger R, Witte M. EnergyPlus testing with building thermal envelope and fabric load tests from ANSI/ASHRAE Standard 140-2007; 2014.

45. Saffari M, de Gracia A, Fernández C, Cabeza LF. Simulation-based optimization of PCM melting temperature to improve the energy performance in buildings. *Appl Energy* 2017;202:420–34. doi:10.1016/j.apenergy.2017.05.107.

46. CEN (European Committee for Standardization). Indoor environmental input parameters for design and assessment of energy performance of buildings addressing indoor air quality, thermal environment, lighting and acoustics; 2007. doi:10.1520/E2019-03R13.

47. Gamrath G, Fischer T, Gally T, Gleixner AM, Hendel G, Koch T, et al. The scip optimization suite 3.2. *ZIB Rep* 2016;60:15–60.

48. Ferreto TC, Netto MaS, Calheiros RN, De Rose CaF. Server consolidation with migration control for virtualized data centers. *Futur Gener Comput Syst* 2011;27:1027–34. doi:10.1016/j.future.2011.04.016.

49. Hart WE, Watson JP, Woodruff DL. Pyomo: modeling and solving mathematical programs in python. *Math Program Comput* 2011;3:219–60. doi:10.1007/s12532-011-0026-8.

50. Prechelt L. Are scripting languages any good? A validation of perl, python, Rexx, and Tcl against C, C++, and Java. *Adv Comput* 2003;57:205–70. doi:10.1016/S0065-2458(03)57005-X.

51. Rao C, Rawlings J. Linear programming and model predictive control. *J Process Control* 2000;10:283–9.

52. Thieblemont H, Haghighat F, Ooka R, Moreau A. Predictive control strategies based on weather forecast in buildings with energy storage system: A review of the state of-the art. *Energy Build* 2017;153:485–500. doi:10.1016/j.enbuild.2017.08.010.

53. Prívara S, Široký J, Ferkl L, Cigler J. Model predictive control of a building heating system: The first experience. *Energy Build* 2011;43:564–72. doi:10.1016/j.enbuild.2010.10.022.

54. Camacho EF, Bordons C. Model predictive. *Control* 2007. doi:10.1007/978-0-85729-398-5.

55. Ellis M, Christofides PD. Performance monitoring of economic model predictive control systems. *Ind Eng Chem Res* 2014;53:15406–13. doi:10.1021/ie403462y.

56. Åström KJ, Hägglund T. The future of PID control. *Control Eng Pract* 2001;9:1163–75. doi:10.1016/S0967-0661(01)00062-4.

57. Kummert M, André P, Nicolas J. Building and HVAC optimal control simulation. Application to an office building. *Proc Third Int Symp HVAC* 1999:857–68.

58. Electricity Authority. NZX Electricity Price Index; 2018. https://www1.electricityinfo. co.nz/price_indexes [accessed 01.03.18].

59. Lefort A, Bourdais R, Ansanay-Alex G, Guéguen H. Hierarchical control method applied to energy management of a residential house. *Energy Build* 2013;64:53–61. doi:10.1016/j.enbuild.2013.04.010.

4.4 A Comparison between Passive and Active PCM Systems Applied to Buildings

Gohar Gholamibozanjani and Mohammed Farid
University of Auckland

4.4.1 INTRODUCTION

Thermal energy storage systems using phase change materials (PCMs) provide an opportunity to effectively store energy for later use. The high energy density of PCMs and their isothermal performance make them suitable for building applications [1]. In general, energy analysis of the PCM-integrated buildings strongly depends on the melting point, thermal properties, location, and the amount of the PCM incorporated into the building, as well as the climatic condition and the design of the building [2]. Usually, thermal comfort is in the range of 22°C–27°C in summer and 18°C–25°C in winter [3]. Therefore, to meet the thermal comfort of the building sector, the chosen PCM should have a melting temperature of18°C–27°C [4].

PCMs can be integrated into buildings passively [5] or actively [6]. In the passive method, PCM is embedded into the construction materials of buildings, such as plasterboard, gypsum or concrete, without requiring any auxiliary equipment to operate. Typical applications of a passive thermal storage system (PTSS) include the integration of PCM into the building fabrics of walls, floors, and roofs (Figure 4.4.1). Moreover, they can also be incorporated into the fenestration, insulation, façade, shutter, and shading systems [7].

A large number of studies show that the application of PTSS into buildings has significantly reduced their energy consumption and enhanced indoor thermal comfort. For instance, using a price-based control method, Barzin et al. [8] investigated the application of underfloor heating in combination with PTSS wallboards for space heating in the city of Auckland, New Zealand. They showed an energy-saving of about 35% and an electricity cost-saving of about 44%, generated from capturing solar radiation and inducing peak load shifting. Also, Saffari et al. [9] studied the effect of the PTSS in buildings using the Fanger comfort model. They performed a numerical study for three different occupancy schedules, including office, residential, and 24-h occupancy over the summer and winter periods in the Madrid climate zone. The energy consumption of the buildings was reduced for different schedules with the exception of office heating.

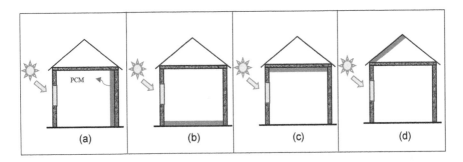

FIGURE 4.4.1 Passive heating by the integration of PCM into the (a) walls, (b) floor, (c) ceiling, and (d) roof. (For interpretation of the references to color in this figure legend, the reader is referred to the Web version of this article.)

Active thermal storage systems (ATSS), on the other hand, require some mechanical or electrical energy for their operation [10]. In circumstances where greater control of the system or greater heat transfer performance is needed, ATSS is preferred [11]. Unlike PTSS, active applications allow the system to store and release heat on demand and ensure more efficient heat transfer and hence reduce the energy consumption of buildings. ATSS integrates PCM into HVAC (heating, cooling, and air conditioning) systems [12] or uses storage containers in the vicinity of buildings (such as aquifers, boreholes, snow storage, and pits or tanks) [13]. As an example, Wang et al. [14] studied the performance of a solar-assisted water-PCM heat storage unit for domestic hot water use and peak load shifting, consequently. By proposing a performance evaluation approach, they confirmed that the mass flow rate of water has an insignificant effect on the thermal performance of the system. On the other hand, solar collector area is a key factor which needs to be optimized based on the size of PCM storage unit. In addition, Prieto et al. [15] showed that, in a flat plate water-PCM heat exchanger, the thickness of PCM, inlet air temperature and arrangement of the plates (vertical/horizontal) are effective in the thermal behavior of the system.

Some researchers also integrated PCM into the hybrid photovoltaic-thermal panels [16,17], which not only increased the electricity efficiency of the panels during day but also provided hot water at higher temperature during night.

Further to active application of PCM, Stathopoulos et al. [18] designed an air-PCM heat exchanger to investigate its potential in creating peak load shifting. In their study, Microtech 37D paraffin was charged using an electrical resistance and tested under artificial environmental condition. Their results confirmed the capability of the active storage unit to be applied in net-zero building designs and shift energy loads from peak hours to off-peak hours. In an experimental study conducted in the UK by Turnpenny et al. [19], a prototype heat storage system was developed by embedding heat pipes into the PCM storage unit. During a summer day, the heat load of the room was absorbed by PCM through the heat pipes, thus reducing the interior temperature. At night the heat pipes released the energy stored in PCM to the environment. The proposed system provided adequate cooling for the room and prevented overheating. Further, the system offered substantial benefits in terms of cost and energy savings compared to conventional air conditioning.

Despite a large number of publications on using PTSS and ATSS in buildings, a comparative study on their relative performance is not available. Our study provides, for the first time, a good comparison of the thermal behavior of two office size buildings, employing PTSS and ATSS under real environmental conditions.

In this experimental study, two identical test huts, each equipped with an advanced control system, were used to investigate the concept. In general, in the ATSS, an air-PCM heat storage unit was installed in one of the huts, while in the PTSS, the PCM was integrated into the wallboards of the hut. The control system was developed using the LabVIEW software to ensure thermal comfort in the huts based on the weather conditions of the city of Auckland, New Zealand. The control strategy was set up to allow storing solar energy provided by the solar air heater in winter and the free night coolness in summer, utilizing the high latent heat of melting of the PCM.

4.4.2 METHODOLOGY

4.4.2.1 EXPERIMENTAL SETUP

Two identical experimental huts were used to demonstrate how to improve the thermal efficiency of buildings utilizing PTSS and ATSS. The huts, with external dimensions of 2.7 m×2.7 m×2.7 m, were located at the Ardmore campus of the University of Auckland. Both huts were constructed using standard lightweight materials and had a 1 m×1 m north-facing, single-glazed window. The walls and ceilings of the huts were insulated with glass wool, while the floor was insulated with polystyrene foam. The insulations were then covered with standard 10 mm gypsum boards. Figure 4.4.2 shows the external view of one of the huts used in this study. The detailed information of the opaque envelope with the associated thermophysical characteristics is found in Ref. [20]. There was no internal heat gain from lighting, occupants, and electric equipment as the huts were empty.

The hut with an ATSS was provided with two identical PCM storage units designed and fabricated at the University of Auckland. Each PCM storage unit was composed of 19 sets of aluminum macroencapsulated PCM panels filled with 9.5 kg RT25HC (manufactured by Rubitherm GmbH). The design of the ATSS is well explained in the research paper published by the authors of the current paper [21]. The thermochemical properties of the PCM used in the ATSS can be found in references [22,23], while the melting temperature range and latent heat were crosschecked using water bath experiment and differential scanning calorimetry, respectively. The wallboards of the hut with a PTSS were impregnated with PT20 PCM (20 wt% PCM impregnation); detailed information regarding the impregnation method is found in Ref. [24]. In the space cooling experiment, the wallboards of all walls were impregnated with about 47 kg PCM, having an energy storage capacity 50% more than that used in ATSS. In a later experiment, only the northern and eastern walls (which gives the maximum benefit of PCM while reduces the capital cost [25]) of the hut were impregnated with 25 kg PCM to have identical energy storage capacity in both PTSS and ATSS. Energy storage capacity was defined through multiplying the mass of PCM by its latent heat.

FIGURE 4.4.2 Experimental hut located at the Ardmore Campus of the University of Auckland, New Zealand.

The room temperature was set between 22°C and 25°C in summer and between 19°C and 25°C in winter. Hence, the maximum amount of sensible heat for 6°C is equivalent to 12 kJ/kg, which is less than 5% of the total energy.

The objective of the experiments was to compare the performance of a PTSS with that of an ATSS for both heating and cooling. To this end, both huts were heated by an electric heater and an air-based solar heater, which information are detailed in Table 4.4.1. The huts were cooled using identical air conditioner (AC) units (Table 4.4.1) and also by the free night cooling stored in the PCM.

In the hut provided with PTSS, heat exchange between PCM and room environment occurred through the natural convection, while in the ATSS, it was through forced convection heat transfer. A fan powered by a photovoltaic module (PV) was used to pump the room air into the solar heater, warm it up, and send it back to the room or the PCM storage unit. Electric valves ($V1$, $V2$, and $V3$) were installed to determine the pathways of airflow for heat exchange. The possible pathways used to supply the heating and cooling demands of the huts are shown in Figure 4.4.3 and discussed below and later in Section 4.4.2.3. Table 4.4.1 summarizes the equipment used to provide comfort in both winter and summer, with some of their specifications.

Pathway 1: Solar air heater sucks the room air into its duct and then sends it back to the room, using a PV-driven fan.

TABLE 4.4.1

List of Equipment Used in the Experimental Huts for both Space Heating and Cooling

Equipment	Model	Operating Conditions	Electric Power (W)
AC	GREE heat pump (Model GWH09MB-K3DNA4H/O)	25°C–22°C	730
Electric heater	SUNAIR, Model OHS1000	19°C–20°C	1000
Solar air heater	1 m² Flat plate	–	Maximum: 550
PCM Storage	RT25HC	Storage capacity: 1.2 kWh	–
Fan powered by PV module	Model AD091 CE2XB-A71GP	Flowrate: 100 m³/h	6.6
Fan powered by electricity	100 mm Plastic Duct Booster Inline Fan	Flowrate: 130 m³/h	12
Valve	PVC	Fully open/fully closed	–

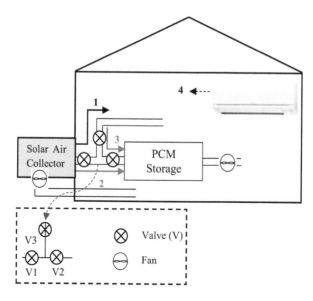

FIGURE 4.4.3 Schematic view of different pathways used to supply heating/cooling energy.

Pathway 2: The indoor air is heated by the solar air heater, sent to the ATSS, and finally is returned to the room. This circulation was done by an electricity-driven fan installed at the outlet of the ATSS.

Pathway 3: The room air is circulated into the ATSS and sent back to the room, using the electricity-driven fan.

Pathway 4: The huts receive heating/cooling from an electric heater/AC unit.

T-type thermocouples, calibrated against a reference thermometer, were used to measure the temperature of PCM, in-door and outdoor temperatures and that of air

TABLE 4.4.2

Measurement Instrumentation

Instrument	Accuracy Range	Operating Range	Model
Thermocouples	±0.45°C	0°C–50°C	T-type
Reference thermometer	0.02°C	0°C–50°C	Ebro TFX430
kWh meter	±0.5%	0–45 A	DRS-45-1 P
Digital anemometer	± (2% + 0.2 m/s)	0.4–30 m/s	AM-4201
Pyranometer	<10 W/m²	0–2000 W/m²	VAEQ08E

outlet from solar air heater. Air velocity was measured manually by a digital anemometer model AM-4201, which was used to calculate air flow rate. The air velocity in the pipes, with a diameter of 10 cm, was constant at 3 m/s. Solar radiation was measured by using a pyranometer. The types of measurement used in this study and their accuracy are listed in Table 4.4.2.

In this experimental study, all parameters including the inlet temperature of PCM storage, room and outdoor temperatures were highly dependent on the environmental conditions. Therefore, it was not possible to duplicate the same experiment and test the uncertainty of the experiment. Nevertheless, several experiments with different inlet air temperatures to the storage system were conducted, and they showed the same trend as discussed in the study published by the authors of this paper [21].

4.4.2.2 DATA ACQUISITION

The data were logged using a Compact Reconfigurable (CompactRIO) Data Acquisition System (NI cRIO-9012, National Instruments, USA) interfaced with LabVIEW software on a dedicated computer. CompactRIO incorporates a real-time processor and reconfigurable field-programmable gate array (FPGA) for embedded machine control and monitoring applications, which requires a precise and fully defined control algorithm [26]. The instrument provides a means for the communication of input and output data. An analog input that is placed in a CompactRIO measures input parameters and then sends corrective commands through an analog output [27]. In the current study, thermocouples were connected to analog inputs to monitor room temperature (T_{room}), PCM temperature (T_{PCM}), and outdoor temperature (T_{out}). A kWh meter was installed to measure the electricity consumed by the electric heater or AC units [8]. The online electricity price was also fed to the system. After processing the data based on the measured comfort temperature, the corrective signals were sent to decide on the operation of the AC unit, solar air heater, electric heater and PCM storage units, where applicable, as shown in Figure 4.4.4.

4.4.2.3 CONTROL SYSTEM

A building automation system was designed to control the air conditioning or heating of the huts and hence ensures indoor thermal comfort. First, the thermal performance

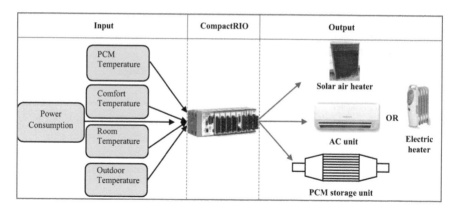

FIGURE 4.4.4 Data acquisition setup for the experimental huts.

of the PTSS in Hut 1 was compared with that of the ATSS in Hut 2 for air conditioning. When the room temperature of the huts exceeded the upper bound of comfort level, the AC units started automatically. Once the AC unit was switched ON, it continued until the room temperature reached the lower bound of comfort level. Then, no cooling energy was provided to Hut 1 until room temperature reached the upper bound of comfort temperature again. In Hut 2, if the PCM temperature was lower than room temperature, the PCM storage units removed heat from the hut while melting the PCM. If the coolness stored in the PCM was insufficient to maintain the indoor comfort and room temperature exceeded the upper bound of comfort level, the AC unit would start automatically.

During the following cooler hours (mainly during the night), the collected/stored heat was automatically released to the room or the outside. Hence, PCM was solidified for use in the following daytime.

Figure 4.4.5 shows the flowchart of the control algorithms implemented for air conditioning in Hut 2, provided with ATSS, with the objective of maintaining the comfort condition. Excluding the left side of the algorithm, leading to pathway 3, from Figure 4.4.5 results in the control algorithm applied to the Hut 1 (no PCM or PTSS). In the flowcharts, "no action" denotes the "OFF" mode of both cooling sources (AC and PCM).

Next, another set of experiments was conducted to investigate the thermal performance of an ATSS for space heating. Electric and solar air heaters were used to satisfy the heating demand of the huts. When room temperature dropped below the lower bound of comfort level, the electric heater sustained comfort until it reached the upper bound of comfort level. Otherwise, the solar air heater in Hut 1, and solar air heater or energy stored in PCM, in Hut 2, assisted in reducing the heating demand. Figure 4.4.6 shows the control algorithm applied to the active PCM-enhanced building for space heating. By removing commands referring to PCM in Figure 4.4.6, the control algorithm would be used for the space heating of the hut provided with PTSS.

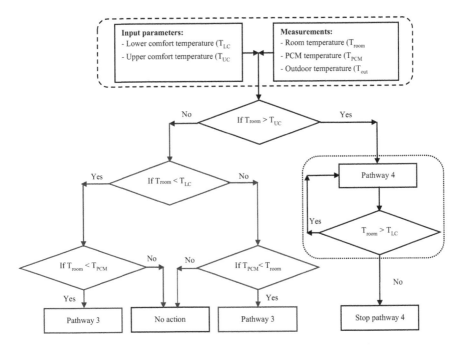

FIGURE 4.4.5 Control algorithm used for air-conditioning of the hut provided with ATSS (The algorithm without red arrows and boxes represent the control method used for the hut with PTSS or no PCM.) (For interpretation of the references to color in this figure legend, the reader is referred to the Web version of this article.)

4.4.3   RESULTS AND DISCUSSION

4.4.3.1   COMPARISON OF PTSS AND ATSS

4.4.3.1.1   Space Cooling

This section illustrates the thermal behavior of PTSS and ATSS compared to their counterparts with no PCM. The comfort level was set between 22°C and 25°C. Following the control algorithm used for air conditioning of the huts (Figures 4.4.5), Figure 4.4.7a shows the thermal performance of Hut 1, in the presence of the PTSS, compared to that of the Hut 2, with no PCM. Immediately after sunrise, the ambient temperature and consequently, the room temperature of the huts, started to increase, but at different rates. At noon, the room temperature of Hut 2 (with no PCM) exceeded that of Hut 1 and then reached the upper bound of comfort level (25° C), at 2 pm. Then, the controller started the AC unit until the room temperature reached the lower bound of comfort (22°C). This cycle was repeated until 7 pm and hence created temperature fluctuations between 22°C and 25°C, as shown in Figure 4.4.7a. However, Hut 1 did not require any air conditioning for almost 4 h, until 5:45 pm, as highlighted in gray. The excess heat was absorbed by the PCM embedded into the wallboards.

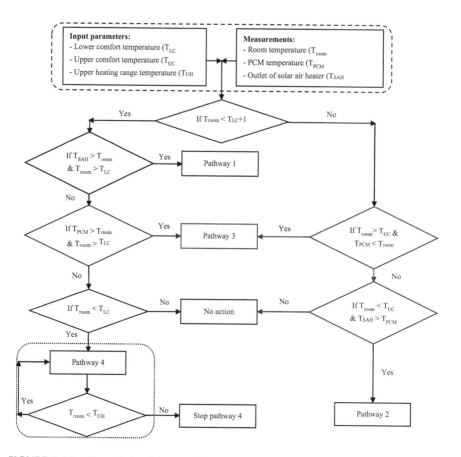

FIGURE 4.4.6 Control algorithm used for space heating of the hut provided with the ATSS (The algorithm without red arrows and boxes represent the control method used for the hut with PTSS or no PCM.) (For interpretation of the references to color in this figure legend, the reader is referred to the Web version of this article.)

Figure 4.4.7b shows the thermal performance of Hut 2, in the presence of an ATSS, compared to that of Hut 1, with no PCM. The room temperature of both huts started increasing from 8 am. Once the room temperature of Hut 1 exceeded 25°C, at 11 am, the AC unit started and then kept operating until the room temperature reached 22°C. This process was repeated until 3:30 pm and created the temperature fluctuations seen in Figure 4.4.7b. Through the high latent heat of melting, the ATSS in Hut 2 maintained the thermal comfort and hence delayed the use of electricity for 2 h, as highlighted in gray. It can be seen from the figure that the PCM temperature increased during this period, which confirms that it has absorbed the surplus heat.

As discussed in Section 4.4.2.3, the scope of this study was to compare the benefit of PTSS with that of the ATSS. The results showed that energy consumption in the hut with PTSS was more susceptible to ambient temperature compared to that in ATSS. For instance, as shown in Figure 4.4.8b, during the days of 4, 5, 6, and Dec 10, 2019, the hut with PTSS required less cooling energy compared to the hut with

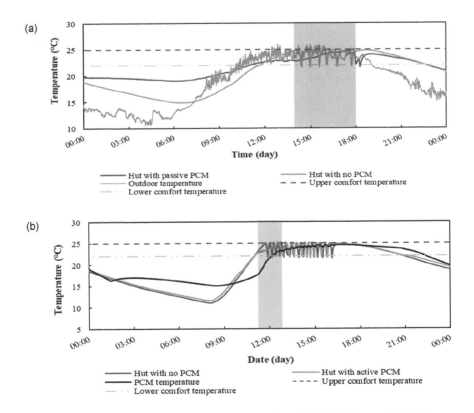

FIGURE 4.4.7 Thermal performance of hut provided with (a) PTSS and (b) ATSS compared to their counterparts with no PCM, for space cooling.

no PCM. However, the opposite result was evident on days of 7, 8, and 9 Dec because the nighttime temperature did not drop enough to solidify the PCM (Figure 4.4.8c). Hence, the extra cooling load was needed to cool the PCM sensibly. In contrast, the ATSS showed the potential of reducing energy consumption (Figure 4.4.9b) in the hut even during the nights of moderate temperatures (Figure 4.4.9c).

As shown in Figure 4.4.10, the comparison between ATSS in Hut 2 with the PTSS in Hut 1, at the same time, showed a slightly higher energy consumption for the ATSS on most of the days. Indeed, the hut equipped with PTSS consumed only 8% less electricity than the hut provided with ATSS. This difference is small remembering that the quantity of PCM used in PTSS is more than twice as much as that used in ATSS (8100 kJ for PTSS versus 4370 kJ for ATSS). This is why it was decided later to reduce the amount of PCM in the PTSS hut so it will have the same energy storage capacity of 4370 kJ as will be discussed later in Section 4.4.3.1.2.

The effect of the differences in melting temperatures of PCMs used in PTSS and ATSS should be noted. The melting temperature of the PCM in the passive configuration (PT20) was within a very narrow temperature range of 19.9°C ± 0.3°C, while the melting range of PCM in the ATSS (RT25HC) was between 22°C and 26°C. The lower melting temperature of PT20 required a lower nighttime temperature

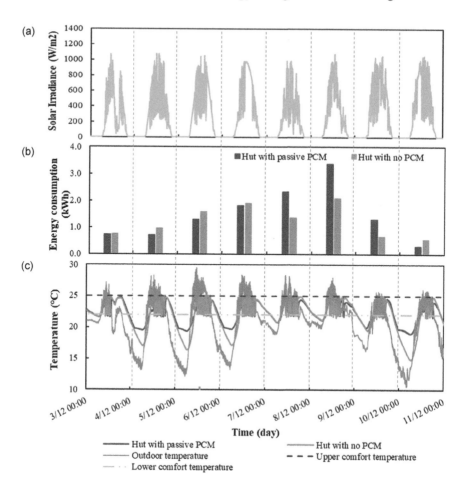

FIGURE 4.4.8   (a) Solar radiation, (b) energy consumption of the AC unit, and (c) the thermal behavior of the hut provided with PTSS, compared to that of t the hut with no PCM, used for air conditioning.

to solidify it; however, it was more efficient in absorbing heat during the daytime. In contrast, the RT25HC could solidify more easily at night while it did not melt entirely during the daytime. Indeed, the upper bound of comfort temperature was set at 25°C while PCM temperature reached up to 24°C only (Figure 4.4.9). It is known that PCM must be selected based on its melting/freezing temperatures suitable for weather conditions in different seasons.

4.4.3.1.2   Space Heating

This section provides a more accurate comparison for the thermal performance of the huts employing PTSS and ATSS for space heating. The amount of PCM (PT20) incorporated into the hut with the PTSS was reduced to 25 kg, compared to 19 kg PCM (RT25HC) in the ATSS unit. These quantities correspond to the same energy storage capacity of 4370 kJ since the two PCMs have different latent heat. In this experiment,

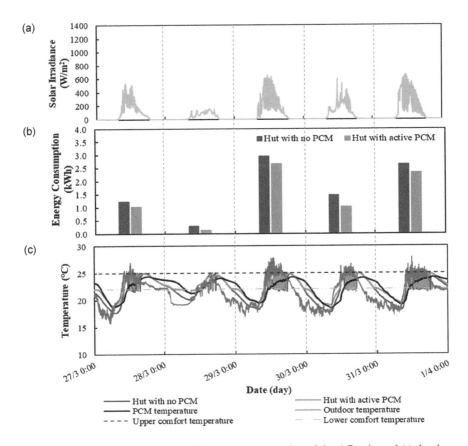

FIGURE 4.4.9   (a) Solar radiation, (b) energy consumption of the AC unit, and (c) the thermal behavior of the hut provided with ATSS, compared to the hut with no PCM.

the setpoint for the electric heater was between 20°C and 22°C during the day and between 19°C and 20°C during night. Based on the algorithm shown in Figure 4.4.6, solar air heaters were used to heat the huts and melt the PCMs, during the day on 27 March. At 8:30 pm, the room temperature of both huts reached the upper bound of the heater's operation range (22°C). At that time, there was no control over the release of energy by PTSS; however, the ATSS operated automatically to sustain the comfort and hence delayed the use of electric heating for a couple of hours (Figure 4.4.11). As a result, the hut provided with the ATSS consumed 0.3 kWh less electricity compared to the hut equipped with PTSS. As shown in Figure 4.4.12, the experiments conducted over 10 days were in favor of the ATSS showing 22% less energy consumption. Under mild environmental conditions (such as 21 Mar), the ATSS showed a better performance compared to the PTSS. Due to a lower heat transfer rate between the walls and room, the full capacity of PCM in PTSS was not exploited. However, under colder environmental conditions (such as 26 Mar), both PTSS and ATSS contributed almost equally. Indeed, the heat transfer mechanism in the PTSS increased due to the existing driving force, i.e. temperature difference between walls and room.

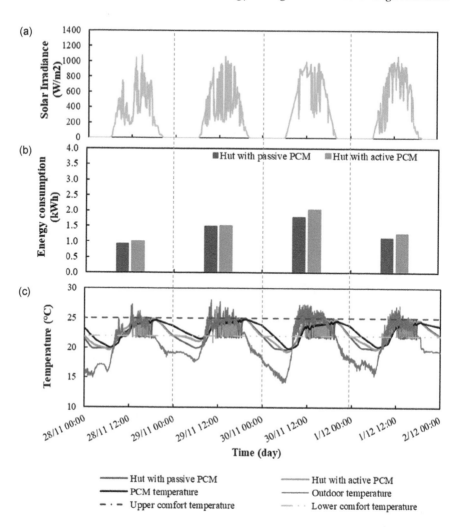

FIGURE 4.4.10 (a) Solar radiation, (b) energy consumption of the AC unit, and (c) the comparison between the thermal behavior of the PTSS and ATSS.

4.4.3.1.3 Peak Load Shifting Using PTSS and ATSS

In this study, an experimental measurement was carried out to compare the benefit of using these two types of storage systems to create heating peak load shifting. To this end, the hut provided with PTSS received energy from the Sun (through solar heater) so long as indoor room temperature remained within comfort level. Otherwise, an electric heater was used to prevent the room temperature from falling below comfort. On the other hand, the hut equipped with ATSS stored solar energy (through solar collector) or the low-rate energy (provided by the electric heater) for use during the high electricity peak period. The electric heater would also start if the room temperature dropped below comfort. For example, Figure 4.4.13 shows the thermal performance of both huts on Sep 7, 2019. The heater's setpoints were between 19°C

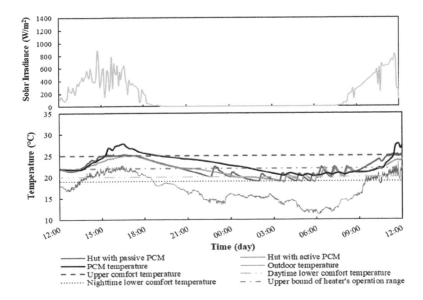

FIGURE 4.4.11   Comparison between PTSS and ATSS, used for space heating on 27 March.

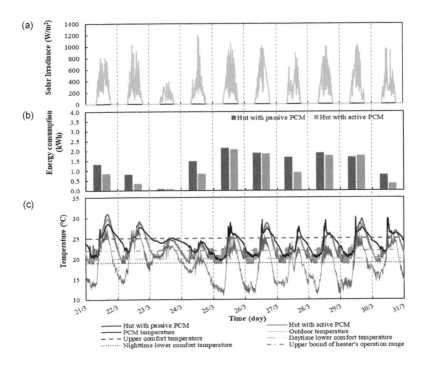

FIGURE 4.4.12   (a) Solar radiation, (b) energy consumption in the huts, and (c) the thermal behavior of the PTSS and ATSS, over 10 days.

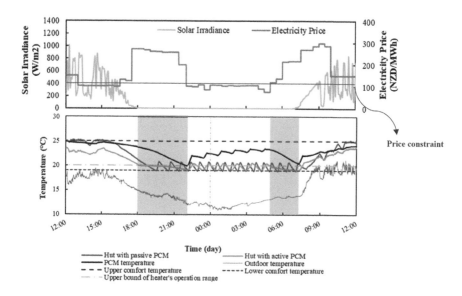

FIGURE 4.4.13 Application of the price-based control for peak load shifting using the ACTSS, versus the use of PTSS.

and 20°C. The daytime solar energy was stored to create evening peak load shifting, from 6 pm to 10:30 pm. Once the room temperature of Hut 2 (with ATSS) dropped below the lower bound of comfort temperature in winter (19°C), the electric heater would have started to sustain comfort. The heater continued operating until the room temperature reached the upper bound of comfort level (20°C). By then, it would have reached an off-peak hour; hence, the electric heater kept heating both the room and the PCM to maintain comfort. The stored heat, produced by the electric heater during off-peak hours, was then utilized for the following morning peak hour, from 5:30 am to 7:30 am and delayed the use of electricity. The application of ATSS in combination with a control strategy led not only to energy-saving but also to electricity cost-saving through peak load shifting.

It is evident that there was a high level of control over the heat transfer from PCM storage to the indoor environment; however, such a level of control cannot be achieved by a PTSS.

The same set of experiments was continued for a couple of days. Figure 4.4.14 shows the thermal performance of both huts provided with PTSS and ATSS, together with energy consumption in each hut, under the same environmental conditions. Results showed that the hut with ATSS always consumed less electricity to maintain comfort compared to the hut with PTSS. This is due to the better control of the heat release from the PCM in the ATSS. In addition, heat is transferred from the thermal energy storage unit to the building indoor environment by forced convection at a high rate [28]. However, the dominant heat transfer mechanism in a PTSS from walls and ceiling is controlled by natural convection at a much lower rate.

On the other hand, when the PCM temperature in a PTSS is higher than room temperature, the stored energy would be released to the room despite it being within

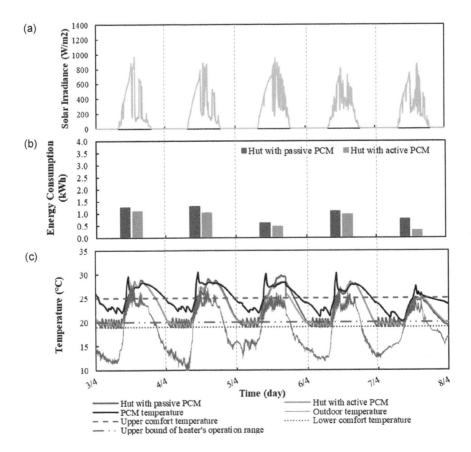

FIGURE 4.4.14   (a) Solar radiation, (b) energy consumption of electric heaters, and (c) thermal performance of both huts for peak load shifting over 5 days.

comfort level or during the off-peak period. Hence, PTSS would work desirably only if the heating demand of a building is within peak hours. Results show that the hut provided with ATSS consumed less energy and saved more cost to maintain comfort. However, the reduction in cost was not proportional to the decrease in energy consumption. For example, the results on the day of 4 Apr, showed 20% less energy consumption with a corresponding 32% less cost when the ATSS was used. This result confirms a more successful peak load shifting through the ATSS. Figure 4.4.15 shows the ratio of the electricity cost reduction to the energy consumption decrease achieved by the ATSS compared to that by PTSS, over 5 days under the similar environmental conditions of Figure 4.4.14.

4.4.3.2 Further Comparison of PTSS and ATSS

ATSS offers some other advantages over PTSS. In PTSS, PCM is integrated into building construction materials. At the end of a building's life, the building will be demolished and the presence of PCM with the demolished materials will cause

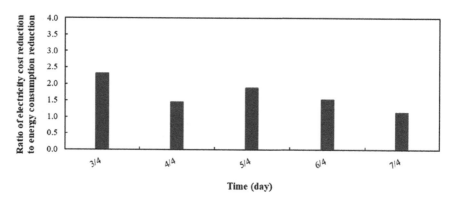

FIGURE 4.4.15 Ratio of the electricity cost reduction to energy consumption reduction of ATSS compared to PTSS.

environmental problems [29]. Indeed, there is no way of recovering microencapsulation PCM from construction materials, while PCM does not degrade easily. More work is needed in this area. In contrast, in the ATSS, PCM is stored in separate containers (macro-encapsulation) allowing it to be recovered as PCMs are known to be chemically and physically stable [30]. Thus, at the end of the macroencapsulation lifetime, PCM can be reused for other purposes. Also, the results showed that ATSS provides better energy-saving in building through better control of heat release. The capital cost of ATSS may be higher but the reduced quantity of PCM, which is expensive, may make ATSS more economically feasible.

4.4.4 CONCLUSION

PCM application not only provides indoor thermal comfort but also contributes to energy-saving and peak load shifting. The comparison of the designed ATSS with a PTSS, having the same energy storage capacity, confirmed a greater energy-saving (22% less energy consumption) and a more efficient peak load shifting (32% less electricity cost) when ATSS was used. PCM is still expensive, and a reduction in the quantity used would improve the economics. Besides, the use of ATSS would provide a better degree of control, higher reliability, and less adverse environmental impact. The final selection of the type of energy storage system should be based on cost analysis, which is the main focus of our future work.

DECLARATION OF COMPETING INTEREST

The authors declare that they have no known competing financial interests or personal relationships that could have appeared to influence the work reported in this paper.

ACKNOWLEDGMENT

The first author thanks the University of Auckland for providing her with an International Doctoral Research Scholarship.

NOMENCLATURE

AC	Air conditioner
ATSS	Active thermal storage system
CompactRIO	Compact Reconfigurable Input/Output
FPGA	Field Programmable Gate Array
HVAC	Heating, cooling, and air conditioning
I/O	Input/Output
PCM	Phase change material
PTSS	Passive thermal storage system
SAH	Solar air heater
T	Temperature (°C)
T_{LC}	Lower bound of comfort temperature (°C)
T_{out}	Outdoor temperature (°C)
T_{PCM}	Phase change material temperature (°C)
T_{room}	Room temperature (°C)
T_{UC}	Upper bound of comfort temperature (°C)
T_{UH}	Upper bound of heater operation range temperature (°C)

REFERENCES

1. A. Sharma, V.V. Tyagi, C.R. Chen, D. Buddhi, Review on thermal energy storage with phase change materials and applications, *Renew. Sustain. Energy Rev.* 13 (2009) 318–345, doi:10.1016/j.rser.2007.10.005.
2. M. Saffari, A. de Gracia, S. Ushak, L.F. Cabeza, Passive cooling of buildings with phase change materials using whole-building energy simulation tools - a review, *Renew. Sustain. Energy Rev.* 80 (2016) 1239–1255, doi:10.1016/j.rser.2017.05.139.
3. CEN (European Committee for Standardization), Indoor Environmental Input Parameters for Design and Assessment of Energy Performance of Buildings Addressing Indoor Air Quality, Thermal Environment, Lighting and Acoustics, 2007, doi:10.1520/E2019-03R13.
4. D. Zhou, C.Y. Zhao, Y. Tian, Review on thermal energy storage with phase change materials (PCMs) in building applications, *Appl. Energy* 92 (2012) 593–605, doi:10.1016/j.apenergy.2011.08.025.
5. G. Li, X. Bi, G. Feng, L. Chi, X. Zheng, X. Liu, Phase change material Chinese Kang: design and experimental performance study, *Renew. Energy* 150 (2020) 821–830, doi:10.1016/J.RENENE.2020.01.004.
6. Y. Guan, T. Wang, R. Tang, W. Hu, J. Guo, H. Yang, Y. Zhang, S. Duan, Numerical study on the heat release capacity of the active-passive phase change wall affected by ventilation velocity, *Renew. Energy* 150 (2020) 1047–1056, doi:10.1016/J.RENENE.2019.11.026.
7. H. Akeiber, P. Nejat, M.Z.A. Majid, M.A. Wahid, F. Jomehzadeh, I. Zeynali Famileh, J.K. Calautit, B.R. Hughes, S.A. Zaki, A review on phase change material (PCM) for sustainable passive cooling in building envelopes, *Renew. Sustain. Energy Rev.* 60 (2016) 1470–1497, doi:10.1016/j.rser.2016.03.036.
8. R. Barzin, J.J.J. Chen, B.R. Young, M.M. Farid, Application of PCM underfloor heating in combination with PCM wallboards for space heating using price- based control system, *Appl. Energy* 148 (2015) 39–48, doi:10.1016/j.apenergy.2015.03.027.
9. M. Saffari, A. De Gracia, S. Ushak, L.F. Cabeza, Economic impact of integrating PCM as passive system in buildings using Fanger comfort model, *Energy Build.* 112 (2016) 159–172, doi:10.1016/j.enbuild.2015.12.006.

10. E. Osterman, V.V. Tyagi, V. Butala, N.A. Rahim, U. Stritih, Review of PCM based cooling technologies for buildings, *Energy Build.* 49 (2012) 37–49, doi:10.1016/j.enbuild.2012.03.022.

11. B.L. Gowreesunker, S.A. Tassou, M. Kolokotroni, Coupled TRNSYS-CFD simulations evaluating the performance of PCM plate heat exchangers in an airport terminal building displacement conditioning system, *Build. Environ.* 65 (2013) 132–145, doi:10.1016/j.buildenv.2013.04.003.

12. G. Gholamibozanjani, J. Tarragona, A.D. Gracia, C. Fernández, L.F. Cabeza, M.M. Farid, Model predictive control strategy applied to different types of building for space heating, *Appl. Energy* 231 (2018), doi:10.1016/j.apenergy.2018.09.181.

13. J. Heier, C. Bales, V. Martin, Combining thermal energy storage with buildings a review, *Renew. Sustain. Energy Rev.* 42 (2015) 1305–1325, doi:10.1016/j.rser.2014.11.031.

14. Y. Wang, X. Yang, T. Xiong, W. Li, K.W. Shah, Performance evaluation approach for solar heat storage systems using phase change material, *Energy Build.* 155 (2017) 115–127, doi:10.1016/j.enbuild.2017.09.015.

15. M.M. Prieto, I. Suárez, B. González, Analysis of the thermal performance of flat plate PCM heat exchangers for heating systems, *Appl. Therm. Eng.* 116 (2017) 11–23, doi:10.1016/j.applthermaleng.2017.01.065.

16. M.C. Browne, B. Norton, S.J. McCormack, Heat retention of a photovoltaic/ thermal collector with PCM, *Sol. Energy* 133 (2016) 533–548, doi:10.1016/j.solener.2016.04.024.

17. A. Gaur, C. Ménézo, S. Giroux–Julien, Numerical studies on thermal and electrical performance of a fully wetted absorber PVT collector with PCM as a storage medium, *Renew. Energy* 109 (2017) 168–187, doi:10.1016/j.renene.2017.01.062.

18. N. Stathopoulos, M. El Mankibi, R. Issoglio, P. Michel, F. Haghighat, Air-PCM heat exchanger for peak load management: experimental and simulation, *Sol. Energy* 132 (2016) 453–466, doi:10.1016/j.solener.2016.03.030.

19. J.R. Turnpenny, D.W. Etheridge, D.A. Reay, Novel ventilation system for reducing air conditioning in buildings. Part II: testing of prototype, *Appl. Therm. Eng.* 21 (2001) 1203–1217, doi:10.1016/S1359-4311(01)00003-5.

20. P. Devaux, M.M. Farid, Benefits of PCM underfloor heating with PCM wallboards for space heating in winter, *Appl. Energy* 191 (2017) 593–602, doi:10.1016/J.APENERGY.2017.01.060.

21. G. Gholamibozanjani, M. Farid, Experimental and mathematical modeling of an air-PCM heat exchanger operating under static and dynamic loads, *Energy Build.* 202 (2019) 109354, doi:10.1016/J.ENBUILD.2019.109354.

22. Rubitherm Technologies GmbH, Macroencapsulation - CSM, n.d. https://www.rubitherm.eu/en/index.php/productcategory/organische-pcm-rt. (Accessed 1 August 2019).

23. S. Khanna, K.S. Reddy, T.K. Mallick, Performance analysis of tilted photovoltaic system integrated with phase change material under varying operating conditions, *Energy* 133 (2017) 887–899, doi:10.1016/j.energy.2017.05.150.

24. R. Barzin, J.J.J. Chen, B.R. Young, M.M. Farid, Application of weather forecast in conjunction with price-based method for PCM solar passive buildings - an experimental study, *Appl. Energy* 163 (2016) 9–18, doi:10.1016/j.apenergy.2015.11.016.

25. K. Biswas, R. Abhari, Low-cost phase change material as an energy storage medium in building envelopes: experimental and numerical analyses, *Energy Convers. Manag.* 88 (2014) 1020–1031, doi:10.1016/J.ENCONMAN.2014.09.003.

26. M. Rosol, A. Pilat, A. Turnau, Real-time controller design based on NI Compact- RIO, *Proc. Int. Multiconference Comput. Sci. Inf. Technol.* (2014) 825–830, doi:10.1109/imcsit.2010.5679917.

27. National Instruments, Getting Started with CompactRIO and LabVIEW, 2009.

28. A. Kasaeian, L. Bahrami, F. Pourfayaz, E. Khodabandeh, W.M. Yan, Experimental studies on the applications of PCMs and nano-PCMs in buildings: a critical review, *Energy Build.* 154 (2017) 96–112, doi:10.1016/j.enbuild.2017.08.037.

29. N. Soares, J.J. Costa, A.R. Gaspar, P. Santos, Review of passive PCM latent heat thermal energy storage systems towards buildings' energy efficiency, *Energy Build.* 59 (2013) 82–103, doi:10.1016/j.enbuild.2012.12.042.

30. A. Karaipekli, A. Sarı, Capricemyristic acid/vermiculite composite as form- stable phase change material for thermal energy storage, *Sol. Energy* 83 (2009) 323–332, doi:10.1016/J.SOLENER.2008.08.012.

Index